Principles of Fishery Science

SECOND EDITION

W. HARRY EVERHART
WILLIAM D. YOUNGS

Comstock Publishing Associates, *a division of*
Cornell University Press, *Ithaca and London*

First edition published 1975 by Cornell University Press.
Second edition published 1981 by Cornell University Press.
Second printing 1989.

International Standard Book Number 0-8014-1334-6
Library of Congress Catalog Card Number 80-15603
Printed in the United States of America

*Librarians: Library of Congress cataloging information
appears on the last page of the book.*

*The paper in this book is acid-free and meets the guidelines for permanence
and durability of the Committee on Production Guidelines for Book Longevity
of the Council on Library Resources.*

Contents

Preface

The problems of managing natural resources grow in magnitude and complexity as human pressures increase needs, wants, and concerns for the quality of life. Our recognition of the intricacy of natural resource utilization has grown with the expanding awareness that nothing in this world exists by itself. Everything is interrelated, and no longer is there any excuse for considering the use of our forests, the farming of our countrysides, the mining of our mineral resources, or the fishing in our lakes, streams, and oceans as a single, simple operation. Successful stewardship of our natural resources demands unity, and increasing acceptance of this principle is progress. University programs and government institutions have been reorganized to implement interdisciplinary studies in resource management by making it easier for the economist, sociologist, political scientist, engineer, and biologist to work together. Appreciation of the whole may seem to make resource management more difficult, but it also makes real and satisfactory solutions more certain.

We define fishery science as the application of scientific knowledge to the problems of providing the optimum yield of fishery products, whether stated in tons of commercial products or in hours of angling pleasure. The definition tells us that fishery science is interdisciplinary, and we expect that fishery principles and technology are developing to meet the challenge. Fishery scientists must build on the traditional foundations of mathematics, chemistry, biometrics, and basic biology while developing an understanding of the economics of natural resources and a sensitivity to sociological problems, and at the same time perceiving the interrelationships of environmental concerns. Each of us must combine an in-depth knowledge of our own discipline with a thorough appreciation for the unity of our world if we value the quality of living and, indeed, survival.

The purpose of our text is to help students and professionals reach a thorough understanding of fishery science. From this base of principles and applications each individual must build his or her own perception and sensitivity to world unity by exposure to the teachings of other disciplines, by experience, and by analysis of environmental problems. Professors and senior administrators have a responsibility to reinforce the holistic concept by their own examples and teachings. Because so many disciplines are strongly interwoven in the understanding and solution of problems of natural resource management, the compilation of the knowledge required to solve these problems in a single text is impossible.

The need for objectiveness in dealing with fishery resource problems generally leads to a necessity for deriving a quantitative solution. Our experience shows this to be an area of difficulty for most fishery students. We have added an Appendix to this edition in an attempt to provide review material in mathematics and statistics. While this is not an integral part of the text, readers may wish to cover this material before reading Chapters 5 through 8.

After publication of the first edition, our former colleague, Dr. Alfred W. Eipper, transferred to the position of Northeast Power Plant Activities Leader in the U.S. Fish and Wildlife Service's Boston Regional Office. These duties prevented Dr. Eipper's active participation in this revision, but his contributions to the first edition remain.

George A. Rounsefell and W. Harry Everhart set forth the principles and concepts in their college textbook *Fishery Science,* published in 1953. We are indebted to Dr. Rounsefell for these and for other major contributions to the 1975 text, *Principles of Fishery Science,* and to its 1981 revision; these contributions are included in the chapters on fishing gear, fishways, screens and guiding devices, fish marking, and regulations.

Alice Klock and Eileen Stanturf prepared the manuscript and helped with the proofreading. General support of the New York State College of Agriculture and Life Sciences at Cornell University is gratefully acknowledged.

W. HARRY EVERHART
Professor and Chairman

WILLIAM D. YOUNGS
Associate Professor

Department of Natural Resources
College of Agriculture and Life Sciences
Cornell University

Principles of
Fishery Science

1

Introduction

Fishery biologists must be prepared to accept variation, dynamic populations, an environment largely uncontrollable, the need for compromise, optimum rather than maximum results in the resolution of many harvest problems, and the conflicting desires of the people we work for. More likely than not there will be no formulas, no handbooks, no set rules to follow in making the numerous decisions fishery management demands whether it be in the writing of an international treaty or deciding what size fish to stock in the local sportsmen's favorite stream. The fishery biologist lives in the demanding world of gray where definite black and white in decision making are an exception. The challenges are obvious, and for those who enjoy such contests, fishery biology is an exciting and satisfying profession.

Fishery research and management should be directed toward specific well-defined objectives. Time was when professors and fishery department chiefs could insist on well-written research proposals and ask no more than that. Now, the introduction of systems analysis requires a more formal recognition of the detailed planning necessary for good management decisions. Systems analysis forces the resource manager to define his objectives, which must be clear if the work is to be productive. Natural resources problems are increasingly more complex, and systems analysis will help the natural resource manager clarify objectives and alternatives. And, too, as society becomes more involved in the decision-making process affecting natural resources, it becomes increasingly important to provide clear, straightforward alternatives and at the same time make known their possible consequences. Research should help define these alternatives.

While definitions of kinds of research are as likely to attain unanimity of opinion as the definition of a species, some explanation of the types and purposes of research is useful. The probability that arbitrary divisions of research will overlap is recognized and is really little cause for concern.

Background Research

With the systematic collection, observation, organization, analysis, and presentation of facts, using known principles to reach objectives that are clearly defined before the study is undertaken, background research can provide a foundation for subsequent research, together with a bank of standard reference data. A large share of the research needed in natural resources is of this character, including most of the habitat surveys, such as the biological inventory of lakes and streams.

No business, particularly environmental management, can operate successfully without a continuing inventory of resources and conditions. Biological inventories are basic to fishery resource management in any area. Information must be available on kinds, numbers, and distributions of species; kinds, conditions, and uses of water bodies; pollution, exploitation, parasites and diseases, and obstructions to fish migration and fishways. In addition, sound management requires continuous monitoring of the effectiveness of current practices such as habitat improvements, stocking, and regulations. Without a continuing inventory, serious problems may go unrecognized. Biological inventories provide a base for determining the relative importance of problems and help in deciding the best solutions. For research-project priorities and the selection of management tools must, most importantly, be based on firm data.

The first biological inventories, called surveys then, began in the early 1900s. The Illinois Natural History Survey was one of the first to consider all the flora and fauna in a state and to devote considerable time to the study of the fishes. The New York State Biological Survey began in 1926 and was the prototype for laying down general techniques and for training biologists who went to other states to direct similar surveys of their own.

Planning and conducting a state-wide biological inventory requires some executive decisions. Four general approaches are possible: (1) by drainage area, (2) by specific problem, (3) by political boundaries (counties), and (4) by a combination. Too many states elected to use the drainage-area approach and thus expended funds and personnel working on data for one drainage-area, while the problems in the remainder of the state went unsolved. Others conducted their inventories around specific problems, such as classification of their lakes by chemical parameters, while the streams were neglected. Following political boundaries usually meant starting in one corner of the state and working to complete the entire state. While work went on in the southern part, northern counties had no information on which to base their fishery programs. A wise approach, in our opinion, would involve a division of funds and personnel on a fifty-fifty basis for both drainage-area studies and relief for specific problems. Regional biologists rightfully want to complete drainage studies, but they need also to consider the requests of the public. Better public relations and service can result from a biological inventory of a lake or stream requested by a county sportsmen's club, a Trout Unlimited chapter, or a group of camp owners who have, or think they have, a fishery problem. With a planned division of funds and personnel it should be possible to satisfy the immediate needs of the public and yet spend time completing the drainage studies.

Publication of the information collected in biological inventories is particularly important. Early surveys emphasized long, detailed tables of various chemical and biological parameters of more value to the regional biologist than to the public. Plans should include arrangements for storage of data, publication of information for the public, and technical publication. Depending on the scope of the inventory, the data should be easily available on keysort cards, IBM cards, or on computer tape. It may be that the data storage will begin on the cards and then through the years be transferred to the computer. Information should be available for immediate reference in the regional office and in the central office. State fishery employees should be able to answer questions regarding fishing regulations and stocking policies on request when the local senator or sportsmen's club stocking committee asks questions. Much of the technical data stored for professional use would have little interest for the general public. These people want to know what fish are present, what access is available, whether there is a launching ramp, or whether there are boats for rent, and the answers to many other questions of concern to recreationists.

The format of the popular publications should provide for distribution of information having to do with single lake and stream areas. Most people fish and boat in a limited area, and there is no need for them to purchase or to be furnished with an expensive publication that provides a complete inventory of all water areas. Separate publication would make it easy to revise an inventory sheet if a lake were reclaimed, the stocking policy were changed, pollution were abated, or whatever. A small, reasonable charge would offer some control on the number of publications distributed and help defray printing costs. Some standard way of identifying lakes and streams is also necessary. Actually, not too much imagination enters into naming lakes, and consequently most states have a number of Round Ponds, Clear Lakes, Mud Ponds, Lily Ponds, and Island Lakes. Lakes and streams should be identified alphabetically by familiar political boundaries, and an up-to-date index should be published each spring prior to the fishing season. Interested persons could easily identify the water areas of their interest and order copies from the Department.

A biological inventory requires careful planning. For the field work, the planning should begin at least a year ahead; ordering of equipment should be started early; maps for field use should be prepared ahead of the actual field season; standard-data collection forms should be available for all field personnel, and they should be instructed at some length on the correct way to complete the forms. The possible different interpretations to the ways of measuring an important parameter make it necessary to train all personnel to make the same kind of field notes. Identification of nursery areas for salmonids is many times a subjective observation, for unless field personnel have worked together and received detailed instructions, a wide variation can be expected in reports from different people. Biological inventories for state fish and game agencies are usually built around the fishery biologists. However, it frequently becomes necessary to employ, perhaps only summers, a parasitologist, a limnologist, an algologist, or a chemistry specialist. These specialists, besides providing necessary information, may also be used to train field personnel and to teach them to collect and to analyze similar information in the future.

Information accumulated from biological inventory studies is useful for making fishing regulations, for recommending habitat improvement, and to develop and implement stocking programs.

Background research in itself does not solve any problems, but it is a necessary part of other research, and often serves to point out problems requiring solution.

Basic, Applied, and Developmental Research

Fundamental or basic research is the theoretical experimentation, analysis, or exploration of the general principles governing natural or social phenomena. Without basic research all other types would eventually perish. Fundamental research has most frequently been associated with universities, but to a greater and greater extent both government agencies and industry are participating. Behavioral studies of fish are examples of basic research.

Applied research is the extension of basic research to a specific application, which may involve devising a specified product, process, technique, or device, or the solution of a particular problem. A major share of fish and wildlife investigation falls into this category, since most research studies rely on already established principles to solve particular questions. A basic fish-behavior study suggests that hatchery-reared Atlantic salmon (*Salmo salar*) have a high level of aggressiveness that may contribute to the mortality of these salmon after their introduction into natural environments, because of problems arising from loss of feeding time, excessive use of energy, and increased exposure to predators. Application of these findings may require complete overhaul of the usual fish-cultural practices to produce lesser numbers of salmon better prepared for survival in the wild. Further, novel hatching and rearing troughs may be devised.

Developmental research is the adaptation of research findings to experimental or demonstration purposes, such as the experimental production and testing of models, devices, equipment, materials, procedures, and processes. Developmental research differs from applied research in that it is done on products, processes, techniques, or devices previously discovered or invented. Thus, work on improving the type and design of a fishway or of a fish-cultural facility is an example of developmental research.

Solution of a problem through research should follow logical steps as suggested by the following: awareness, exploration, hypothesis or explanation, design, test, communication, trial, and use. There must first be an awareness that a problem exists. In basic research the problem is usually suggested by a scientist. In applied or developmental fishery research, ideas for programs or projects often come from commercial or sport fishermen (for example, gear research and regulation justification).

Exploration of the problem includes a diagnosis of the elements involved and the collection of data and their analysis. Exploration is a "before" study, such as might be dictated in research associated with a river-basin program. Exploration, then, should provide enough information about the problem to permit development of a hypothesis, which is an unproved conjecture or supposition postulated to explain, provisionally, certain facts and to guide in the investigation of others. Sometimes it is called a working hypothesis to indicate it is merely assumed to aid in the search for facts.

Further, there should be a properly designed research project that includes every possible contingency—from field logistics and laboratory work to statistical records and analysis—to test the hypothesis and to insure meaningful interpretation of results. The assimilation of acquired information into useful statistical records is a part of the concept of systems analysis. Dr. Donald McCaughran, Center for Quantitative Sciences in

Forestry, Fisheries, and Wildlife at the University of Washington in Seattle, has given permission to use the following material that explains something about systems analysis.

The study of systems is by no means new: the Phoenician astronomers studied a system of stars, and Plato devised a system of government. However, during the last twenty years systems analysis has become a major new approach in scientific investigation. Today, there are systems analysts in all major fields of human endeavor. It would seem reasonable, then, that fishery biologists should look into this ''new'' approach to see what it has to offer the field of fishery management. In doing this, it is necessary to learn some of the basic concepts of systems study, some approaches to analyzing systems, and finally the difficult task of framing fishery problems in the systems approach. First, let us look at what is called general systems theory.

General Systems Theory

It is difficult to define a system; supposedly one can say that it is some conceptual ''whole unit of interest.'' What makes this whole unit a system is the idea of reductiveness: that is, the breaking down of the whole into well-defined interconnected parts. A few examples will clarify this idea: (1) a road map of a logging claim is a system; the parts of the system are the roads and spurs; (2) the digestive system of a fish is a system; one can recognize the various parts; (3) the circulatory system in a body is a system; here the arteries, veins, and capillaries are its parts; (4) a steam engine is a system; in this instance, the cylinders pistons, valves, and so forth, are its parts; (5) the heating plant in a house is a system; here the furnace, hot-air ducts, thermostat, etc., are it parts; (6) a population of trout is a system; the parts can be thought of as age and sex categories of the fish, that is, young-of-the-year, mature males, and so on.

Some workers see general systems theory as a revolution in the quest for knowledge; they see a similar phenomenon in many fields of study—for example, in the growth of crystals, growth of cells, growth of individuals, growth of the economy; and they look at building systems which transcend the various disciplines to explain this phenomenon in general. They see this approach as a scientific revolution. Many who work with systems do not completely agree with such thinking, but feel that the systems approach is, in simpler terms, the method that rational people follow; that is, *it is the common-sense approach to large problems*. One ordinarily formulates a problem in some total manner, recognizes relevant parts to this whole and the interrelationships and interactions between the parts, rearranges these interrelationships into whatever models seem appropriate, and attempts to derive the logical outcome of the whole model. The computer provides a quick and easy way to sort out alternatives and hence has contributed much to this new approach. The complexities one gets into in even the simplest system most often result in models for which no analytical solutions are obtainable. With modern-day computers, however, one can frequently turn to a simulation of the systems and obtain empirical outcomes. For some purposes these empirical outcomes may suffice; at least one gets a clue to the functioning of the system.

Definition of Systems

In defining a system one must be careful to include only the parts that are relevant to the system for the purpose of solving the problem at hand, for the possible number of interrelationships gets extremely large as the number of system elements increases; hence one can very quickly have an unmanageable model to contend with. The first step in defining a system is to specify the purpose and restrictions under which the abstraction of the system will be created.

Systems Diagrams

It may be useful to represent systems in some pictorial fashion, and many schemes have been used for this purpose. Generally, what is done is to represent the parts of the system as points, circles, or boxes, and represent the interrelationships between the boxes as arrows or lines running between the boxes. An example, shown in Figure 1-1, taken from Regier and Henderson (1973), identifies a series of variables and interrelationships for use in analyzing fisheries problems.

Systems Analysis

Since there are no general methods of solution to cover all systems— because of the complexity of the problems—a start can be made by analyzing the systems approach. This is done by first asking the right questions, by cycling through a procedure of: (1) defining objectives; (2) designing alternative systems to achieve these objectives; (3) evaluating the alternatives in terms of their effectiveness and costs; (4) questioning the objectives and other assumptions underlying the analysis; (5) opening new alterna-

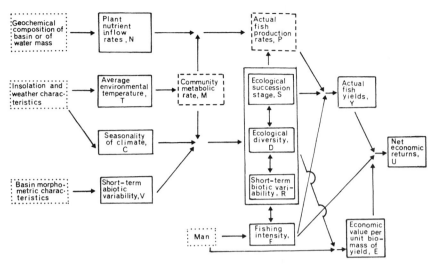

Figure 1-1. Schematic representation of variables and interrelationships to provide a basis on which to examine ecological aspects of fisheries problems in general. Courtesy of H. A. Regier and H. F. Henderson, 1973, Trans. Amer. Fish. Soc. 102(1):58.

tives; and (6) establishing new objectives. At step (6) the analyst re-enters at step (1) and continues until a satisfactory formulation of the problem is obtained, at which point it is apparent what are the right questions to ask. However, objectives will differ, depending on the role the individual plays in the organizational structure. If he is a fisherman, his goal would be to manage the species in such a way as to maximize his own chances of quality fishing; if he is a fish manager, his goal might be to manage the population to obtain maximum numbers harvested; the director of the management agency wants the fish managed in a way that yields the fewest complaints from the fishermen and from the politicians above him; and, the politician who has control over the department has as his management goal the method that keeps the public happy so they will re-elect him to office. In any case, depending on who he is, the goals are frequently determined for him. There is a rigid set of constraints placed upon every individual.

Once the goals are defined, and the formulation of the problem is clear, the next step in the systems approach is to construct a mathematical model to represent the system under study. From the model it should be possible to derive optimal solutions, which should then be tested thoroughly, and a study made of methods for implementation of the solution. The recommended solution should be tried in a limited or controlled situation to confirm its correctness, after which the solution should be publicized and recommendations made for its general use.

Some disadvantages to systems analysis are the tendency to model away the unknown, the tendency to ignore the probabilistic aspects of the system, and ignoring or not accounting for real uncertainty because of lack of knowledge.

A most important step in the solution of a problem involves comunication of research results to interested parties. Unprocessed data, mimeographed reports, and intra-agency reports frequently result in wasting of effort. Unless research findings are communicated in a quality scientific publication, much of their value will be lost. Of particular importance is the response of other scientists to the work done. The close scrutiny by other scientists is invaluable in obtaining criticism of, and reinforcement for, your methods and conclusions. Wide distribution of research findings is stimulating and provides the catalyst for a profitable exchange of ideas. Further, in fishery science the responsibility of the biologist should usually include a popular explanation of his findings, so fishermen and other citizens interested in natural resources can understand and appreciate management recommendations. This is especially true in those areas where public funding is required; it is important for concerned citizens to know what is being done, for an understood and appreciated program is more likely to be a supported program.

Proper conduct of research may require years of work and study. Too frequently today a biologist is considered fortunate if his proposal is funded for a three-year period. But in the variable world of natural resources this is simply too short a time to provide for proper replication and confirmation of results. Surely the lesson is plain. Only long-term research results can plug the holes still remaining in the protective dike of fishery management.

References

Davis, H. S. 1938. Instructions for conducting stream and lake surveys. U.S. Fish Wildl. Serv., Bur. Fish., Fish. Circ. 26. 55 pp.

Embody, G. C. 1927. An outline of stream study and the development of a stocking policy. Cornell Univ., Aquicultural Lab. 21 pp.

Hazzard, A. S. 1935. Instruction for lake and stream work. U.S. Bur. Fish., Cir. 1935:1–34. Mimeogr.

Huet, M. 1959. Profiles and biology of western European streams as related to fish management. Trans. Amer. Fish. Soc. 88(3):155–163.

Leger, L. 1910. Principes de la methode rationnelle du peuplement des cours d'eau à salmonides. Univ. Grenoble, Trav. Lab. Piscic., fasc. 1:531.

Maine Department of Inland Fisheries and Game. 1939-1946. Biological surveys of Maine lakes and ponds. Augusta, Maine, Fish Surv. Rep. 1-7.

_____. 1953-1977. Maine lakes: a sportsman's inventory. Augusta, Maine.

Moyle, J. B. 1956. Relationships between the chemistry of Minnesota surface waters and wildlife management. J. Wildl. Manage. 20(3):303-320.

New Hampshire Fish and Game Department. 1937-1939. Biological surveys of New Hampshire, by watersheds, 1936-1938. Surv. Rep. 1-4.

New York State Conservation Department. 1927-1940. Biological surveys of New York, by watersheds, 1926-1939. Supplements to the Annual Reports.

Regier, H. A., and H. F. Henderson. 1973. Towards a broad ecological model of fish communities and fisheries. Trans. Amer. Fish. Soc. 102(1):56-72.

Smith, L. L., Jr., and J. B. Moyle. 1944. Stream improvement on the north shore watershed. A biological survey and fishery management plan for the streams of the Lake Superior north shore watershed. Minnesota Dep. Conserv., Tech. Bull. 1:147-164.

Stroud, R. H. 1953. Spot-poisoning applied to the Massachusetts lake and pond fisheries survey. Prog. Fish-Cult. 15(1):3-10.

Thorpe, L. M., D. A. Webster, G. W. Hunter, III. 1942. A fishery survey of important Connecticut lakes. Connecticut Geol. Natur. Hist. Surv., Bull. 63. 339 pp.

Van Deusen, R. D. 1953. A simplified technique for classifying streams useful in fishery and related resource management. Prog. Fish-Cult. 15(1):14-19.

2

Characteristics of Fisheries

A fishery is the complex of interactions within and between the population(s) of fish being harvested, the population(s) of fishermen, and the environment of each. Fishery management ranges from an individual fisherman's concern to problems of international magnitude.

Fishes, of all the animals and plants in the sea, are the most important food source for man. We need not preserve fish by not using them, but rather must develop ways to use them so populations continue to provide optimum sustained yield. Unless some stress or limiting factor should upset the usual conditions, fish populations tend to remain in balance with their environment. We've come a long way since early biologists impressed by the vastness of the oceans assumed marine resources were inexhaustible. Today the disappearance or near extinction of the flora and fauna of our planet are an all too familiar concern.

We must continually recognize that fisheries and economics are inseparable whether we are dealing with the commercial or sport fishery industries.

Fish Population Data

Most of the fish species sought by recreational or commercial fishermen occupy a relatively high position in the food pyramid. For example, elements in the food chain leading to largemouth bass production are likely to include phytoplankton and/or bacteria, zooplankton, aquatic insects, and smaller fish species. Production at each of these levels is influenced by such processes as predation, competition, reproduction, growth, and all the environmental factors that affect each. So there are a vast number of physical, chemical, and biological interactions that ultimately determine

21

the production of a particular fish species, but which are difficult or impossible for the fishery biologist to observe or measure directly. Because the biologist—a terrestrial animal—is working in a foreign (aquatic) environment, he must use the approach of periodically taking "blind" samples of the fish population, from which he tries to deduce what processes have operated to produce the apparent differences between one sample and the next. In the more restricted aquatic environments (bays, small lakes, ponds, and streams) the fishery biologist may be able to collect larger samples more easily than can the biologist working with terrestrial animals, but the terrestrial biologist usually can observe his subjects more readily than can the aquatic biologist. Thus, the fishery biologist's approach to research tends to be more deductive, that of the wildlife biologist more observational.

Fish-population data developed by this sampling process, unfortunately, show great variability between perceived effects and apparent causes—and hence uncertain conclusions—and require a long time, indeed years, to determine true causes of perceived effects. Such barriers to assembling adequate research data make it clear that a blind, unreasoned "trial and error" approach to managing a fishery is usually a poor one, for two reasons: it often takes years to determine the true results of the trial management measure, and errors, once ascertained, may take more years to correct. At the same time, fishery biologists are increasingly confronted with situations involving conflict between competing uses of aquatic resources, where quick decisions are required. But decisions cannot be delayed for the five or ten years needed to obtain the research data needed for a decision based on firm biological evidence.

When faced with the choice of making a quick decision or having the decision made without his input, clearly the fishery biologist must choose the first alternative. In such cases he must rely heavily on judgment, based on past experience, and most especially on prudence, recognizing the frequency with which ecosystems have been damaged in ways that were at the time unforeseen and unforeseeable.

Recreational Fisheries

Much reliable information on trends in recreational fishing in the United States has been produced by the comprehensive National Survey of Fishing and Hunting which has been made for the U.S. Bureau of Sport Fisheries and Wildlife in 1955, 1960, 1965, and 1970. We can expect additional reliable data when the 1980 survey data are published.

Fishing is one of the most popular forms of outdoor recreation. In 1970 one of every three U.S. men and one of every nine women fished, and the country's total of 33,158,000 anglers spent nearly 5 billion dollars on recreational fishing. Comparing data from the 1960, 1965, and 1970 surveys, the number of fishermen has increased steadily (up 31 percent since 1960). In 1970 about 113.7 million angler-days were spent in salt-water fishing and 592.5 million angler-days in freshwater fishing. The study of sport fishing in the United States conducted by the Outdoor Recreation Resources Review Commission (1962) indicated that species other than salmonids attract about 70 percent of the freshwater fishing effort, and make up 75 percent (by weight) of the freshwater catch. Reservoirs and impoundments provide nearly 40 percent of the United States' freshwater fishing.

While the number of anglers in this country is increasing five times faster than the population, the total area of fishing waters is at best static, with additions in the form of impoundments being offset by subtractions in the form of waters degraded by pollution and others rendered inaccessible by posting of privately owned lands. The critical problem for public fishing is to find the funds to maintain and improve aquatic environments, and to purchase access to them. The 1970 survey revealed that money spent for licenses is the smallest category of fishing expenditures. On the other hand, expenditures for special privileges to fish on private waters appear to have increased at least threefold in the past five years.

Commercial Fisheries

In contrast to the continued rapid growth of recreational fishing, commercial fishing in the United States, overall, has remained nearly static for the past ten years or longer. Indeed, the sport-fishing catch is reported to greatly exceed the commercial catch of many ocean fishes, including mackerel, bluefish, striped bass, croakers, snappers, mullet, groupers, and weakfish. Estimates indicate that sport fishermen take as much as seven times the weight of striped bass that commercial fishermen do. Estimated saltwater catch by U.S. sport fishermen is 1.6 billion pounds, and freshwater catches would bring the total to over 2 billion pounds annually or about 30 percent of the total American catch. Saltwater anglers take from three to nine times as much fish as the freshwater angler.

One of the reasons our commercial fishery has not grown is that per capita fish consumption in the United States has remained fairly stable at a low average of 5 kg as compared to 39 kg in Iceland and 74 kg in Japan.

There are indications of a rising consumption in United States which we are meeting by becoming the largest importer of fish in the world. But a nonexpanding market is by no means the only problem faced by the fishing industry. Other problems are rooted in the basic characteristics of many commercial fisheries.

One large subset of problems in managing commercial fisheries consists of those directly associated with the fishing process itself:

1. It is difficult or impossible to forecast catch, because:
   (a) Abundance of fish stocks in a particular locale can vary in many unpredictable ways (such as migrations or undetected year-class failures associated with unpredicted and/or unrecognized environmental changes).
   (b) Catchability may vary with changes in weather, and with the characteristics of the locality fished.
2. There are frequently large discrepancies between the distribution of fish supplies (stocks) and the distribution of demands (markets):
   (a) Geographically—great distances occur between the fishing ground and the market, and
   (b) Temporally—fish supplies are too often abundant when demand is low, or the converse.
3. A further complication is that fish is a highly perishable product. (Chemical changes often can be measured within a few minutes following death.) This adds to the difficulties caused by the discrepancies noted above and penalizes the small operator lacking equipment for sophisticated fish-processing techniques such as onboard freezing or canning.
4. Fisheries are common-property resources (open-access resources). That is, a commercial fishery resource is usually open to exploitation by all (unrestricted entry). Such resources tend to become overexploited, chiefly because no one individual can control either the resource or its exploitation. As exploitation increases with more fishermen entering the fishery, the catch per unit of fishing effort starts to decline. But it is obvious to each individual fisherman that if he were to reduce his own fishing efforts in the interests of conserving the resource, he would only be penalizing himself. Any reduction in his catch would produce a commensurate increase in his competitors' catches, but no increase in the fish stock. Since each operator is bound by the same reasoning, exploitation can only continue to increase. This problem, fundamental to all resource management, is excellently described in Garrett Hardin's

"Tragedy of the Commons" (1968). The larger the common-property resource and the more diverse its exploiters (culturally, economically, politically), the more acute the problem and the more difficult to solve. An international high-seas fishery exemplifies this situation.

Removal of fisheries as common-property resources is discussed by Francis T. Christy, Jr., in J. C. Mundt (see references). Direct or indirect entry controls administered by a public agency would remove the common-property characteristics by limiting the number of fishing licenses, vessels, or gear, or by reducing the incentive to fish by imposing taxes or user fees, or by dividing up the resource among fishermen with quotas.

Christy lists four important criteria in evaluating entry controls: (1) economic efficiency, (2) biological effectiveness, (3) equitable distribution of values, and (4) political feasibility. Careful analysis of the alternatives available for removing commercial fisheries as common property supports the combination of entry controls that most closely upholds a free market for the resource. The problem of limited entry make excellent illustrations of the dilemma natural resource managers so frequently face. Advantages and disadvantages of general areas of control are summarized:

	Advantages	Disadvantages

License limitations	Most direct
	Most closely related
	  to present methods
	Gradual reduction possible
	Politically most feasible
		What should be licensed?
		  a. Number?
		  b. Size?
		  c. Gear?

Taxes and fees	Extracts economic rent Equitable	Fluctuating fisheries High tax necessary Difficult to tax already depressed industry

Fisherman quotas	Individual share of annual
	  yield
	Biologically effective
	Own quota but not actively
	  fishing
		Difficult to adopt initial
		  allocation

Other commercial fishery problems are associated with marketing and with other aspects of the fisherman's livelihood:

1. The total proportion of the retail price of fish that is preempted by processors and distributors tends to be relatively high and inflexible. Some of the reasons for this are perishability and the uncertainty and geographical separation of both fish stocks and markets, mentioned earlier.

2. Most commercial fishing operations, like most farming operations, require such a major investment in equipment that, on having made the investment, the average operator cannot change jobs. Moreover, since fishermen typically harvest a publicly owned resource they have no control and must many times accomodate to the regulations imposed by the towns, state governments, or the federal government.
3. Fishing is a high-risk occupation in which the chances for damage or loss of equipment or a catch are many and large.
4. The above three factors combined to produce a fourth: very small changes in the retail price of fish can drastically affect the fisherman's income. Consider the following reasonably realistic example: a particular fish retails for 86¢/pound. Of this the fisherman gets 4¢/pound, and the remaining 82¢ goes to the various processors and distributors. Now, due to some slight economic shift, the retail price drops by 2¢/pound, and the fisherman absorbs half of that shift. This means his earnings have dropped from 4¢/pound to 3¢/pound—a cut of 25 percent.

Too frequently, the net result of all the preceding characteristics of fisheries and the fisherman's livelihood is to place the commercial fisherman in a cost-price squeeze, with fishing costs rising but the prices received for fish remaining relatively stable, because of:

1. Low demand
2. Rising distribution/processing costs
3. Increasing foreign competition
4. Increasing competition from agriculture (notably, poultry)

Since the fisherman cannot change jobs often, his only recourse for economic survival is to increase his catch by increasing his fishing effort. These economic factors aggravate the problems inherent in the "commons" situation, with the result that many commercial fisheries tend to be pushed to the point of economic marginality, and then to biological marginality (overexploitation). The northwest Atlantic haddock fishery is a case in point.

Increasing competition in commercial fisheries puts a premium on fishing efficiency, and hence on gear technology. There is now a tendency for the smaller operators with more conventional gear to be replaced by very large operators with, for example, a fleet of highly technological fishing vessels servicing a mother ship equipped to process and store the catch.

Although recreational fishing in the United States is growing rapidly and is increasingly limited by the supply of fishing waters, by contrast, our commercial fishery is not growing because of a combination of factors including basic physical, biological, and sociological characteristics of the commercial fishing process, fish consumption patterns in the United States, and economic characteristics of the fishery.

The Fisheries Conservation and Management Act of 1976

The Fishery Conservation and Management Act of 1976 was implemented on March 1, 1977. This act under the administration of the National Oceanographic and Atmospheric Administration and the National Marine Fisheries Service by establishing the 200-mile fisheries conservation zone put 10 percent of the world's marine fisheries and 20 percent of the worldwide offshore groundfish resources under national management. Quotas in fisheries open to foreign fishermen will be controlled by quotas, permit fees, poundage fees, and observer fees. Final implementation of the Act will be the responsibility of eight Regional Fishery Management Councils. The 108 voting members of these councils will be made up of federal, state, and private citizens. They are expected to produce about 75 management plans which should include the following fishery assessment information:

1. Reasonably accurate inventory
   a. License all fishermen
   b. Adequate systems of collecting effort and catch statistics
2. Identification of a unit or population
   a. Distribution
   b. Spawning areas
   c. Age composition
   d. Morphological and physiological characteristics
3. Population parameters
   a. Growth
   b. Food
   c. Natural mortality
   d. Recruitment
4. Effects of "natural" environmental factors (limiting)
   a. Temperature
   b. Bottom conditions
   c. Currents
   d. Seasonal changes
   e. Pollutants

5. Effects of human activities
   a. Catch/effort
   b. Yield
   c. Characteristics of fishery
6. Interrelationship among species
   a. Food chains
   b. Trophic levels
7. Enforcement of regulations
8. Fishery management policy

These management plans will represent an ambitious but long overdue assessment of our commercial fisheries. Further, they will focus attention on the combination of factors including basic physical, biological, and social characteristics of the commercial fishing process, fish consumption patterns in the United States, and the economic characteristics that have affected the growth and health of our commercial fisheries.

Further impact on the consideration of commerical fisheries will be the effect of the principle of optimum sustained yield replacing the long-standing maximum sustained yield which concentrated largely on fish rather than fishermen and largely avoided the more complex issues of society, economics, and politics. More exposure of fishery problems will result in more support.

References

Acheson, J. W. 1975. Fisheries management and social context: The case of the Maine lobster fishery. Trans. Amer. Fish. Soc. 104(4):653–668.

Anderson, L. G. 1980. Necessary components of economic surplus in fisheries economics. Can. J. Fish. Aquat. Sci. 37(5):858–870.

Borgstrom, G. 1964. Japan's world success in fishing. Fishing News (Books), Ltd., London. 312 pp.

Borgstrom, G., and A. J. Heighway, eds. 1961. Atlantic Ocean fisheries. Fishing News (Books), Ltd., London. 335 pp.

Clark, C. W. 1976. Mathematical Bioeconomics—The optimal management of renewable resources. John Wiley and Sons, New York. 352 pp.

Crutchfield, J. A., ed. 1965. The fisheries: problems in resource management. Univ. Washington Press, Seattle. 136 pp.

Donaldson, E. M. 1976. Aquaculture Symposium, 13th Pacific Science Congress. J. Fish. Res. Bd. Can. 33(4):875–1119.

Hardin, G. 1968. The tragedy of the commons. Science 162:1243–1248.

Hickling, C. F. 1961. Tropical inland fisheries. John Wiley and Sons, New York. 287 pp.

McHugh, J. L. 1975–1976. Does fishing have a future? Search 2:20–27.

Mundt, J. C., ed. 1974. Limited entry into the commercial fisheries. Institute Marine Studies, Univ. Washington, Pub. Ser. IMS-UW-75-1. 143 pp.

Pearse, Peter H. 1979. Symposium on policies for economic rationalization of commercial fisheries. J. Fish. Res. Bd. Can. 36(7):711–866.

Regier, H. A. 1976. Science for the scattered fisheries of the Canadian interior. J. Fish. Res. Bd. Can. 33(5):1214–1232.

Ricker, W. E. 1969. Food from the sea. In Committee on Resources and Man. Nat. Acad. Sci.-Natur. Resour. Counc. Resources and Man. W. H. Freeman & Co., San Francisco, pp. 87–108.

Sinclair, W. F. 1978. Management alternatives and strategic planning for Canada's fisheries. J. Fish. Res. Bd. Can. 35(7):1017–1030.

Skud, B. E. 1976. Jurisdictional and administrative limitations affecting management of the halibut fishery. International Pacific Halibut Commission. Sci. Dep. 59. 24 pp.

Tont, S., and D. A. Delistraty. 1977. Food resources of the oceans: An outline of status and potentials. Environmental Conservation 4(4):243–251.

U.S. Fish and Wildlife Service. 1962. Sport fishing—today and tomorrow. Bur. Sport Fish. Wildlf., Outdoor Recreation Resour. Rev. Comm. Stud. Rep. 7. 127 pp.

_____. 1964. America goes fishing. Bur. Sport Fish. Wildlf., Conserv. Note 14. 8 pp.

_____. 1972. National survey of fishing and hunting, 1970. Bur. Sport Fish. Wildlf., Resour. Publ. 95. 108 pp.

Walford, L. 1965. Research needs for salt-water sport fisheries. In Dep. Fish. Canadian Fish. Rep. 4. Ottawa, Ont., pp. 81–90.

3

Fishing Gear

Fishery management requires a knowledge of fishing gear. There is great divergence in the efficiency of different forms of gear, in their adaptability to certain conditions, and in their desirability for specific jobs. Although it would take several volumes to describe the various kinds of fishing gear and their modifications, a comparatively few types take the major share of the catch. This chapter is necessarily a mere outline, and no attempt has been made to include all minor types of fishing gear. A general classification includes: (1) impaling gear; (2) hook-and-line gear; (3) the maze; (4) gill nets; (5) encircling nets; (6) towed nets. In addition, many types of research gear have been developed as modifications of the basic types described in this chapter.

We can be certain that more research and experience will produce more sophisticated methods of taking fish in response to world demands for increased protein. Generally, fishing gear of the future will require less energy per unit of effort, specificity for species, selectivity of catch as to size and age, and increased survival of the incidental and accidental catch.

IMPALING GEAR is any form of gear by means of which a fish is impaled on a sharp shaft.

The spear or gaff is very ancient gear used where fish, especially large ones, are concentrated in a small area. It is used to capture freshwater and anadromous fish on their spawning beds or when they are handicapped while negotiating stretches of swift water. This type of gear is often used to spear fish through holes in the ice or where fish are resting in stream eddies. It may consist of a metal head of one-to-several tines or points, usually barbed, on a long handle. Against larger fish the head usually slips into a socket on the end of the handle from which it becomes detached as soon as the fish is struck. The head may be attached to the handle by a short line, or it may become entirely detached but have a long line with which the fish is landed. The gaff is a modified form of spear, consisting usually, but not invariably, of a single recurved tine. The tine is seldom barbed, as dependence is placed on the recurved angle to prevent escape. As used by the northwest Indian tribes, it is fished with a long handle and detachable head as just described for the spear. It is very effective against salmon held up by obstructions, but many injured fish escape. A gaff attached firmly to a short handle is widely used to land rod-and-reel fish as they are brought to boat or to the shore.

Spears are used extensively in taking fish among coral reefs and are employed in the popular sport of shallow diving, as with scuba gear or snorkeling.

The harpoon is a spear modified for throwing (Figure 3–1). The rear end of the harpoon or "lily iron" has two broad thin "barbs" or blades. These differ from actual barbs in that the rear edges are dull and are not only broad but also curve at a slight angle from the head. The line is attached in front of these rear blades. Thus when a fisherman, standing in the "pulpit" protruding past the prow of his vessel, strikes deeply into a tuna or swordfish, the backward pull of the line turns the back end of the lily iron

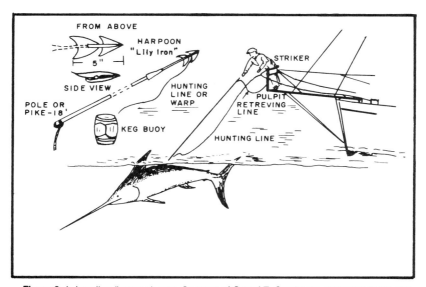

Figure 3-1. Impaling (harpoon) gear. Courtesy of Gustaf T. Sundstrom, illustrator, 1957. U.S. Fish Wildl. Serv., Bur. Commer. Fish., Circ. 48.

into the flesh and anchors it securely. Large fish are often played by throwing overboard a keg or other float to which the end of the line is attached. This tires the fish but does not pull with enough force to dislodge the iron. The harpoon, for example, accounts for the largest share of the swordfish catch. While lying motionless or playing slowly about the surface, the swordfish is vulnerable to approach by the harpoon fisherman.

The bow and arrow deserve mention as an impaling type of sports gear. HOOK-AND-LINE GEAR, used in many modern fisheries, is ancient in origin and was probably used by practically all primitive peoples. Anthropological studies indicate that primitive hooks were made of pieces of shell, hardwood, or bone, or combinations of these items.

The jig consists of one-to-several bare hooks attached, usually by short lines or ''gangings,'' to a weighted line. Sometimes the hooks are attached to a wire frame, The jig is fished by jerking it up and down in the water. Some species, for instance mackerel, are snagged as they strike at the bare hooks; some are caught because the jig passes through a dense school and the fish do not avoid the hooks.

The baited hook consists essentially of one or more simple hooks attached to various types and arrangements of lines to which bait is fastened as an attraction. The long line—trot line, flag line, or line trawl—is an adaptation of the older hand line, in which the main line or ground line is horizontal and carries many short vertical gangings, each with a baited hook (Figure 3–2). In fishing, this gear may be anchored or drifted. For bottom-feeding species it is usually anchored. First a float is thrown overboard and sufficient anchor line is payed out to reach the bottom, allowing some extra anchor line for changes in tide levels and to prevent dragging. A small anchor or grapnel is then attached. The float is usually a keg, a cork, or a mesh-enclosed glass float, although some floats are very elaborate. The float often carries a tall bamboo pole bearing a red or white flag, a light, and a radar target. A main line is then attached to the anchor and the line payed out, with its baited hooks spaced on short gangings, as the vessel moves ahead. As the end of each ''line'' is reached, another is attached until a long string of gear, often several miles in length and holding thousands of baited hooks, has been set. The lines composing a string are usually coiled on the boat deck in half barrels or tubs, sometimes in shallow baskets, and in the halibut fisheries on pieces of canvas called a ''skate,'' and are payed out as the vessel moves. For such long strings an extra float on a buoy line is attached at intervals depending on tide, weather, and the roughness of the bottom, to aid in finding the gear and retrieving it if the main line is broken in hauling. The end of the string is

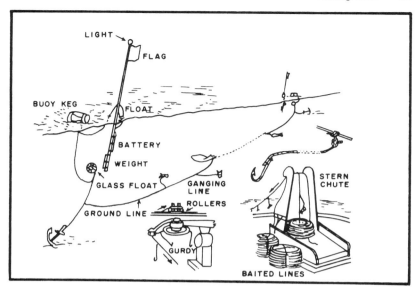

Figure 3-2. Hook-and-line (long line) gear. Courtesy of Gustaf T. Sundstrom, illustrator, Circ. 48, above.

also anchored and buoyed. By shortening the buoy lines of the floats, the line may be fished at any desired depth.

Such a string of gear is usually called a long line on the Pacific coast and a line trawl in the North Atlantic. It is also called a set line, and in inland waters the term trot line is more usual. The long line may be fished without being anchored, in which case it is called a drift line. This is extensively employed by the Japanese for fishing tuna; when the catch is collected, one end of the line is attached to the vessel, and smaller boats are used to overhaul the line.

Snap-on gear has provided certain advantages to the more traditional set-line gear. The ground line is one continuous line and can be stored efficiently on the same drum used for storing salmon gill nets, allowing more flexibility in fishing. Snap-on gear allows the baited hooks to be attached to the ground line as it is set with an opportunity to vary the distances between hooks. Retrieval is easier as the snap-on gangions are removed and stored on racks as the gear is brought in. Hook space varies with species and fisherman preferences. Hooks may be as close as 5 to 6 feet for pelagic species such as the densely schooled mackerel or salmon. For tuna, which are faster moving and less dense, the spacing may be as much as 600 feet, although the usual is between 80 and 200 feet.

Trolling gear involves towing hook-and-line gear through the water. The hook may be baited with live or artificial lures (Figure 3–3). In trolling for salmon or albacore the boats use several lines. One or two lines may be fished over the stern of the vessel, and others are attached to outrigger poles. In a regular trolling vessel each line is led through small blocks to its own small "gurdy" or reel. These reels are arranged along a shaft which is kept turning by the vessel's engine. Each gurdy has its own clutch so that each line may be reeled in individually. The depth of trolling is controlled by the vessel's speed and by the weight of the lead sinker.

Usually a boat fishes with lines at different depths and with different lures until fish start striking. All lines are then fished as close to the depth at which the fish are biting as can be accomplished without fouling the lines, and the lure is used that the fish are striking.

The gear is rigged in various ways to suit the individual fisherman. Usually a heavy spring or a short length of heavy rubber is attached to the outrigger. To the end of this is attached a small running block through which the heavy cotton line is run to a gurdy. To the end of the line are fastened a heavy swivel and then a sinker. The sinker varies in shape and may be keeled to keep it from turning in the water. Next follows a split ring

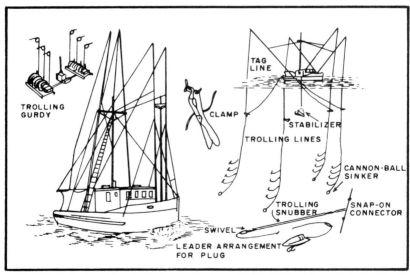

Figure 3–3. Hook-and-line (trolling) gear. Courtesy of Gustaf T. Sundstrom, illustrator, Cir. 48, above.

or another swivel and then a length of bronze, stainless steel, or piano wire to which the lure is attached.

Hooks used in trolling are rarely offset, and the point is usually long. Lures of many types are used; hooks baited with a whole small fish, hooks baited with a fillet from the side of a fish, metal spoons, metal spinners, feathered jigs, plugs, and so on. The type of lure used varies with the species, the season, the locality, and the whim of the fisherman. Artificial lures are preferred when the fish will strike them well, since they can be more quickly taken from the fish, they save time in rebaiting, and are less costly than bait which is sometimes difficult to obtain in suitable variety or quantity.

For small boats with only one or two lines, a ''spreader'' is sometimes used on the end of the line to prevent fouling. This is merely a piece of very stiff wire (at the least no. 10) bent to form two unequal-length arms and fastened at the apex of the arms to a swivel in the end of the line. To one arm is attached the lead followed by a short line, leader, and lure. The second arm is fitted in the same manner but with no lead.

Trolling is used extensively in taking chinook and silver salmon and albacore (*Germo alalunga*) on the Pacific Coast and is universally popular for sport purposes in taking many species of fish, such as tuna, mackerel, and trout.

Pole and line gear is merely the elaboration of the original hand line. Gear ranges from the bamboo pole to the expensive fly rod and reel of the trout and salmon angler.

MAZE GEAR consists essentially of equipment for leading fish into a situation or enclosure from which they cannot escape, or from which the avenue of escape is not readily apparent. It includes many varieties of fish pots, fyke nets, pound nets, and tidal weirs.

Pot gear are portable traps that fish enter, usually through a small opening, and with or without enticement by bait; they are used to capture lobsters, spiny lobsters, crabs, sea bass, eels, catfish, king crabs (Figure 3-4), octopi, carp, and live bait for angling. Pots are most effective in capturing slow-moving creatures that exist for the most part about, on, or just above the sea or lake bottom. The pots are usually small enough so that a large number can be piled on the deck of a small boat, and light enough so that they can be readily hauled aboard or set out in choppy weather.

For lobsters, fishermen prefer either the half-round or the rectangular pot. The half-round pot is usually about 70 cm long by 70 cm wide and about 50 cm high. A half-round bow is used at each end, and in the center. The trap is built of laths, usually about 3 cm wide by 1 cm thick, spaced

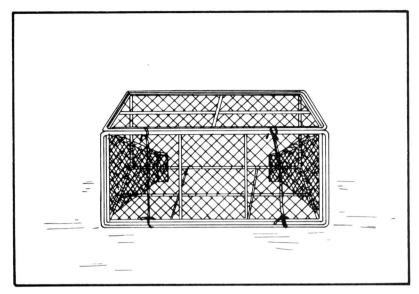

Figure 3-4. King crab pot

about the width of a lath apart. A curved door to permit removal of the catch is made of three laths running the length of the trap, and nailed to three curved cleats at one end and hinged on the other. The door is hinged with strips of leather or rubber, and fastens down with two wooden buttons. The center bow is placed a little off center from the middle of the trap, making two compartments called the "chamber" and the "parlor." The shorter section, the chamber, usually has an entrance on both sides of the trap. Each opening between the laths is usually about 20 cm high by 33 cm wide, and the bottom of the opening is about 6 cm above the floor of the pot. Each has a tunnel of netting about 18 cm deep ending in a ring 10 cm to 15 cm in diameter. The tunnels from opposite sides are tied together with brace lines to keep the netting taut. From this chamber the inner portion, or parlor, is entered by a longer tunnel or "head" of netting that has its base lashed to the center bow and extends lengthwise of the pot into the parlor. These traps may vary from 71 cm to 122 cm in length. Tunnels may be made of laths, and sometimes the parlor tunnel is replaced by a vertical self-closing lath door so that lobsters cannot escape.

The rectangular lobster pot is easier to make, to store, and to repair and is therefore becoming more popular. It is made about 76 to 102 cm in length, with the top usually about 10 cm narrower than the bottom.

In either case, the completed trap is weighted with a few bricks, concrete, or flat stones, and the buoy line is attached to a lower corner at the chamber end. Bait is kept on a hook or in a mesh bag attached to the center bow so that it hangs in the parlor entrance. The buoy line has a small float attached about a fathom from the pot to keep the lines from becoming cut or entangled on the rocky bottom. The surface end of the line is attached to a small buoy that usually has a spindle through it, to aid both in finding it and in retrieving the buoy. Pots are often fished in long strings, trawl fashion. About 10 to 15 traps are fastened by ganging lines about two fathoms long to a heavier trawl line. Pots are spaced about 5 to 10 fathoms apart.

Crabs are often fished with pots, and although they can be taken in lobster pots, special types are also used. Thus in Chesapeake Bay a trap 0.2 m square is made of 3.8-cm mesh, of doubly galvanized 18-gauge poultry wire. If 2.5-cm mesh is used, the pots are also effective for large eels. A horizontal wire partition 20 cm off the bottom divides the trap into an upper and a lower chamber. The lower chamber is entered from all four sides through small wire tunnels. The partition bulges upward in a fold about 20 cm high for about one third of its width. In the top of this fold are two small openings that give access to the upper chamber.

Spiny lobsters can be taken in pots similar to those used for lobsters, but they are also taken by leaving old ice-making molds on the sea bottom into which they retreat to hide. When the mold is hauled, they cannot climb the sides to escape. Similarly, the Japanese use large terra-cotta vases which octopi enter in seeking a convenient shelter.

Cylindrical pots are used extensively for capturing eels, and in Chesapeake Bay they are used to take a large share of the catfish. Eel pots can be made of almost any material, but wire is the most easily obtainable. A typical pot is 76 cm long and 23 to 28 cm in diameter. Three 10-mm stock metal rings are covered with 16-gauge 13-mm galvanized wire netting. Two flat iron bars, each 76 cm long, 3 cm wide, and 6 mm thick are used for reinforcement. The center ring is 25 cm from one end, so that the first chamber is 25 cm long, the second 50 cm long. The first chamber is entered through a knit web tunnel about 8 cm in depth with an opening of 10 cm attached to a metal ring. The second tunnel is about 18 cm deep, with an opening only 5 cm in diameter. This opening is sometimes made larger and then stretched into an oval shape. The catch is removed through an end opening in the larger chamber. The pots are fished in a manner similar to that of lobster pots. Bait is suspended in front of the second tunnel.

Many eel pots are much smaller for ease in handling, and some are made of oak splints. These lighter pots are often fished trawl fashion, about 25 pots on 3-meter gangings fastened about 10 m apart. Rectangular and half-round wire eel pots are also in use. One form of rectangular pot is about 76 cm long, 30 cm wide, and 20 cm high and is fished with only one long funnel in one end. Catfish traps are usually larger, about 1 m in length and 45 to 50 cm in diameter; they are similar to the cylindrical eel pot. Various types of pots of all shapes and sizes are used in different localities. In the tropics, where pots are used extensively for taking reef fishes, they are called basket traps. These are usually made of split bamboo, wood, or vines, but wire is also used.

Fyke nets are essentially shallow-water gear, for they are difficult to set effectively in deep water (Figures 3–5, 3–6). The fyke net is used extensively in river fisheries, and it is best adapted for use in a fair current, and under such conditions it is sometimes used without any wings or leader.

The simplest form of fyke is merely a long net bag with a rectangular opening, each side of the opening being lashed to a stake. If there is a current flowing into the opening, such a net may take a number of different species, especially shrimps and prawns. A more common type of fyke net is a long bag mounted on one-to-several hoops which serve a double purpose: they keep the net from collapsing, and they form an attachment for the base of net funnels to prevent the fish from escaping. The catch is

Figure 3–5. Maze (Maine fyke net) gear

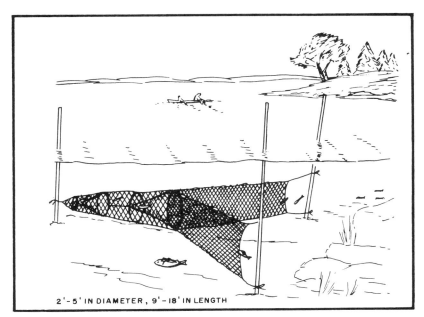

2'-5' IN DIAMETER , 9'-18' IN LENGTH

Figure 3-6. Maze (fyke net) gear. Courtesy of Gustaf T. Sundstrom, illustrator, Circ. 48, above.

removed from the last pocket. The fyke usually has short, vertical, net wings set obliquely on either side of the mouth of the bag. As fish moving with the current strike these wings they are deflected toward the mouth of the net. In the catfish (*Ictalurus catus*) fishery in the Sacramento River, fykes 7½ m long with hoops 1½ m in diameter are used.

In swift currents the wings are necessarily short and the bag very long so as not to offer too much resistance to the water. However, fykes are also used in many shore fisheries, especially for taking flounders and other bottom fishes. In such situations very long wings can be used, since the currents are fairly slow. These wings are heavily leaded on the bottom and have cork or glass floats on top. Instead of stakes the net is set with anchors.

Pound or trap gear is very effective for migratory species that tend to follow a shore line. There are scores of designs and variations (Figure 3–7), but the basic principle is the same for all. Fish moving along the shore encounter a lead of brush, netting, or wire. They follow this lead in an attempt to get past it and are led into one or more enclosures from which they find it difficult to escape. Sometimes the lead is placed on a fishing

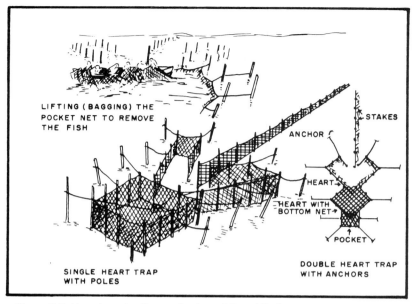

LIFTING (BAGGING) THE
POCKET NET TO REMOVE
THE FISH

STAKES

ANCHOR

HEART

HEART WITH
BOTTOM NET

POCKET

SINGLE HEART TRAP
WITH POLES

DOUBLE HEART TRAP
WITH ANCHORS

Figure 3-7. Maze (pound net) gear. Courtesy of Gustaf T. Sundstrom, illustrator, Circ. 48, above.

bank offshore, and either an enclosure is placed at each end, or one end terminates in a huge rounded hook or ''jigger'' which tends to shunt fish back in the direction from whence they came.

The simplest maze gear is the brush weir used in the Maine sardine fishery. These are built of wooden stakes and saplings driven into the bottom in shallow waters. The young herring encounter the lead which they follow toward deeper water, finally passing into an enclosure of brush or netting. The passage into the enclosure may be provided with a net apron that the fisherman lowers after a school has entered the enclosure or pound to prevent its escape. The fish concentrated there may then be removed with a small seine.

A variation of this type is the tidal weir, which takes advantage of shallow bars or flats. Fish enter the trap over a considerable area at high stages of the tide. As the tide falls, the shallow bars or flats are exposed first, but the fish are concentrated in a deeper portion, and escape is cut off by a barricade. Such weirs, built of stone or bamboo, are also used in the islands of the Pacific.

The typical pound net used in the Great Lakes and along the Atlantic seaboard is built of wire, cotton, or nylon netting (the mesh size varying with locality and species sought), except for the final enclosure which is usually a tarred cotton webbing. Piling is driven into the bottom in depths up to 25 m or more, and the netting is hung from these piles.

The most elaborate traps or pound nets are those used in the Pacific salmon fisheries. They are of two kinds, pile and floating. Pile traps often have a lead of half a mile or more. Schools of salmon moving with the tide encounter the wire lead. They hesitate and sometimes sound, looking for a deep opening. Usually they start toward deep water following the wire at a distance of 3 to 6 m. This leads the school into the huge funnel formed at the end of the lead and the top of the "heart." The entrance to the heart is a wide opening usually 3 to 5 m across. Traps located where the flood and ebb tides flow in opposite directions usually have an opening into the heart from both sides of the lead, but if the traps are in a tidal eddy, the single opening is on the uptide side. The heart has no bottom, but the tunnel from the point of the heart into the next chamber or "pot" extends only part of the way to the bottom and has no floor. The wire tunnel from the heart to the pot has a narrow mouth about 46 to 76 cm in width, and the pot itself is about 4 m square, is usually made of wire, and has a bottom.

The salmon are now led from the pot into a similar square chamber called a "spiller," usually made of heavily tarred webbing, although some have web sides and a wire bottom. The spiller is placed alongside the pot and is connected by a web tunnel with a narrow opening usually 15 to 30 cm in width. The opening is adjustable and may be entirely closed. The salmon enter the trap with the tide, but after being confined they commence to swim continually against the tide. Thus in a tidal eddy the spiller must be located on the uptide side of the pot, and in other situations the connecting tunnel is alternately opened and closed as the current changes direction. In exceptionally favorable locations traps are provided with two spillers to prevent overcrowding during the peak of the run.

The four corners of the spiller are held down by ropes running through the ring of a "downhaul" which is a chain that slips down the corner piling. To remove the fish, the four corners of the bottom are hauled close to the surface by hand windlasses. The web is overhauled by men on a large skiff or small scow inside the spiller until the salmon are confined in a bag made by one side of the spiller. They are then removed by lowering the webbing on the outside of the spiller low enough to fasten it to the side of the transporting scow or vessel. A "brail" or apron, consisting of a rectangular piece of webbing, has one end fastened to the scow. The other end is fastened to a heavy pipe that is lowered into the "bag" of heavily concentrated fish. The pipe end of the brail is then hoisted by a power winch. Each time it is hoisted, hundreds of struggling salmon are dumped into the scow.

Because of the difficulty or impossibility of driving piling in many favorable fishing locations which have hard or rocky bottoms, the *floating* trap was evolved. In this trap the heart, pot, and spiller are hung from a floating framework of huge logs anchored in position by heavy wire cables to huge anchors made of concrete. The lead is supported by a cable stretched over floats.

The deep trap net, also sometimes termed a "submarine pound net," was developed in the Great Lakes. It differs from the typical pound net held by stakes or pilings, and from the floating trap, which has a rigid floating framework, by being held in position and in shape entirely by anchors and buoys, and by having a top of webbing on all but the heart. It can thus be fished in water of considerable depth, regardless of the hardness of the bottom, and can be readily taken up and shifted to a new location in order to follow the concentrations of fish. Because of this ability to follow the fish concentrations, it has proved very effective for whitefish and can be used to catch them during periods when the fish are at levels too deep to be available to pound nets.

Deep trap nets may have leaders 8 to 10 m long and 6 to 15 m deep, and hearts of the same depth as the leaders with a spread of about 30 m at the tips and a length of about 14 m. In some nets the outside walls of the hearts are extended about 7 m, using single thicknesses of netting known as wings. The "hood" or "breast" connecting the heart with the tunnel has both a top and bottom and a length varying from 7 to 8 m. The tunnel, varying from 14 to 23 m in length, is tapered from the same depth as the heart to form a 0.23-meter-square opening extending inside the pot or crib which is roughly cubical, varying from 5 to 15 m in depth and 9 to 12 m in length. In lifting the nets, the main anchor line is brought to the surface, and the vessel is worked under or alongside the pot from which the fish are removed through laced openings in the bottom.

GILL NETS are used to take a great variety of fishes: salmon, cod, mackerel, pollock, herring, whitefish, smelt, crabs, sharks, tuna, sardines, and spiny lobsters. In the gill net the fish becomes caught by the mesh of the net in trying to swim through (Figure 3–8). Usually, if the mesh is the correct size for the size of fish sought, the fish is able to get its head through a mesh but its body is too large, and when the fish attempts to free itself the twine slips under the gill cover, preventing escape. Many fish will be caught around the middle of the body, and some may be caught by the twine entangling the maxillary bone of the jaw or the teeth.

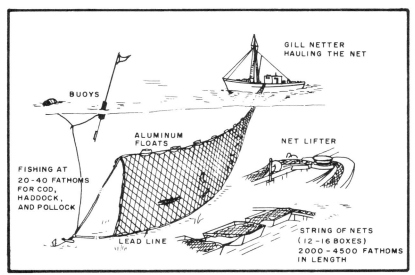

Figure 3–8. Entangling (gill net) gear. Courtesy of Gustaf T. Sundstrom, illustrator, Cir. 48, above.

Gill nets vary in size of mesh in accordance with the size of the individual or species sought. The twine also varies in size and strength, although in general the larger-sized mesh is made of heavier twine. Gill nets are usually made of nylon. The ordinary gill net consists of a single wall of webbing kept vertical in the water with sinkers and floats. The webbing is hung to a cork line at the top and a lead line at the bottom. At least 50 percent more net is used than the length of these lines so that the net will hang full. A loosely hung net is more efficient.

A drift net may be fished in either of two ways: at the surface or in midwater, always suspended vertically with a cork line at the top edge and a lead line on the bottom. In fishing below the surface, the size and number of floats and weights are adjusted so that the net will sink; the desired fishing depth is maintained by lines fastened at intervals from the cork line to large surface buoys.

A stake net is fished by attaching the net to driven stakes. It is used principally in shallow bays and estuaries. A diver net, used often in salmon rivers, drifts on the bottom, the weights applied being a little less than needed to stop the net from dragging. A sink or anchor gill net is set in a stationary position on the bottom by anchors. It is also referred to as a set net. Circle gill netting consists of setting the gill net around a school of fish usually in water sufficiently shallow for the net to reach the bottom, and then scaring the fish so that they rush the net and become gilled.

As gill nets are not so effective when the fish can see the net they are usually fished at night, but in rivers the turbidity of the water often renders them effective in daylight. When there are many phosphorescent organisms in the water, gill nets sometimes will not fish effectively when it is dark but may take fish in twilight. When the species sought may be off the bottom, the fishermen sometimes set a gill net obliquely from the bottom toward the surface until they discover the depth being frequented.

Drift gill nets are used principally for pelagic species, such as herring and salmon. When used on large bodies of water or at sea they are set in long strings or gangs with lighted buoys at intervals. One string may be several miles long, and modern vessels using this system employ a special net puller which consists essentially of a smooth curved metal chute and a large revolving metal drum set vertically. When drifted in midwater, gill nets are used for lake trout, whitefish, cod, and other species that, though usually dwelling somewhat below the surface, are often found some distance above the bottom.

Anchored gill nets may also be fished off the bottom merely by employing sufficiently light weights on the lead line, with heavier weights on lines at intervals. Anchor or sink gill nets are used in the North Atlantic for capturing cod and other bottom fish.

In fish surveys, gill nets, made up of gangs, 30 m in length, with different size mesh, are used to provide qualitative samples of different species and different sizes. Also used, experimentally, are the vertical gill nets set from surface to bottom to provide some idea of depth distribution.

ENCIRCLING NETS capture fish by placing a mobile net either completely around the schooling fish (Figure 3-9), or so as to block their escape. The nets are classified according to the method of fishing them as shore landing, vessel landing, and blocking.

Haul seine or beach seine was undoubtedly the forerunner of both the purse seine and the otter trawl. Once used extensively, it became unprofitable for general use and is now employed only in special fisheries. Thus it was at first one of the principal types of gear used in the Pacific salmon fisheries but is now used in only a few localities. A few large haul seines operate along the southern Atlantic coast, and they are used for shad in the lower reaches of the Connecticut River. Largely though, the haul seine is used sparingly.

Essentially a haul seine is a strip of strong netting hung to a stout cork line at the top and to a strong very heavily weighted lead line on the

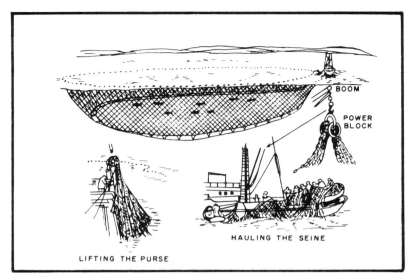

BOOM

POWER
BLOCK

HAULING THE SEINE

LIFTING THE PURSE

Figure 3-9. Encircling (purse seine) gear. Courtesy of Gustaf T. Sundstrom, illustrator, Cir. 48, above.

bottom. On muddy bottoms the lead line may require a little weight. The wings of the net are often of larger mesh than the middle portion or bunt. The wings may taper so that they are shallower on the ends. The bunt sometimes has the center portion of the netting formed into a bag to aid in confining the fish. At the ends of the wings the cork and lead lines are often fastened onto each end of a short stout pole or a brail. The hauling lines are then attached to the top and bottom of the brail by a short bridle.

The usual method of fishing is to draw the net onto a boat. One end of the hauling lines is left on the shore, and the boat leaves the shore, paying out the line until the other end of the line is reached. The boat changes direction and lays out the net parallel to the beach. When the net is all in the water, the boat brings the end of the second hauling line ashore.

Sometimes, especially when fishing for pelagic or schooling species, the net is left in this position until fish are sighted between the net and the shore; then hauling commences. When it is known that fish are traveling along the beach in a certain direction, one end of the net may be hauled in first so as to form a hook against the beach. As soon as a school of fish enters the area, the second line is hauled. In fishing for nonschooling fish both ends of the net are usually hauled in at once.

Stop seines are used to close off the opening into a small cove or bight.

When the bay is large it may require several seines lashed together to close the opening. Once the fish are thus impounded they are removed as needed with a small seine. This method has been used for centuries in the Norwegian spring fishery for adult herring and is one method used in Maine to take young herring for the sardine canneries.

Purse seines account for a major share of our fish harvest. The purse seine is very efficient for taking pelagic schooling species, and the recent use of various types of modern sounding devices, such as echo sounding, radar, and sonar in locating schools of fish beneath the surface is already increasing its effectiveness.

The purse seine has certain advantages. Because of the strength imparted by its fine mesh and weight, it is more suitable in rough weather or when the catch is large. The fine mesh prevents the escape of fish while the net is being pursed, and it can be pursed more rapidly than the other forms. The purse seine is a continuous deep ribbon of web with corks on one side and leads on the other. Rings are fastened at intervals, each by two short ropes, to the lead line. A purse line runs completely around the net through these rings. The net is piled on to a turntable on the stern of the vessel. The end of the net is fastened by a short rope to a skiff that is towed behind the vessel. As soon as a school of fish of the proper size and species is located, the net is set around it, an attempt being made to set the net across the path of the fish if they are moving. As the skiff is fastened to the end of the net, it acts as a drag as soon as it is cast off from the vessel. The vessel sets the net in a circle while cruising at full speed and, as soon as both ends are brought together, the purse line is hauled aboard. Usually the turntable is swung so that the net is taken in from one side. Most vessels use a steel cable for a purse line. It is hauled in by taking a turn around a "gypsy" powered by a shaft from the vessel's main engine. The roller on the margin of the turntable also may be powered to aid in hauling in the heavy net, but today the net is usually hauled in from the boom with a power block or on a net drum. When all of the purse line and most of the net is aboard, the fish are safely confined in a bag formed by the fine-meshed net swung from the boom.

Drum seining is practiced in Puget Sound. The drum is usually about 8 feet long with 8-foot-diameter flanges and provided with a fairlead to level wind the net onto the drum. A ring stripper holds the purse rings at the surface after pursing is complete. Major advantages to the drum over the power block are a reduction in crew to four, overall speed in setting and retrieving, and fewer accidents to the crew since nothing is lifted overhead.

Some analyses of catch records indicate higher annual earnings for the drum seiners.

Purse seines vary in size according to the size of the vessel, the size of the mesh, the species sought, and the depth to be fished. The length runs from 30 to 40 fathoms up to 360 fathoms. Tuna seines are nearly a kilometer. The webbing is manufactured in strips 3½ fathoms deep. The number of strips used is governed by the depth. In deep water 10 to 11 strips are about the maximum used, but in fishing in shallow bays, where the net may take bottom and tear, less depth is desirable, and as few as 4 to 5 strips may be used. Some of the species taken are salmon, herring, sardines, mackerel, bluefin tuna, barracuda, jack mackerel, and anchovies.

Purse seining for tuna has created a serious environmental controversy. Tuna fishermen have long used the porpoise to spot the schools of large (average 100 pounds) yellowfin tuna. Similar feeding habits of the porpoises and large yellowfin seem the most likely explanation for their close association, although other theories, including using each other for directional sense, have been advanced. Prior to the late 1950s the tuna fishery used hook-and-line gear which caught the yellowfin without affecting the porpoises. When the tuna fishery switched to purse seining for more efficiency, the problems of porpoise mortality began as the air-breathing mammals were entangled in the mesh or otherwise held under water until many drowned.

Commercial fishermen themselves, in an effort to reduce porpoise mortality, developed a technique of "backing down" as they retrieved the net, combining it with modifications to the purse itself. As the seiner is backed down, the tension on the net causes the far end to dip below the surface and the porpoises are free to swim over to freedom. Further protection is the addition of fine mesh panels at the far end and sides of the net which helps to prevent entanglement. Some of the porpoises go to the bottom of the net instead of staying near the surface. Careful captains can save these porpoises by simply waiting until they come to the surface and guiding them from the net. Finally, the use of crew members in rubber rafts or small boats to herd the porpoises from the net also increases survival.

Quotas have been set under the Marine Mammal Protection Act of 1972 limiting the kill of porpoises to 51,945 in 1978, 41,610 in 1979, and 31,150 in 1980. The objective is to reach a 50 percent reduction in porpoise mortality by 1980 from the 1977 quota of 62,429. Problems still exist on how to treat the different species of porpoises when some are rare and others are abundant.

Gear modifications and other research continues in an effort to ensure that fishing for this valuable resource can continue without endangering the porpoise populations.

TOWED NETS strain the water through the meshes of a moving net, leaving the fish in the bag. The main types of towed nets are otter trawls (Figure 3–10) and midwater or pelagic trawls.

Otter trawls derive their name from the otter boards attached to the forward end of each wing to keep the mouth of the net open. The otter board is a large, rectangular board usually 1 to 2.5 m long, 1 to 1.5 m high, and 5 to 10 cm thick. One edge is heavily shod with an iron runner to slide over the bottom. The front edge is usually rounded on the bottom to aid in bouncing over obstructions. The towing bridle or warp is attached to the board by four heavy chains or short heavy rods. The two forward rods are shorter, so that when towed the board sheers to the side. As the two boards sheer in opposite directions, they keep the mouth of the net open.

At first the otter boards were attached directly to each wing of the net. In the Vigneron-Dahl modification, known as the V-D trawl, each wing was attached to a short brail or spreader, the headrope to the upper end and the footrope to the lower end. The spreader was attached by a short bridle to one long ground cable which led to the lower after-edge of the otter board.

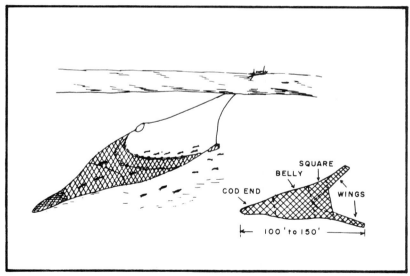

Figure 3-10. Towed net (otter trawl) gear. Courtesy of Gustaf T. Sundstrom, illustrator, Cir. 48, above.

This arrangement kept the otter boards farther apart, and it would appear that many of the fish disturbed by the ground line eventually were caught in the net, since this form of trawl was considerably more effective than its predecessor.

The original V-D gear, like its predecessor, did not fish far off the bottom, for in spite of floats (chiefly glass or metal balls, rubber or cork floats) along the headrope, the outward pressure kept the headrope low. This was a big disadvantage in fishing for many species that are not always close to the bottom. The New England trawlers discarded the spreader and used two long ground cables to each otter board. One went from the lower edge of the otter board to the footrope and the other from the upper edge of the otter board to the headrope. The wing was made higher, and the new method of attachment permitted the headrope to arch upward, greatly increasing the height of the net. Later the cables from the upper and lower edges of the wing, called ''leg wires,'' were joined a little distance in front of the net into a single ground wire that was attached to the upper edge of the otter board. This tended to keep the net from digging into the bottom.

The footrope, which constitutes the lower edge of the net, is made in various fashions. For small trawls it may be merely a heavy manila hawser, weighted with lead and perhaps, for chafing gear, wrapped in burlap lashed on with rope yarn. In large trawls it is usually a heavy steel cable. In New England the trawlers fished at first only on smooth bottom to avoid tearing the net or having the footrope hang up on heavy boulders or ledges. However, the best fishing was often on rough ground where tears were frequent. Soon the fishermen largely overcame this difficulty by using ''rollers.''

Rollers are wooden disks 20 cm thick and either 20 or 50 cm in diameter. They are strung like beads on a cable, first a 50-cm roller, then two 20-cm rollers, then a 50-cm roller, and so on. This cable is attached to a footrope by short roller chains, one chain between each pair of small rollers. On a typical large trawl there are 16 rollers for the belly section, 6 large ones, with 5 pairs of small rollers between them. Sometimes only these ''bosom'' rollers are used, but for rougher bottom they are sometimes used along the half of each lower wing next to the belly. Each side takes 13 rollers, 4 large ones, with 3 sets of 3 small rollers between them.

All but the very early otter trawls are towed by two warps, one to each otter board, as this permits the net to spread properly. The winches on most trawlers are provided with two drums, one for each cable.

The small inshore otter-trawl vessels, known in New England as ''draggers'' often use small trawls 9 to 15 m or more in width. The trawl nets used by the larger vessels usually measure 25 m along the headrope and 35 m along the footrope for groundfish. The smaller-meshed trawls used for ocean perch are 18 m on the headrope and 25 m on the footrope.

Trawlers usually land their catch amidships. As the otter boards come in, they are hauled up and attached to heavy metal frames shaped like an inverted V called "gallows frames." One is placed well forward, and the other just aft of amidships. Practically all New England trawlers have two gallows frames on each side of the vessel so that they can fish on either the port or starboard side. If a net is sufficiently torn to require some time for mending, a different net is set on the opposite side of the vessel so that fishing can continue while the first net is being repaired. Most Pacific Coast trawlers set the nets off the stern and have one gallows frame well aft on each side. Modern otter trawlers set and pull the trawl over the stern rather than the port or starboard.

Midwater trawls using subsidiary otter boards to spread the net mouth vertically as well as horizontally have been tried experimentally but are not in commercial use in North America. Otter trawls, one of the principal forms of modern gear, are in use throughout the world. They take any species that customarily lives on or near the sea bottom. Midwater trawls or pelagic trawls are the object of much developmental gear research. Engineering designs for the otter boards of the midwater trawls are as complicated as for the wing of an airplane. Other research is concerned with designing trawls that will sort out the catch, as is done in the shrimp fishery where two sizes of nets, one within the other, for example, permit the shrimp to be collected in the bag of the net, and hold the large animals in front.

The use of electricity, as with electrified trawl nets, to attract and hold the fish is also receiving much attention. Electrofishing is widely used by freshwater fishery biologists to collect and to estimate fish populations. Most every state and federal fishery installation uses some modification of the electrofishing gear. Generally, alternating current is used where electronarcosis will produce the sample. Direct current is used in areas where it is necessary to draw out the fish from underneath banks or brush piles or some other cover. Galvanotaxis results in the fish being attracted to the positive electrode. Electrofishing units range from back pack models with small gasoline generators or batteries providing the power to larger units with sizable generators operating from the banks or from boats. Electrofishing has proved to be an efficient method for sampling.

Some idea of the design and performance studies necessary for developing fishing gear can be obtained by examining the 79-page booklet Otter

Board Design and Performance, published by the Food and Agriculture Organization of the United Nations in 1974.

References

Alverson, D. L. 1959. Trends in trawling methods and gear on the west coast of United States. *In* H. Kristjonsson, ed. Modern fishing gear of the world. Fishing News (Books), Ltd., London, pp. 317–320.

Atton, F. M. 1955. The relative effectiveness of nylon and cotton gill nets. Can. Fish Cult. 17:18–26.

Baldwin, W. J. 1961. Construction and operation of a small-boat trawling apparatus. California Fish Game J. 47(1):87–95.

Barraclough, W. E., and W. E. Johnson. 1956. A new midwater trawl for herring. Fish. Res. Bd. Can. Bull. 104. 25 pp.

Berst, A. H. 1961. Selectivity and efficiency of experimental gill nets in South Bay and Georgian Bay of Lake Huron. Trans. Amer. Fish. Soc. 90(4):413–418.

Brandt, A. von 1964. Fish-catching methods of the world. Fishing News (Books), Ltd., London. 191 pp.

Bullis, H. R. 1951. Gulf of Mexico shrimp trawl designs. U.S. Fish Wildl. Serv., Fish. Leafl. 394. 16 pp.

Burnet, A. M. R. 1959. Electric fishing with pulsatory electric current. New Zealand J. Sci. 2(1):48–56.

Collins, J. J. 1979. Relative efficiency of multifilament and monofilament nylon gill net towards lake whitefish (*Coregonus clupeaformis*) in Lake Huron. J. Fish. Res. Bd. Can. 36(10):1180–1185.

Dale, P., and S. Moller. 1964. The development of a midwater trawl. *In* H. Kristjonsson, ed. Modern fishing gear of the world. Fishing News (Books), Ltd., London, pp. 482–489.

Devereux, R. F. 1953. Isaacs-Kidd midwater trawl. Final report. Univ. California, Scripps Inst. Oceanogr., Oceanogr. Equip. Rep. 1. 18 pp.

Donaldson, I. J., and F. K. Cramer. 1971. Fishwheels of the Columbia. Binfords & Mort, Portland, Ore. 124 pp.

Haskell, D. C. 1939. Electrical fields as applied to the operation of electric fish shockers. New York Fish Game J. 1(2):130–170.

Haskell, D. C., and W. F. Adelman, Jr. 1955. Effects of rapid direct-current pulsations on fish. New York Fish Game J. 2(1):95–105.

High, W. L., I. E. Ellis, and L. D. Lusz. 1969. Progress report on the development of a shrimp trawl to separate shrimp from fish and bottom-dwelling animals. U.S. Fish Wildl. Serv., Commer. Fish. Rev. 31(3):20–30.

High, W. L. 1974. Puget sound drum seining. National Marine Fish. Serv., Marine Fish. Rev., 36(12):5–11.

Horak, D. L., and H. A. Tanner. 1964. The use of vertical gill nets in studying depth distribution. Horsetooth Reservoir, Colorado. Trans. Amer. Fish. Soc. 93(2):137–145.

Klima, E. F. 1968. Shrimp behavior studies underlying the development of the electric shrimp-trawl system. U.S. Fish Wildl. Serv., Fish. Ind. Res. 4(5):165–205.

Lackey, R. T. 1968. Vertical gill nets for studying depth distribution of small fish. Trans. Amer. Fish. Soc. 97(3):296–299.

McCombie, A. M., and F. E. J. Fry. 1960. Selectivity of gill nets for lake whitefish, *Coregonus clupeaformis*. Trans. Amer. Fish. Soc. 89(2):176–184.

Meyer-Waarden, P. F. 1957. Electrical fishing. FAO, U.N., Fish. Stud. 7. 77 pp.

Monan, G. E., and D. E. Engstrom. 1962. Development of a mathematical relationship between electric-field parameters and the electrical characteristics of fish. U.S. Fish Wildl. Serv., Fish. Bull. 63(1):123–136.

Nedelec, C. ed. 1975. Catalogue of small-scale fishing gear. FAO. Fishing News (Books) Ltd. England. 191 pp.

Rupp, R. S., and S. E. DeRoche, 1960. Use of a small otter trawl to sample deep-water fishes in Maine lakes. Prog. Fish-Cult. 22(3):134–137.

Schwartz, F. J. 1961. Effects of external forces on aquatic organisms. Maryland Dep. Resour. and Educ., Chesapeake Biol. Lab. 168:3–26.

Scofield, W. L. 1951. Purse seines and other roundhaul nets in California. California Dep. Fish Game, Fish Bull. 81. 83 pp.

Todd, I. S. P., and P. A. Larkin. 1971. Gill selectivity on sockeye (*Oncorhynchus nerka*) and pink salmon (*O. gorbuscha*) of the Skeena River System, British Columbia. J. Fish. Res. Bd. Can. 28(6):821–841.

U.S. Fish and Wildlife Service. 1957. Commercial fishing vessels and gear. Bur. Commer. Fish., Circ. 48. 48 pp.

Vibert, R., ed. 1967. Fishing and electricity. FAO, U.N. Fishing News (Books), Ltd., London. 276 pp.

Webster, D. A., J. L. Forney, R. H. Gibbs, Jr., J. H. Severns, and W. F. Van Woert. 1955. A comparison of alternating and direct electric currents in fishery work. New York Fish Game J. 2(1):106–113.

4

Population Management

A population of fish is a biological entity. These are populations that have characteristics not necessarily applicable to single organisms, including size, density, mortality, longevity, growth rates, sex ratios, and behavior patterns. Fishermen and fishery biologists refer to stocks of fish. Others define populations as a group of organisms of the same species occupying space at a particular time and then refer to a deme as subdivided from the populations as a "local" population in a lake. Whatever we call them and however we recognize them are not as important as the fact that we do recognize them and can manage them. The traditional taxons of species, subspecies, and even races may not be discriminating enough for the fishery resource manager who finds he must recognize subpopulations and strains to reach the subtle level demanded by present and future fishery management. The species is simply too gross a classification. Consider the domesticated strains of dogs, cattle, sheep, and swine that have been developed for human benefit. How, then, can one expect a single strain of hatchery-reared trout to be equally successful in large and small lakes, in small streams, or in larger rivers, or in various waters with various chemical, physical, and biological characteristics?

Populations become adapted to ecological niches through adjustments in characters interwoven with natural selection. There is little preadaptation, but we are able to use the plasticity of some species to utilize them in a range of habitat parameters. The population required for a special fishery problem may already occur naturally and needs only to be recognized and utilized. A recent note on the steelhead populations in the Province of British Columbia makes reference to the possibility of 800–1000 specific "races" exhibiting different characteristics of size, age, times of river entry, distances migrated, and spawning times. A promise for the future may exist in the construction of the strain needed by careful genetics selection for the characteristics that best adapt them for the habitat for which they are intended.

When a fishery depends wholly on one stock of fish it will be affected by the quantities caught in any one locality. If the stocks of fish are local in distribution, each must be treated as a separate unit, and it becomes possible to reduce the numbers of fish in one locality without affecting them elsewhere. An excellent example of the critical importance of recognizing fish stocks is the Pacific halibut, an international fishery for which biological and political questions must be solved. The United States Fishery Conservation and Management Act of 1976 specifies that, "to the extent practicable, an individual stock of fish shall be managed as a unit throughout its range, and interrelated stocks of fish shall be managed as a unit in close coordination." A careful analysis of the fishery records and research supports the premise that the stocks between Canada and Alaska intermingle and should, in compliance with U.S. law and basic management, be considered as a whole in formulating international treaties and fishery regulations.

Washington State has utilized intraspecific diversity of the steelhead by concentrating its program on the summer steelhead, which enter fresh water from May to late summer in prime condition as their spawning may be up to a year away. Thus, steelhead anglers have an extended, pleasant time of year to fish not provided by the winter runs.

Dwight Webster, Professor of Fisheries at Cornell University, has been directing research programs to study the potentials of wild brook trout as compared with domesticated stocks. He is also experimenting with two Quebec strains and with intraspecific hybrids. Results of this New York work confirm that wild trout exhibit higher survival, longer life, higher return to anglers, and produce five times as much poundage in nature as their domestic counterparts.

California has reported on the harvest results of four strains of rainbow trout planted as fingerlings. Two of these strains, the Virginia and Whitney, have been domesticated for many years. The Virginia strain dates back to the 1880s and the Whitney strain to the early 1900s. The Shasta strain was developed by crosses made in California in 1951 and 1952. The fourth strain, Kamloops, is native to the interior waters of British Columbia and is characteristically a lake fish spawning in tributaries where the young spend variable amounts of time before migrating to the lake. Results of stocking fish from these four strains in a coldwater reservoir were as follows: Kamloops were harvested at rates ranging from 17 to 33 percent;

the Shasta strain had varying success with a best harvest of 22 percent; in contrast, the Whitney strain gave a 2.5 percent harvest; and the Virginia strain only 4.2 percent. The Kamloops gave a higher return to the fishermen, even though they were more inclined to migrate from the reservoir and seemed to prefer the open-water areas of the reservoir, thereby decreasing their availability to shore anglers.

Two distinct groups of kokanee (*Oncorhynchus nerka*) have been reported from Oregon.

Group I	Group II
Spawned mid-September until November 10 (entered spawning area as early as August 22)	Spawned December 6 until mid-January
Chose streams and shoreline spawning areas	Chose shoreline spawning areas
Eighty-one percent hatchery-origin fish	No hatchery-reared fish

Fish transplants to supplement existing populations or to introduce a new fishery often fail. More and more evidence is being accumulated to show that these failures are more likely to result when innate behavior of the donor population is not correctly matched with the requirements of their new environment. Stream populations of trout are not usually successfully transplanted to lake environments and vice versa. Inlet-spawning salmon and trout are not successful in establishing populations in lake environments where only outlet spawning is available. Years ago, as the result of a biological survey indicating a limited inlet-spawning area in a Maine lake, the logging dam obstructing downstream salmon movement out of and back into the lake was removed. Resident Atlantic salmon did not use the now available outlet, nor did the transplanted fish from an inlet-spawning population. Not until salmon from an outlet-spawning population were transplanted into the lake was the outlet utilized and the salmon population built up for a successful fishery. Similar results have been reported for genetic control in lakeward migrations of cutthroat trout fry in Yellowstone Lake. Fertilized eggs from cutthroat in an inlet and in the outlet of Yellowstone Lake were collected and incubated under identical conditions. When the resultant fry were introduced into three simulated stream sections under carefully controlled conditions the highly significant results indicated that fry from the inlet-spawning populations migrated downstream, and fry from the outlet-spawning populations moved upstream—exactly the directions necessary for return to Yellowstone Lake.

Exploitation of the Atlantic salmon from North America and Europe in the oceanic and coastal areas off West Greenland makes stock identification critical to the management of commercial fisheries. Consistent differences in electrophoretic behavior of serum proteins and liver esterases in salmon from the two continents, plus supporting evidence from life histories, lengths, and abundance of two parasites are reported to separate the stocks.

If fishery biologists are to work with these various populations, frequently at levels of classification below the traditionally accepted taxonomic separations, then how can the populations be recognized? What methods are available?

Marking fish is the most direct means of studying populations. A hypothetical example follows where 1,000 marked fish were released near the center of a 500-kilometer stretch of coastline several months in advance of the regular fishing season. Actual recoveries made during the fishing season were compared to the number of recoveries that might have been expected in the area if the catch were all from one population and if the marked fish were randomly distributed through such a population. Recoveries do not support this hypothesis, as shown in the accompanying table (from Rounsfell and Everhart, 1953). Catch figures are given in hundred thousands.

Area in kilometers	Number marked	Numbers in catch	Number of observed recoveries	Number of expected recoveries
0–100		150	0	84
100–200		200	50	113
200–300	1000	120	200	68
300–400		50	40	28
400–500		30	20	17
Total	1000	550	310	310

Another example of the usefulness of marking fish might be in a situation where a sports fishery concentrates its activity at the mouth of a tributary stream to a large lake. To find out whether the sport fishery is exploiting a general lake population or a discrete population from its tributary stream, fish descending the stream could be marked. A check of subsequent recaptures throughout the entire lake would indicate where the fish had come from; if the fish came from the general population, the fishery would be considered less likely to reduce the population. In another example, at the present time Atlantic salmon are being marked in several countries to help identify stocks in the Greenland fishery.

Meristic characteristics, such as vertebral counts, fin-ray counts, scale counts, and the enumeration of gillrakers, branchiostegals, caeca, and ova, may be used to separate populations. Racial investigations of the striped bass (*Morone saxatilis*) made use of fin-ray counts in young striped bass to determine if the Chesapeake Bay system and Hudson River populations were differentiated and at what level. Dorsal-ray counts, anal-ray counts, pectoral-ray counts, and a character-index count obtained by adding the soft rays of the dorsal, anal, and pectoral (both sides) fins were studied. The best results are obtained from the character index which permitted the separation of a high percentage of the individuals sampled with Hudson origin from those spawned in the several tributaries of Chesapeake Bay. Although the separation is below the species level, it is possible thus to refer to the Hudson race and the Chesapeake-Delaware race.

Diseases and parasites have been used to separate fish populations, particularly in the marine environment. Their appearance in a fish population is sometimes referred to as a "natural tag." If a disease is known to occur throughout a fish species, as with the Atlantic herring, *Clupea harengus,* then one indication of population separation would be the incidence of the disease in one locality or another. Thus, in some years Canadian populations will have the disease in epidemic proportions with a very low incidence in northeastern United States: populations of ocean perch (*Sebastes marinus*) can be separated by their copepod parasites. Two studies progressing at present are supplying additional information for population separation through identification by way of diseases and parasites; one study is on the American lobster, and another is to provide some indication of the origin of Atlantic salmon now fished off the west coast of Greenland. Atlantic salmon could originate in United States, Canada, Scotland, Ireland, England, Norway, Sweden, France, and Spain.

Age composition, too, can be used to separate populations. Growth rate is another possibility and has been used in the separation of Pacific salmon. Otoliths, one described as an S-type characterizing spring-spawning herring and another as an A-type characterizing fall-spawning herring, are used in the southern Gulf of St. Lawrence and adjacent waters to separate Atlantic herring into spring-spawning and fall-spawning stocks. Ninety-seven percent separation has been reported, with the distinct advantage that the fish can be identified individually.

Periodic abundance and scarcity may also be indicators of population separation. If the spawning run in one tributary is low, while across the bay or lake another equally suitable tributary is teeming with fish, you might suspect separate populations.

Cytotaxonomic studies, presently not definitive at the population or strain level, may eventually provide us with cytogenetic parameters, particularly with respect to the karyotypes. The reduction in diploid number through time via Robertsonian or centric fusion is important in separating evolutionary trends. Work with the coregonine fishes indicates higher numbers of chromosome arms with the more primitive species. Recent species have fewer. Some of these studies are in progress, but at least one researcher refers to an Atlantic salmon chromosome and to a cutthroat chromosome. Of value too is simply the count of chromosomes as well as their configuration.

X-ray fluorescence spectrometry has been used by Canadian workers to characterize the chemical composition of samples of fry and adult sockeye salmon (*Oncorhynchus nerka*). The method depends on the possibility that fish living in different environments acquire a chemical composition that is quantifiable. This "chemoprint" would thus serve as a natural tag for identifying geographical origins and separating populations. Seventy-seven percent of the fry in the Canadian study were correctly assigned to their area of origin and 73 percent of the adults. Increased sample sizes could lead to as high as 95 percent correct classification and separation.

Electrophoretic mobility of protein molecules is also of value in separating populations. Soluble lens proteins of inbred strains and their hybrids of brook trout were examined by acrylamide gel electrophoresis. Differences at the molecular level can be directly related to genetic variability. These differences can be used to characterize and identify separate breeding populations: electrophoretic analysis of serum of striped bass populations from rivers flowing into the Chesapeake Bay indicate distinct populations in the Elk, Choptank, and Nanticoke. Strengthening of the distinct population hypothesis would support the necessity for individual river management rather than for an overall area or state.

Serology has been used increasingly as a method of separating populations. Serological methods have been of interest since the first demonstration of individual differences in human erythrocyte antigens in 1901. Many animal groups have been studied, including particularly marine fish, although some work has been done in fresh water too. This work has demonstrated the existence of individual variations of erythrocyte antigens of certain fishes and in some cases quantitative differences in the frequency of occurrence of antigens. Thus, a major blood-group system in immature Atlantic sea herring has been distinguished from normal lobster serum and rabbit antiherring serum. Eastern and western herring have been tentatively separated.

The use of behavior patterns as criteria for separating populations promises to be of particular value to fishery management. Spawning times, spawning migrations, territorial defense, and general migration may be used.

References

Averett, R. C., and F. A. Espinosa, Jr. 1968. Site selection and time of spawning of two groups of kokanee in Odell Lake, Oregon. J. Wildl. Manage. 32(1):76–81.

Behnke, R. J. 1972. The systematics of salmonid fishes of recently glaciated lakes. J. Fish. Res. Bd. Can. 29(6):639–671.

Bodaly, R. A. 1979. Morphological and ecological divergence within the lake Whitefish (*Coregonus clupeaformis*) species complex in Yukon Territory. J. Fish. Res. Bd. Can. 36(10):1214–1222.

Booke, H. E. 1968. Cytotaxonomic studies of the coregonine fishes of the Great Lakes, USA:DNA and karyotype analysis. J. Fish. Res. Bd. Can. 25(8):1667–1687.

Calaprice, J. R. 1971. X-ray spectrometric and multivariate analysis of sockeye salmon (*Oncorhynchus nerka*) from different geographic regions. J. Fish. Res. Bd. Can. 28(3):369–377.

Calaprice, J. R., and J. E. Cushing. 1967. A serological analysis of three populations of golden trout, *Salmo aquabonito* Jordan. California Fish Game J. 53(4):273–281.

Carscadden, J. E., and W. C. Leggett. 1975. Meristic differences in spawning populations of American shad, *Alosa sapidissima:* Evidence for homing to tributaries in the St. John River, New Brunswick. J. Fish. Res. Bd. Can. 32(5):653–660.

Cordone, A. J., and S. J. Nicola. 1970. Harvest of four strains of rainbow trout, *Salmo gairdnerii,* from Beardsley Reservoir, California. California Fish Game J. 56(4):271–287.

Dalziel, J. A., and K. J. Shillington. 1961. Development of a fast-growing strain of Atlantic salmon (*Salmo salar*). Can. Fish. Cult. 30:57–59.

Donaldson, L. R., D. D. Hansler, and T. N. Buckridge. 1957. Interracial hybridization of cutthroat trout, *Salmo clarkii,* and its use in fisheries management. Trans. Amer. Fish. Soc. 86:350–360.

Dymond, J. R. 1957. Artificial propagation in the management of Great Lake fisheries. Trans. Amer. Fish. Soc. 86:384–391.

Eckroat, L. R., and J. R. Wright. 1969. Genetic analysis of soluble lens protein polymorphism in brook trout, *Salvelinus fontinalis.* Copeia 1969(3):466–473.

Fenderson, O. C. 1964. Evidence of subpopulations of lake whitefish, *Coregonus clupeaformis,* involving a dwarfed form. Trans. Amer. Fish. Soc. 93(1):77–94.

Flick, W. A., and D. A. Webster. 1976. Production of wild, domestic, and interstrain hybrids of brook trout (*Salvelinus fontinalis*) in natural ponds. J. Fish. Res. Bd. Can. 33(7):1525–1239.

Henricson, J., and L. Nyman. 1976. The ecological and genetical segregation of two sympatric species of dwarf char (*Salvelinus alpinus* (L.) species complex). Inst. Freshwater Res. Rept. 55. Drottningholm: 15–37.

Imhof, M., R. Leary, and H. E. Booke. 1980. Population or stock structure of lake whitefish, *Coregonus clupeaformis,* in northern Lake Michigan as assessed by isozyme electrophoresis. Can. J. Aquat. Sci. 37(5):783–793.

Lux, F. E., A. E. Peterson, Jr., and R. F. Hutton. 1970. Geographical variation in fin ray number in winter flounder, *Pseudopleuronectes americanus* (Walbaum) of Massachusetts. Trans. Amer. Fish. Soc. 99(3):483–488.

Messieh, S. N. 1972. Use of otoliths in identifying herring stocks in the southern Gulf of St. Lawrence and adjacent waters. J. Fish. Res. Bd. Can. 29(8):1113–1118.

Messieh, S. N., and S. N. Tibbo. 1971. Discreteness of Atlantic herring (*Clupea harengus harengus*) populations in spring and autumn fisheries in the southern Gulf of St. Lawrence. J. Fish. Res. Bd. Can. 28(7):1009–1014.

Morgan, R. P., II, T. S. Y. Koo, and G. E. Krantz. 1973. Electrophoretic determination of populations of striped bass, *Morone saxatilis,* in the Upper Chesapeake Bay. Trans. Amer. Fish. Soc. 102(1):21–32.

Northcote, T. G., S. N. Williscroft, and H. Tsuyuki, 1970. Meristic and lactate dehydrogenase genotype differences in stream populations of rainbow trout below and above a waterfall. J. Fish. Res. Bd. Can. 27(11):1987–1995.

Nyman, O. L., and J. H. C. Pippy. 1972. Differences in Atlantic salmon, *Salmo salar,* from North America and Europe. J. Fish. Res. Bd. Can. 29(2):179–185.

Plosita, D. S. 1977. Relationship of strain and size at stocking to survival of lake trout in Adirondack Lakes. N.Y. Fish and Game Jour. 24(1):1–24.

Raleigh, R. F. 1967. Genetic control in the lakeward migrations of sockeye salmon (*Oncorhynchus nerka*) fry. J. Fish. Res. Bd. Can. 24(12):2613–2622.

Raleigh, R. F., and D. W. Chapman. 1971. Genetic control in lakeward migrations of cutthroat trout fry. Trans. Amer. Fish. Soc. 100(1):33–40.

Raney, E. C., and D. P. DeSylva. 1953. Racial investigations of the striped bass, *Roccus saxatilis* (Walbaum). J. Wildl. Manage. 17(4):495–509.

Sidell, B. D., and R. G. Otto. 1980. Apparent genetic homogeneity of spawning striped bass in the upper Chesapeake Bay. Trans. Amer. Fish. Soc. 109(1):99–107.

Sindermann, C. J., and D. F. Mairs. 1959. A major blood group system in Atlantic sea herring. Copeia 1959(3):228–232.

Skud, B. E. 1977. Drift, migration, and intermingling of Pacific halibut stocks. Internat. Pacific Halibut Commission, Sci. Rept. 63. 42 pp.

Trojnar, J. R., and R. J. Behnke. 1974. Management implications of ecological segregation between two introduced populations of cutthroat trout in a small Colorado lake. Trans. Amer. Fish. Soc. 103(3):423–430.

Vrooman, A. M. 1964. Serologically differentiated subpopulations of Pacific sardine, *Sardinops caerula.* J. Fish. Res. Bd. Can. 21(4):691–701.

Webster, D. A., and W. A. Flick. 1975. Species management. Proc. Wild Trout Symposium, U.S.D.I. and Trout Unlimited:40–47.

Wilkins, N. P. 1968. Multiple haemoglobins of the Atlantic salmon (*Salmo salar*). J. Fish. Res. Bd. Can. 25(12):2651–2663.

5

Age and Growth

Age

Knowledge about the age composition of a population is essential to good resource management. Whether a fish lives to over thirty years, as with the striped bass and the lake trout, about eleven years for the longnose gar, or that the brook trout in our streams average about three years of age tells us much we need to know for proper utilization of the resource. Correct age information is necessary for longevity predictions, to establish records for rates of growth, to know the age at maturity, and age during important migrations, and to learn what periods in the life history of a fish population represent critical stages, or when specific habitat requirements change.

Three methods used for determining the age of fishes are comparison of length-frequency distribution, recovery of marked fish, and interpretation of layers laid down in the hard parts of the fish.

Length-frequency distribution has been used to estimate fish age since the latter part of the nineteenth century. This method depends on knowing that the lengths of fish of one age tend to form a normal distribution pattern. Ages are determined by counting the peaks, as shown in Figure 5-1.

Determination of age distribution by length-frequency studies is often adequate for the first two to four years, but usually fails in reliable separation of the older age groups because of increasing overlap in length distribution (see Figure 5-1). Overlap results from increased dispersion which is measured by a larger standard deviation. Furthermore, increased overlap is due also to the lessened distance between modes. Other major disadvantages of the length-frequency method are: (1) fish of a size tend to school

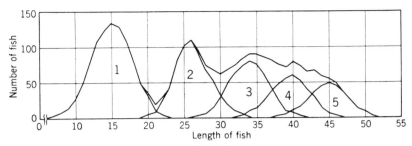

Figure 5-1. Overlapping size frequencies in fish of different ages.

together; (2) hatching may occur at irregular times yielding size groupings not indicative of year classes; (3) parts of a given year class may develop under variable conditions resulting in groupings of different size but of the same age; and (4) one or more of the year classes may be poorly represented or lacking in the sample. To be successful, this method requires large random samples of the population. If fish are scaleless or if the scales and other hard structures are impossible to interpret, length-frequency may be the only way. Frequently this method is used as a check on the scale method of age determination, at least for the younger age groups.

A direct and positive way of determining fish age is by the liberation and *recovery of marked fish* of known age. This method is also the most costly and time consuming. Its real value lies in providing an accurate base on which to compare ages determined by other methods. Scales from fish of known age can be included in a series of scales, and a check is provided as ages are determined. Scales from known-age fish are particularly valuable when more subtle interpretations of the sculpturing are used to determine when anadromous fish leave the stream, the length of time the fish spend in the estuary, and the exact length of its marine life.

Interpreting annual layers deposited in the hard parts of fish is the most generally used method for determining age. Scales are used most frequently, with otoliths, spines and rays, and bony structures such as the vertebra, dentary bone, and cross sections of other bony structures following in that order. Success with this method depends on recognition of changes in the rate of growth or metabolism during certain periods as the changes are reflected as layers in the hard parts—accuracy of the age determination depends on the ability to interpret the layers correctly. Clarity of the layers varies with species and structure, although increased age progressively obscures age interpretation in all structures and for all fishes.

Otoliths

Otoliths are used to determine the age of fish which are either entirely scaleless or where the scale sculpturing cannot be interpreted accurately. The otolith may serve as a check on scale interpretations or vice versa. Although each teleost has 6 otoliths, 3 on each side, the one ordinarily used is the sagitta or sacculolith laid down in the sacculus of the inner ear. A calcareous concretion, the otolith is laid down in concentric layers, a process which is probably continuous. Unknown factors, likely dependent on food or seasons, cause slight color or density variations which produce definite laminations in most species.

Preparation of the otoliths for study varies, and appears to depend on the species and on the preferences of the biologist. Otoliths are examined fresh, sectioned, and in various liquid preparations. Obtaining the otolith necessitates sacrificing the fish, occasionally a disadvantage where a small population is involved or when valuable sport or commercial fish are concerned.

Spines and Rays

Spines and rays have been found serviceable for age identification, since growth is regular with an annular addition of material so differentiated it can be accurately interpreted. Spines and rays must be sectioned near the base if all annular rings are to be included, and sections must be cut exactly transversely. A fine jeweler's saw is usually used to cut the sections. Preparation varies, from no special technique to mounting in glycerin. The use of spines and rays does not require killing the fish.

Although some authors make certain claims for using spines and rays, we can see no advantage over scales.

Other Bony Structures

Technical papers describe utilization of various bony structures, depending on species and author. Vertebrae and scales have the added advantage of possible use in back-calculation of length. Which structure—fin rays, spines, otoliths, scales, vertebrae—serves best may be a matter of personal choice.

The Teleost Scale

The scale remains still the most popular means of estimating the age and calculating the growth of fishes. Understanding the process of teleost scale formation may help in interpreting ages and other life-history events. Laying down of bone is associated with the presence of cells known as osteoblasts. In the teleost scale these osteoblasts are concentrated in primary papillae which first appear along the lateral line in the posterior region of the body. These primary papillae send out oblique outgrowths above and below the lateral line from which, at regular intervals, arise secondary papillae. Scales first appear in the primary papillae and then in the secondary papillae.

The usual teleost scale consists of an outer bony layer and an inner fibrillary plate consisting of closely applied fibrous lamellae. The scale first appears as osteoid tissue within a close investment of osteoblasts. The osteoid tissue is collagenous (containing collagen, an organic material) and does not become definitive bony tissue until the process of calcification is complete. The bony layer is formed only at the margin of the scale, through the activity of the osteoblasts lying at the periphery of the scale pocket. As long as the scale continues to grow, osteoid tissue is present at the outer margins. The fibrillary plate appears as a very thin layer in immediate contact with the bony layer. It increases in thickness through the addition of broad, thin sheets of connective tissue from the floor of the scale pocket. The fibrillary plate, like the bony layer, is collagenous at first, becoming infiltrated later with ichthylepidin, an organic substance recognized in teleost scales. The fibrillary plate is laid down behind the bony layer.

Ridges or circuli used in determining the age of fish are continuous and homogeneous with the general bony layer and are the result of elevations of the osteoid marginal area. The circuli increase in height during the process of calcification. Their formation probably depends on the presence of more bone-forming materials in the intercellular fluid than can be utilized at the growing margins.

Radii represent the lines of flexibility in the scale. The flexible condition is attained by the absence of the bony layer and through a special condition whereby the underlying fibrillary plate does not become impregnated with ichthylepidin beneath the radii or grooves.

Ctenoid and cycloid scales are commonly used in age determinations and growth calculations. The most obvious distinction between these two types is the presence of the ctenii, small spines, on the posterior of the ctenoid scale. In general, fishes with soft-rayed fins have cycloid scales, and spiny fishes have ctenoid scales. There are interesting combinations of the two. Some sea perches have ctenoid scales above the lateral line and cycloid below. Some flatfishes have ctenoid scales on the upper side and cycloid on the underside. Scales may be weakly ctenoid with small ctenii or strongly ctenoid with long, coarse spines. All scales begin as cycloid, and even in spiny fishes, some most commonly located on the head, cheek, and opercle, may never develop ctenii. One of the arguments for the continual growth of the fish scale is the presence of long, sharp ctenii on the posterior margin of scales of even the oldest fish.

There are certain structures common to both types of scales. The *focus* near the center of the scale is a small, clear area which presumably represents the original scale platelet of the young fish. *Ridges* or *circuli* are numerous more or less concentric striations around the focus. The ridges or circuli provide the sculpturing that is interpreted in determining the age. *Radii* represent lines of flexibility in the scale, and examination of the scales tends to show more radii present where a greater body movement is necessary. In the absence of the ctenii, as in the cycloid scale of the white sucker, radii may be found in the posterior field.

Several attempts have been made to develop taxonomic keys to fishes on the basis of scale characteristics and shape. The future shape of a scale depends on the amount of space available between it and the scales around it. In the mirror carp, a mutant form of *Cyprinus carpio,* the scales, few and isolated, grow to an immense size and have rounded margins. The scales of the burbot and the brook trout retain their juvenile, embedded condition as regular, circular cycloid scales. In this type of scale the focus remains in the approximate center surrounded by equal growth. Ctenoid scales with overlapping or imbrication have an unequal growth of the anterior and/or the posterior fields which shifts the relative position of the focus and serves to change the general outline of the scale.

In examining series of scales, irregularities are frequently observed. The most common irregularity is *regenerated* (latinucleate) *scales* in which the clear, well-defined focus is replaced by an expanded central area, devoid of circuli, rough or granular in appearance and irregular in outline. The relative size of this regenerated center depends on the size of the scale at time regeneration began. Although growth beyond the regenerated portion is normal, these scales are for the most part of no value in estimating age. Small scars or patches commonly found on scales are presumably repaired injuries which occurred earlier on the former margins of the scale. An interesting irregularity occurs when a yound scale becomes dislocated and turns slightly in the scale pocket. This results in the appearance of a smaller scale off center in a larger scale. In some instances two papillae may grow together so that one scale may develop with two foci.

Three primary conditions must be met if scales are to be accepted for accurate determinations of age and for back-calculating lengths of fish at the time of earlier annulus formation:

1. The scales must remain constant in number and identity throughout the life of the fish. That they retain their identity is proved when one examines the nuclear area or focus of the scales from young fish and finds them structurally identical with the scales of the older fish. All regenerated scales have a characteristic central portion. Scales increase in size with growth of the body of the fish. That the number of scales remains reasonably constant in all but a few species is determined by use of the lateral-line scale count, and other scale counts, to distinguish between closely related species and subspecies of fish.

2. Growth of the scale must be proportional to the growth of the fish, but scale growth rarely bears an exact linear relation to body growth. However, agreement has been found among the growth histories of the same year class at different ages and among different year classes as to the relative amount of growth in certain calendar years. Supposing a sample is taken annually to follow the survival of fish of a certain species hatched in 1940 in one area. Environmental conditions during 1943, for example, made it a very poor growing year. This poor growth will be recorded on all fish of the 1940-year class surviving during and subsequent to that time whether they be taken in the 4-year age group or in the 8-year age group. On the other hand, fish hatched in 1938, or 1942, though belonging to other year classes, should still reflect the poor growing season of 1943 on their scales.

3. The annulus must be formed yearly and at the same approximate time each year. The majority of experiments designed to test this have shown that the annulus may be used in determining the age of fish: annuli increase in number as the fish grows. Modes in length-frequency distributions of younger fish coincide with modal lengths of age groups based on scale interpretation. The persistent abundance or scarcity of certain year classes are further proof.

Although lists of characteristics are available for recognizing the annulus, there is no substitute for experience in actually reading scales. Scales of different species have peculiarities that can be learned only by observation. One of the best and surest characteristics, especially with ctenoid scales, is the so-called ''cutting over'' or ''crossing over.'' In the fall, or with the cessation of scale development, the outer ridges or circuli tend to flare outward so that several of them may end on the side of the scale. The greater the eccentricity of the focus, the greater the discontinuity of the circuli. The first circulus laid down in the spring follows regularly around the entire scale margin and thus must "cut across" or "cross over" the incomplete ridges which did not grow completely around the scale at the cessation of scale development the previous fall. Another recognition characteristic is the presence of discontinuous circuli, commonly observed in the sculpturing of cycloid scales. Individual circuli arising at different loci around the scale never grow completely together as a result of cessation of scale growth. Crowding of the circuli prior to the resumption of growth, and, in some species, the erosion of the scale during the spawning period may be clues to the exact location of the annulus.

The irregularities in the sculpturing of the scale are undoubtedly reflections of some environmental change acting on the physiological processes of the fish. Temperature is thought to be the most effective single factor. The most apparent effect caused by a drop in temperature is the reduction of the metabolic rate resulting in the cessation of active feeding. An unusually high temperature may produce the same effect. In general, then, the time of annulus formation can be readily correlated with water temperature.

The fishery biologist must be alert to detect accessory year marks or false annuli, which are usually less distinct or less continuous than the true annulus. These accessory marks or false annuli may occur as the result of changes in the rate of growth during the spawning period, seasonal variation in amount of food, extremes of climatic conditions, periods of floods, or times of drought. In many cases the close proximity of the false annulus to a true annulus will aid in its detection. An interesting annulus formation is that of a "natal annulus" formed at birth in some members of the family Embiotocidae, presumably because the embryonic food supply is cut off and a temporary reduction in the food supply occurs until the young fish becomes adapted to capturing food in its new environment. Just as the fishery scientist must be on the alert for accessory annuli, so must he be watching for the omission of annuli, not usual in this hemisphere. Omission would more likely occur in species occupying an environment with uniform temperature condition.

Ctenoid scales are taken from the side in the region near the tip of the pectoral fin, whereas cycloid scales are generally removed from an area between the dorsal fin and the lateral line. The scales in these areas occur in regular order and with a minimum of regeneration. The practice of taking "key" scales, those determined by actual count, cannot be recommended. This is time consuming, and better results can be obtained by taking several scales from a general area. Removing as much mucus, dirt, and epidermis as possible prior to scale removal saves having to clean the scales later. Scales are usually removed with a knife or stout forceps.

Temporary mounts of scales can be made by simply placing the scales in water between two glass slides. This serves to flatten out the scales and is quick and easy to do. Permanent mounts can be made in mounting medium, but for most permanent records plastic impressions are the most popular. The impression records only the scale sculpturing, making it especially useful in interpreting thick scales through which the transmission of light is difficult. Generally, the impression method consists of softening the surface of the plastic slide, usually with heat, and then pressing the scale, sculptured side down, firmly into the softened area. The scale is removed, and the impression is ready for study. Early presses were simple screw-down types, and they worked well. More sophisticated developments include thermostatically controlled heat, hydraulic pressure, and rollers.

Age determinations and measurements of scales for back-calculations of lengths can be accomplished most efficiently with a projector such as the Bausch and Lomb Tri-Simplex. This inexpensive projector can be cheaply equipped with three lenses and used to project scales down on the table or onto a screen. Experience of the authors would indicate that the best scale readings can be accomplished by making temporary mounts, projecting them, and having them interpreted by two or three competent biologists. This method saves remounting, projection time, and takes advantage of multiple consulting. The biologists should have full knowledge of the fish from which the scales were taken.

Some authors have followed the custom of reporting annuli in Roman numerals and growing seasons in Arabic numerals. In either case it is important to establish the criterion. For example: determining the age of a smallmouth black bass in October would mean that the age, if decided merely on the basis of the number of annuli present, would be one less than the growing seasons the fish has been exposed to, since the last annulus on the scale in this case would have been laid down in late spring or early summer. After October or November little growth would take place. This point is important where comparisons of growth are made, and sometimes necessitates an annulus being postulated at the edge of each scale.

If one uses the terminology in which fish up to one year of age are called the 0-age group, the age averages 6 months low. A better method perhaps is to use only Arabic numerals and describe a fish as being in the first year (group 1) when it is 0 to 1 year old and in the second year (group 2) when it is 1 to 2 years old, and so forth. In stating the age of anadromous fish it is customary to show the length of time spent in both the freshwater and sea-life periods. Thus a salmon aged 5, migrated to the sea in its third year (always reckoned from the time the eggs are spawned), and is returning to spawn in its fifth year. Since such a fish is almost exactly 5 years of age at spawning time, the use of the 5 for the fifth year is a very convenient and meaningful designation. The terminology occasionally employed, especially in freshwater studies, of using only the annulus would result in this fish being called a 4-year-old, even though it belongs in a 5-year cycle.

Growth

A fishery manager frequently has as a management objective the production of the greatest harvest from a given population, while at the same time he has to keep costs to a minimum. The harvest may be of many forms. For example, a manager might wish to have the greatest number of fish caught, or to catch the greatest number of fish weighing over 2 pounds, or the greatest total weight of fish, or he may wish the harvest to be the greatest number of rainbow trout jumping up over small falls for people to watch. He may then have a variety of objectives depending upon the actual situation he is faced with at the moment. Growth and mortality studies of fish populations can be observed and interpreted to help the manager achieve his objective.

Growth is the process of increase or the progressive development of an organism. Although in actuality a very complex process, we might measure growth by the change in length or weight of an individual fish or a group of fish between two sampling times. Growth of a population might refer to a change in the number of fish observed at different times. Changes in population number are considered in the discussion of mortality in Chapter 7. The basis for growth in our text is therefore the observation or measurement of an attribute (that is, length, weight, number) at different times.

Many factors may influence the growth of fish. Among the more common determinants are the amount and size of food available; the number of fish using the same food resource; temperature, oxygen, and other water-quality factors; the size, age, and sexual maturity of the fish. If we could accurately measure all the factors influencing growth at every instant of time, we would be able to describe the process of growth. This of course is not practical or for that matter, necessary. What is necessary is to have some representation, that is, a model, that describes the process sufficiently for the purpose at hand. A graph of length at age may be all that is necessary for some purposes, while much more complex representation such as von Bertalanffy's equation might be necessary for calculating yield. The model chosen may be in part determined by the intended use.

Length is the more frequently measured attribute of growth, perhaps because of ease of measurement and the relationship that has been developed between length and weight, and length and scale measurements. These relationships will be developed later, after a consideration of length as a measurement of growth.

A simple representation of growth could be obtained by securing a sample of fish from a population and recording length observations from each individual fish. A convenient way of recording information and at the same time securing a scale sample is to use a printed coin envelope: recording the necessary information on the outside and putting the scale inside to ensure that the scale will be correctly associated with the information. After the age of each fish is determined, length may then be summarized for each age and a mean length calculated. From a graph of mean length and age, which provides a representation of growth, it is then possible to obtain some idea of growth from a single population of fish. A discussion about the worth of such a growth curve follows.

Estimates of mean length are, in the first place, based upon the *sample*. Therefore, if the sample is not representative of the population, the estimates of length will be in error. In this connection, the possibility of gear selectivity should always be kept in mind—younger age groups may not be sampled at all. When possible, it is a good idea to compare samples taken at the same time but with different gear. More confidence would be associated with mean-lengths-at-age that were statistically equal.

Another problem with a single sample is that it is composed of different year classes of fish. Not all year classes have the same size at any given age, since growing conditions are subject to many variations. Age-3 fish this year might be quite different in mean length from age-3 fish last year. Generally, year classes that have a smaller mean length at early ages tend to the long-term average at older ages. This is called growth compensation. It is also true that the year classes with larger than average mean length at younger ages again tend to the long-term average for mean length at older ages. A procedure called back-calculation, to be discussed below, may be used to help overcome the problem of different growth of year classes when using a single sample.

Comparison of fish growth between different bodies of water using single samples for each water has an additional problem. Growth is influenced by many factors which would seldom be the same at the same time for any two bodies of water. Therefore, when lengths are observed for a sample of fish, the mean-length-at-age relationship from the observed data would not necessarily be the same even if in fact fish size was equal in both bodies of water. For example, assume a simple situation in which two populations are growing equally, but from which samples are obtained at two different times. One way to partly correct for this would be to collect samples after growth had slowed down, if such a period occurred. A better way would be to compare lengths at a specified annulus. This can be done by back-calculation.

Reference has already been made to the use of the fish scale (measurements to annuli on the vertebrae have also been used) in back-calculating body lengths. The back-calculation method allows determination of past body lengths if certain growth relationships exist between the scales of fishes and body length. Many formulas and many modifications have been proposed. This discussion will be concerned with the basic direct proportion straight-line relationship and one transformation if a curvilinear relationship should exist.

The easiest application of the back-calculation method would be to assume that a direct proportion exists between the scale and body growth throughout the entire life of the fish. The basis for this assumption comes from the general knowledge that although fish may vary in size, they possess approximately the same body contour, that the scale counts in both small and large fish are approximately the same. The direct-proportion formula is also based on the assumption that isogonic growth is demonstrated by the scale and body length or that their growth is at the same proportional rate. A final assumption is that the body and scale growth demonstrate a straight-line relationship, with the origin of the line passing through zero as illustrated by line *A* in Figure 5–2 below.

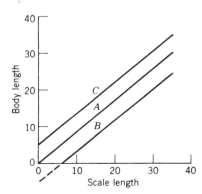

Figure 5–2. Relation of body length and scale length

The formula for back-calculation of body lengths, if a direct proportion between the scale and body growth is assumed, is

$$\frac{\text{Length of scale radius to annulus } X \ (S')}{\text{Length of total scale radius } (S)} =$$

$$\frac{\text{Length of fish when annulus } X \text{ was formed } (L')}{\text{Length of fish at time scale sample was obtained } (L)}$$

as in $S'/S = L'/L$, or $L' = (S'/S)L$.

However, biologists have found that in almost all studies the results from direct proportion give calculated lengths lower than empirical lengths. This discrepancy results because the fish has attained some length before the scales are laid down, and thus line C in the above figure is the more likely. For example, smallmouth bass are approximately 2 cm, brook trout 3.6 cm, king salmon about 5 cm, and the cod 3 to 4 cm in body length before the scales appear. The easiest way to handle this is to determine the correction factor (C) by extrapolating the regression line represented by plotting the body length against the scale measurement. The value on the y axis where it is cut by the regression line is the correction factor which may be either plus or minus (lines C and B). Fortunately, it is not often necessary to know the length of a fish before the first annulus has been formed, so the intercept as calculated from the regression of body length on scale length is valid. When the intercept is plus, as in line C, then the correction factor C would be subtracted from the body lengths L and L' in the direct-proportion formula as:

or:

$$L' - C = \frac{S'}{S} (L - C)$$

$$L' = C + \frac{S'}{S} (L - C)$$

Up to this point only a linear relationship between body length and scale length has been considered. There are, however, small differences in the body proportions between small and large individuals which make heterogonic rather than isogonic growth the more correct assumption describing the growth relationship of the scale and fish. Usually the relationship between scale and body length when plotted graphically is found to be curvilinear. In some cases the relationship may be represented even by a sigmoid curve.

Usually the greatest curvilinearity is in the youngest ages back-calculated from scales of older fish, assuming a straight- or nearly straight-line relationship. The problem of curvilinearity can then be solved by omitting the earliest years and back-calculating only those ages that do not deviate appreciably from a straight line. The formula for a straight-line relationship with the intercept as the correction factor may be used to calculate past lengths. If the *entire* range of the curve between scale and body length deviates significantly from a straight line, then it may be necessary to transform the data. A logarithmic transformation may straighten out the regression. The formula for the back-calculation of lengths with the logarithmic transformation is $L' = cS'^n$, or, changed to logarithms, it becomes $\log L' = \log c + n (\log S')$, where $L' =$ length to be calculated; $S' =$ measurement of scale; and c and $n =$ constants derived from the data.

Where a sigmoid curve best fits the body length-scale length relation it may be necessary to derive two formulas; one to fit the data below the inflection point and the other to fit the data above the inflection. Possibly only the data above the inflection are pertinent, in which case the formula as suggested above for the curvilinear relationship would be employed.

Once information is obtained describing length at age, the way in which the relationship is expressed will depend upon the purpose for which the study is being made. Absolute growth is generally considered to be the average size of fish at each age. The curve for growth may be quite different from the line connecting average size at age. In many populations seasonal growth is more of a sigmoid-shaped curve expressing periods of rapid growth with increase in food supply and optimum temperatures, and periods of slow growth often associated with cold temperature and scarce food. The absolute growth curve may also be sigmoid in shape due to slow growth of very young and old fish with a period of years of faster growth for in-between ages. To show the rate of absolute growth in the razor clam in a slightly different manner, the annual increments, or differences between successive total lengths, are plotted against the age. This type of curve is known as the first differential of the absolute growth curve (Figure 5–3). The curve shows an increase of growth rate which reaches a maximum at the point of inflection of the total length curve. After the maximum, the rate declines throughout life and for a time closely approximates a descending geometric series, which is another way of saying that each yearly growth is a certain percentage of the preceding. This relationship breaks down in extreme old age when growth is actually greater than predicted.

Relative growth is the gain in size for some time period in relation to the

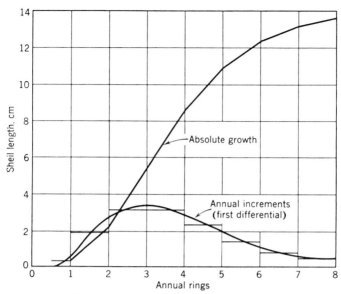

Figure 5-3. Curve of absolute growth, showing growth of the razor clam (*Siliqua patula*) at Hallo Bay, Alaska. Courtesy of F. W. Weymouth and H. C. McMillin, 1931, U.S. Bur. Fish., Bull. XLVI(1099):543.

size at the beginning of the time period. Letting subscripts designate age, relative growth, h, in length would be expressed as:

$$h = \frac{l_6 - l_5}{l_5} = \frac{100 - 75}{75} = 1/3$$

or:

$$l_6 = l_5 + hl_5$$

and:

$$l_5 = \frac{l_6}{1 + h}$$

Notice that relative growth is here an expression of the rate of growth, that is, for the change in size for some specified period of time. The relative growth rate is constant for the period being considered, and as such assumes a straight line connecting the two points as the representation of growth. This may be adequate for short time periods but is seldom accurate when considering an entire year.

This type of growth may also be shown on either of two curves (Figure

5-4), by plotting the logarithms of the total size against time (line *A*), or by plotting percentage yearly increases against time (line *B*). When the logarithm of the total dimension is used, the curve rises steeply at first, but the slope continually declines throughout life. When the percentage yearly gains are plotted against time, the continually decreasing increment is obvious.

The greatest difference in the absolute and relative treatment of growth comes in early life, since the slow growth of old age differs little, whether regarded from the absolute or the relative viewpoint. The absolute growth of young organisms is slow, but it increases constantly up to the maximum at the point of inflection, after which it declines regularly. Relative growth

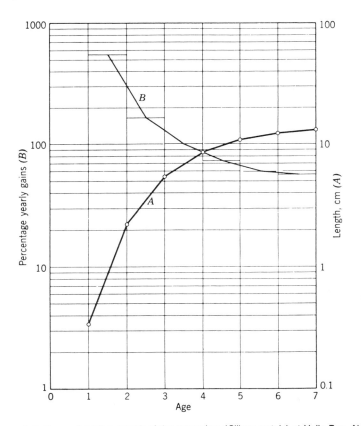

Figure 5-4. Curve of relative growth of the razor clam (*Siliqua patula*) at Hallo Bay, Alaska. Courtesy of F. W. Weymouth and H. C. McMillin, 1931, U.S. Bur. Fish., Bull. XLVI(1099):567.

on the other hand is most rapid at the youngest ages and declines constantly. The most significant difference between the two viewpoints relates to the rate of growth.

An estimate of midyear size made from relative growth information would generally be in poor agreement with actual data. This is one reason why other models of growth were developed. Other reasons were to attempt to have a biologically sound basis for the model and to obtain a growth model that would be tractable in the mathematical calculations involving production and yield. The use of these models has been mainly in computation of production and yield on a weight basis. Before developing the growth models, and since weight of fish produced and harvested is necessary, some consideration should be given to the relationship between length and weight.

Length and weight are attributes that can be measured. Length is generally more easily and more accurately measured than weight. Weight is a criterion of harvest in many commercial fisheries and is frequently the form in which harvest is reported as the weight of the total catch. Length and weight are highly correlated, and therefore by knowing one, the other may be predicted. Length-weight relationships have been of two forms, one relating to isometric growth, the other to allometric growth. All body parts grow at the same rate with isometric growth, whereas with allometric growth it is assumed that different parts of the body grow at different rates. The length-weight relationship for isometric growth is of the form $W = CL^3$, where W is the weight, L is the length, and C is a constant to be empirically determined. This formula implies that the growth rate for length, breadth, and depth of a fish is equal.

The coefficient C in the above formula when multiplied by 10^5 is often referred to as the condition factor, or coefficient of condition. The coefficient of condition is a convenient way of comparing the relative well-being of fish populations. The larger the coefficient, the heavier the fish for a given length. However, there are many variables, such as sex, season of year, stage of maturity, and size of the fish, that influence the condition factor, and when comparing condition these should be equal.

The second expression of the length-weight relationship results from curve fitting when allometric growth is occurring. The form of the model is the same as above, but two constants are determined empirically. The model is $W = aL^b$, where a and b are constants. These constants cannot be interpreted in the same manner as for isometric growth; they are determined by linear regression, using logarithms of the length and weight data. Both isometric and allometric models have been used for computation of production and yield, with the isometric model having been most commonly used.

Two growth models, used extensively in fishery work, have been developed. The first is more intuitively applied to weight, since weight, as shown above, tends to increase exponentially. The model is:

$$w_t = w_o e^{gt}$$

where w_t is weight at time t, w_o is initial weight, g is the growth coefficient, and e is the base of natural logarithms. This model is a sufficient approximation if short intervals of time are considered, as for example for a year or still better for a growing season. The main advantage for this model is the simplicity of the calculation for production and yield. The main disadvantage is that the entire life span cannot be treated by one growth coefficient, and many short time intervals have to be used. The growth coefficient g for a time interval is estimated by the natural logarithm of the ratio of final weight to initial weight for that time period, as follows:

$$g_t = ln \ \frac{w_t}{w_o}$$

Length gives a somewhat poorer fit to the model that does weight, suggesting that if the length is used, a shorter time interval should be used.

The second growth model is ascribed to von Bertalanffy. This model seems to fit most data on observed fish growth, at least for that period after the point of inflection in the absolute growth curve has been reached. The model, which has a biological basis for its development, is as follows:

$$l_t = L_\infty \, (1 - e^{-K(t - t_o)})$$

where l_t is length at age t, L_∞ is the ultimate length for the population, K is a growth coefficient, and t_o is a time when length would theoretically be zero. An equivalent and alternate form of the von Bertalanffy equation is:

$$l_t = L_\infty \, (1 - e^{-Kt}) + l_o e^{-Kt}$$

with symbols as in the above formula and l_o equal to length at time zero.

Obtaining estimates for the parameters L_∞, K, and t_o or l_o requires more calculation than for the estimate of g above. There are computer programs which iteratively solve for the parameters and if at hand will give better results than the alternate procedure. The alternate procedure is not difficult but does sacrifice some accuracy. The estimates for the parameters are arrived at through another procedure for handling length data.

Walford noted that if the length at one age was plotted against the length at the next younger age, the result was a straight line for many populations. In fact, if the resulting Walford plot is a straight line with a slope less than one, a von Bertalanffy growth curve will fit the data. The two are very closely related. The slope of a Walford line is equal to e^{-K}, so therefore the natural logarithm of the Walford slope—with sign changed—provides an estimate of the growth coefficient K. The ultimate length, L_∞, may also be estimated from the Walford plot. This estimate is taken as the point where the growth line intersects a 45° line drawn through the origin. These relations are shown in the illustration (Figure 5-5).

Mathematical solution is more accurate than graphic solution and should therefore be done when possible. The t_o parameter may be estimated from

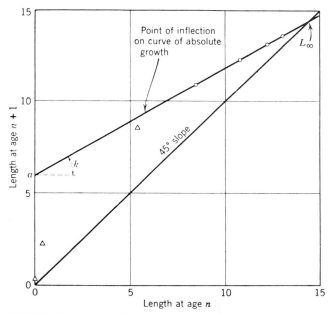

Figure 5-5. Walford's growth transformation using data of Weymouth and McMillin (1931) for razor clams at Hallo Bay, Alaska. Data fit straight line after inflection point on the absolute growth curve.

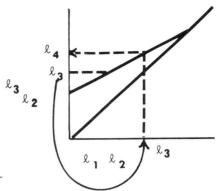

Figure 5-6. A Walford line used to generate growth information

the growth equation after estimates of K and L_∞ have been made, or an iterative procedure may be used to solve for the three parameters simultaneously.

There are several other uses for the Walford plot in addition to providing means for estimating growth parameters. It is possible, for example, to construct growth curves from data on animals which cannot be aged directly, by tagging the animal and observing the size at recapture. The Walford plot can also be used to determine the effect of tagging. In this case, size at tagging against size at recapture may be compared to the regular Walford line. If the point for tagged fish is below the line, growth is considered impaired. The Walford line may also be used to extend or generate growth information to ages which may not be readily determined from scales. The procedure is illustrated in Figure 5-6 as follows. If, for example, lengths at age 1, 2, and 3 are known, but scales for fish older than 3 years are not readable, a Walford line may be drawn through the two points (l_1, l_2) and (l_2, l_3). Length at age 3 may then be taken as the point on the x axis, extended to the Walford line and then to the y axis. This point on the y axis is length at age 4. The procedure is repeated, thereby generating the expected size for ages beyond what may be read from scales. This procedure might also be used to check aging, with the assumption, of course, that the growth curve is the von Bertalanffy form.

The following example will demonstrate the various growth representations discussed above, using data from a lake trout population. The accompanying list shows the average length and weight for ages I through VII.

Age	Length (mm.)	Length increment	Weight (g.)	Weight increment
I	159		68	
		87		132
II	246		200	
		114		317
III	360		517	
		121		544
IV	481		1061	
		82		513
V	563		1574	
		53		395
VI	616		1969	
		47		399
VII	663		2368	
		7		59
VIII	670		2427	

The absolute growth curve is shown in Figure 5-7 as line A. This is simply a line connecting the mean length for each age. The growth increments are plotted as line B with the scale for increments given on the right side ordinate.

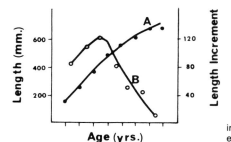

Figure 5-7. Average length and length increments plotted against age for lake trout example

Relative growth is calculated as the increment between two age groups divided by the length at the younger age. Thus relative growth (h) between age I and II is:

$$h = \frac{l_2 - l_1}{l_1} = \frac{246 - 159}{159} = 87 \div 159 = 0.55$$

Between ages V and VI, relative growth is:

$$h = \frac{53}{563} = 0.09,$$

or applied to weight for the same ages, it is:

$$h = \frac{395}{1574} = 0.25$$

The coefficient of growth or instantaneous rate of growth of Ricker would be calculated as the difference between natural logarithms of weight for consecutive age groups; for example, between ages III and IV;

$$g = ln\ w_4 - ln\ w_3 = 0.72$$

The Walford growth transformation of plotting length at age $n + 1$ against length at age n is shown in Figure 5–8. The points used for this relationship are:

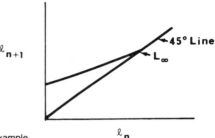

Figure 5–8. Walford line for the lake trout example

Length at age n + 1	(ordinate)	246	360	481	563	616	663	670
Length at age n	(abscissa)	159	246	360	481	563	616	663

Least-squares linear regression using the lake-trout data provides a relationship for the Walford line:

$$l_{n+1} = 147.25 + 0.83\ l_n$$

We may determine the ultimate length, L_∞, for this population as follows:

$$L_\infty = \frac{\text{intercept}}{1 - \text{slope}} = \frac{147.25}{1 - 0.83} = 875$$

The growth coefficient in the von Bertalanffy equation, K, was shown to be equal to minus the natural logarithm of slope of the Walford line. The estimate for the lake-trout data above would be:

$$K = -ln\ \text{slope} = -ln\ 0.83 = 0.186$$

The equation for the von Bertalanffy relationship was:

$$l_t = L_\infty (1 - e^{-K(t - t_o)})$$

Thus far we have preliminary estimates for L_∞ and K, with t_o being the remaining parameter to estimate. If we take the natural logarithm of the von Bertalanffy equation, we find that:

$$ln (L_\infty - l_t) = ln L_\infty + Kt_o - Kt$$

which is a straight line. Therefore, plotting the natural logarithm of $L_\infty - l_t$ against age t should result in a straight line if the data conform to von Bertalanffy's equation and if our preliminary estimate of L_∞ is accurate. The slope of the plot will be equal to $-K$ with the intercept value equal to $ln L_\infty + Kt_o$. We may iteratively choose the value of L_∞ which gives the best linear fit and from this determine a better estimate of K as well. The intercept value of this regression may then be used to solve for t_o:

$$t_o = \frac{\text{Intercept} - ln L_\infty}{K}$$

For our lake-trout example we find that the regression of $ln (L_\infty - l_t)$ against t is an excellent linear fit ($r = 0.99$). The slope of this regression, 0.196, is taken as the estimate of K. The estimate of t_o is, from the above equation:

$$t_o = \frac{6.785 - ln\ 875}{0.196} = 0.057$$

The von Bertalanffy equation as estimated for the lake-trout population is:

$$l_t = 875 (1 - e^{-0.196(t - 0.057)})$$

This equation could then be used for comparison with other populations of lake trout to compare growth or ultimate size, and for calculation of yield to determine optimum fishing by changing to weight. This is discussed in more detail in Chapter 8.

References

Abramson, N. J. 1963. Computer programs for fisheries problems. Trans. Amer. Fish. Soc. 92(3):310.

Adams, L. A. 1931. Determination of age in fishes. Trans. Illinois Acad. Sci. 23:219–226.

_____. 1940. Some characteristic otoliths of American Ostariophysi. J. Morphol. 66(3):497–527.

_____. 1942. Age determination and rate of growth in the *Polydon spatula*, by means of the growth rings of the otoliths and dentary bone. Amer. Midl. Nat. 28:617–630.

Aikawa, H. 1937. Age determination of chub-mackerel, *Scomber japonicus* (Houttyn). Jap. Soc. Sci. Fish., Tokyo, Bull. 6:9–12.

Alexander, M. 1958. The place of aging in wildlife management. Amer. Sci. 46(2):123–137.

Alvord, W. 1954. Validity of age determinations from scales of brown trout, rainbow trout and brook trout. Trans. Amer. Fish. Soc. 83:91–103.

Appleget, J., and L. L. Smith, Jr. 1951. The determination of age and rate of growth from vertebrae of the channel catfish, *Ictalurus lacustris punctatus*. Trans. Amer. Fish. Soc. 80:119–139.

Bagenal, T. B., ed. 1974. Proceedings of an international symposium on the ageing of fish. Unwin Brothers Limited, The Gresham Press, Old Working, Surrey, England. 234 pp.

Bardach, J. E. 1955. The opercular bone of the yellow perch, *Perca flavescens*, as a tool for age and growth studies. Copeia 1955(2):107–109.

Beamish, R. J., and D. Chilton. 1977. Age determination of lingcod (*Ophiodon elongatus*) using dorsal fin rays and scales. J. Fish. Res. Bd. Can. 34(9):1305–1313.

Beckman, W. C. 1943a. Annulus formation on the scales of certain Michigan game fishes. Pap. Michigan Acad. Sci. Arts Lett. 28:281–312.

_____. 1943b. Further studies on the increased growth rate of the rock bass, *Ambloptites rupestris* (Rafinesque), following the reduction in density of population. Trans. Amer. Fish. Soc. 74:72–78.

_____. 1948. The length-weight relationship factors for conversions between standard and total lengths, and coefficients of condition for seven Michigan fishes. Trans. Amer. Fish. Soc. 75:237–256.

Bertalanffy, L. von 1938. A quantitative theory of organic growth. Hum. Biol. 10(2):181–213.

Bilton, H. T., and D. W. Jenkinson. 1968. Comparison of the otolith and scale methods for aging sockeye (*Oncorhynchus nerka*) and chum (*O. keta*) salmon. J. Fish. Res. Bd. Can. 25(5):1067–1069.

_____. 1969. Age determination of sockeye (*Oncorhynchus nerka*) and chum (*O. keta*) salmon from examination of pectoral fin rays. J. Fish. Res. Bd. Can. 26(5):1199–1203.

Bilton, H. T., D. W. Jenkinson, and M. P. Shepherd. 1964. A key to five species of Pacific salmon (*Genus oncorhynchus*) based on scale characters. J. Fish. Res. Bd. Can. 21(5):1267–1288.

Bilton, H. T., and W. E. Ricker. 1966. Supplementary checks on the scales of pink salmon (*Oncorhynchus gorbuscha*) and chum salmon (*O. keta*). J. Fish. Res. Bd. Can. 22(6):1477–1489.

Boyko, E. G. 1946. Age determination of fishes based on cross sections of fin rays. C. R. Acad. Sci. Moscow, New Ser. 53(5):483–484; English reprint in Prog. Fish-Cult. 12(1):47–48.

Bulkley, R. V. 1960. Use of brachiostegal rays to determine age of lake trout, *Salvelinus namaycush* (Walbaum). Trans. Amer. Fish. Soc. 89(4):344–350.

Butler, R. L., and L. L. Smith, Jr. 1953. A method for cellulose acetate impressions of fish scales with a measurement of its reliability. Prog. Fish-Cult. 15(4):175–178.

Cable, L. E. 1956. Validity of age determination from scales, and growth of marked Lake Michigan lake trout. U.S. Fish Wildl. Serv., Fish. Bull. 107. 59 pp.

Campbell, R. S., and A. Witt, Jr. 1953. Impressions of fish scales in plastic. J. Wildl. Manage. 17(2):218–219.

Clemens, H. P. 1951. The growth of the burbot, *Lota lota maculosa* (LeSueur) in Lake Erie. Trans. Amer. Fish. Soc. 80:163–173.

Coble, D. W. 1966a. Dependence of total annual growth in yellow perch on temperatures. J. Fish. Res. Bd. Can. 23(1):15–20.

_____. 1966b. Alkaline phosphate in fish scale. J. Fish. Res. Bd. Can. 23(1):149–152.

Cooper, R. A. 1967. Age and growth of the tautog. *Tautoga onitis* (Linnaeus), from Rhode Island. Trans. Amer. Fish. Soc. 96(2):134–142.

Creaser, C. W. 1926. Structure and growth of scales of fishes in relation to the interpretation of their life-history, with special reference to the sunfish *Eupomotis gibbosus*. Univ. Michigan, Misc. Publ. 17. 82 pp.

Eschmeyer, R. W. 1937. Some characteristics of a population of stunted perch. Pap. Michigan Acad. Sci. Arts Lett. 22:613–628.

_____. 1938. Further studies of perch populations. Pap. Michigan Acad. Sci. Arts Lett. 23:611–613.

_____. 1940. Growth of fishes in Norris Lake, Tennessee. J. Tennessee Acad. Sci. 15(3):329–341.

Everhart, W. H. 1949. Body length of the smallmouth bass at scale formation. Copeia 1949(2):110–115.

_____. 1950. A critical study of the relation between body length and several scale measurements in the smallmouth bass, *Micropterus dolomieu* Lacépède. J. Wildl. Manage. 14(3):266–276.

Fitch, J. E., and R. L. Brownell, Jr. 1968. Fish otoliths in cetacean stomachs and their importance in interpreting feeding habits. J. Fish. Res. Bd. Can. 25(12):2561–2574.

Ford, E. 1933. An account of the herring investigations conducted at Plymouth during the years from 1924–1933. J. Mar. Biol. Assoc. 19:305–384.

Fry, F. E. J. 1939. A comparative study of lake trout fisheries in Algonquin Park, Ontario. Ontario Fish. Res. Lab., Publ. 46. 69 pp.

_____. 1943. A method of calculation of the growth of fishes from scale measurements. Ontario Fish. Res. Lab., Publ. 61:7–18.

Galtsoff, P. S. 1952. Staining of growth rings in the vertebrae of tuna (*Thunnus thynnus*). Copeia 1952(2):103–105.

Gerking, S. P. 1965. Two computer programs for age and growth sutdies. Prog. Fish-Cult. 27(2):59–66.

Graham, M. 1929. Studies of age determination in fish. Part II. A survey of the literature. Min. Agric. Fish and Food (U.K.), Fish. Invest. Ser. II, 11(3):1–50. Reviewed in J. Conseil Int. Explor. Mer 5(1):117–121.

Greenbank, J., and D. J. O'Donnell. 1950. Hydraulic presses for making impressions of fish scales. Trans. Amer. Fish. Soc. 78:32–37.

Harrison, E. J., and W. F. Hadley. 1979. A comparison of the use of cleithra to the use of scales for age and growth studies. Trans. Amer. Fish. Soc. 108(5):452–456.

Hart, J. S. 1941. The availability of smallmouth black bass, *Micropterus dolemieu* Lacépède, in lakes of Algonquin Park, Ontario. Trans. Amer. Fish. Soc. 70:172–179.

Hatch, R. W. 1961. Regular occurrence of false annuli in four brook trout populations. Trans. Amer. Fish. Soc. 90(1):6–12.

Havey, K. A. 1959. Validity of the scale method for aging hatchery-reared Atlantic salmon. Trans. Amer. Fish. Soc. 88(3):193–196.

Hickling, C. F. 1931. The structure of the otolith of the hake. Q. J. Microsc. Sci. 74:547–562.

Hile, R. 1941. Age and growth of the rock bass, *Ambloplites rupestris* (Rafinesque), in Nebish Lake, Wisconsin. Trans. Wisconsin Acad. Sci. Arts Lett. 22:189–337.

———. 1943. Mathematical relationship between the length and the age of rock bass (*Ambloplites rupestris*). Pap. Michigan Acad. Sci. Arts Lett. 28:331–341.

———. 1970. Body scale relationship and calculation of growth in fishes. Trans. Amer. Fish. Soc. 99(3):468–474.

Hile, R., and F. W. Jobes. 1942. Age, growth and production of the yellow perch. *Perca flavescens* (Mitchill), of Saginaw Bay. Trans. Amer. Fish. Soc. 70:102–122.

Hubbs, C. L. 1921. The ecology and life-history of *Amphigonopterus aurora* and of other viviparous perches of California. Biol. Bull. 40(4):181–209.

Irie, T. 1957. On the forming of annual rings in the otolighs of several teleosts. J. Fac. Fish. Anim. Husb., Hiroshima Univ., Fukuyama 1(3):311–317.

———. 1960. The growth of the fish otolith. J. Fac. Fish. Anim. Husb. Hiroshima Univ., Fukuyama 3(1):203–221.

Jenkinson, J. W. 1912. Growth, variability and correlation in young trout. Biometrika VIII:444–466.

Kennedy, W. A. 1943. The whitefish, *Coregonus clupeaformis* (Mitchill) of Opeongo, Algonquin Park, Ontario. Ontario Fish. Res. Lab., Publ. 62:23–66.

———. 1954. Growth, maturity and mortality in the relatively unexploited lake trout, *Cristivomer namaycush,* of Great Slave Lake. J. Fish. Res. Bd. Can. 11(6):827–852.

Koo, T. S. Y., and A. Isaranrua. 1967. Objective studies of scales of Columbia River chinook salmon. *Oncorhynchus tshawytscha* (Walbaum). U.S. Fish. Wildl. Serv., Fish. Bull. 66(2):165–180.

LaLanne, J. J., and G. Safsten. 1969. Age determination from scales of chum salmon (*Oncorhynchus keta*). J. Fish. Res. Bd. Can. 26(3):671–681.

Lane, C. E., Jr. 1954. Age and growth of the bluegill, *Lepomis m. macrochirus* (Rafinesque) in a new Missouri impoundment. J. Wildl. Manage. 18(3):358–365.

Larkin, P. A., J. G. Terpenning, and R. R. Parker. 1957. Size as a determinant of growth rate in rainbow trout *Salmo gairdneri.* Trans. Amer. Fish. Soc. 86:84–96.

Linhart, S. B., and F. F. Knowlton. 1967. Determining age of coyotes by tooth cementum layers. J. Wildl. Manage. 31(2):362–365.

Marcy, B. C. 1969. Age determination from sclaes of *Alosa pseudoharengus* (Wilson) and *Alosa aestivalis* (Mitchill) in Connecticut waters. Trans. Amer. Fish. Soc. 98(4):622–630.

Marzolf, R. C. 1955. Use of pectoral spines and vertebrae for determining age and rate of growth of the channel catfish. J. Wild. Mange. 19(2):243–249.

McConnell, W. J. 1952. The opercular bone as an indicator of age and growth of the carp, *Cyprinus carpio* Linnaeus. Trans. Amer. Fish. Soc. 81:138–149.

McEachran, J. D., and J. Davis. 1970. Age and growth of the striped sea robin. Trans. Amer. Fish. Soc. 99(2):343–352.

McKern, J. L., and H. F. Horton. 1970. A punch to facilitate the removal of salmonid otoliths. California Fish Game J. 56(1):65–68.

Miller, R. B. 1946. Notes on the arctic grayling. *Thymallus signifer* Richardson, from Great Bear Lake. Copeia 1946(4):227–236.

Mosher, K. H., and H. H. Eckles. 1954. Age determination of Pacific sardines from otoliths. U.S. Fish Wildl. Serv., Res. Rep. 37. 40 pp.

Mottley, C. M. 1942a. The effect of increasing the stock in a lake on the size and condition of rainbow trout. Trans. Amer. Fish. Soc. 70:414–420.

———. 1942b. The use of the scales of rainbow trout (*Salmo gairdnerii*) to make direct comparisons of growth. Trans. Amer. Fish. Soc. 71:74–79.

Neave, F. 1936. The development of the scales of *Salmo*. Trans R. Soc. Can. 30 Sect. 5:55–72.

———. 1940. On the histology and regeneration of the Teleost scale. Q. J. Microsc. Sci. 81(Part IV):541–568.

———. 1943. Scale pattern and scale counting methods in relation to certain trout and other salmonids. Trans. R. Soc. Can. 37 Sect. 5:79–91.

Nesbit, R. A. 1934. A convenient method of preparing celluloid impressions of fish scales. J. Conseil Int. Explor. Mer 9(3):373–376.

Neuhold, J. M. 1957. Age and growth of the Utah chub, *Gila atraria* (Girard), in Paguitch Lake and Navajo Lake, Utah, from scales and opercular bones. Trans. Amer. Fish. Soc. 85:217–233.

Nordeng, H. 1961. On the biology of char (*Salmo alpinus L.*) in Salengen, North Norway. Nytt Magsin Zool. 10:67–123.

Paloheimo, J. E., and L. M. Dickie. 1965. Food and growth of fishes. 1. A growth curve derived from experimental data. J. Fish. Res. Bd. Can. 22(2):521–542.

———. 1966. Food and growth of fishes. 2. Effects of food and temperature on the relation between metabolism and body weight. J. Fish. Res. Bd. Can. 23(6):869–908.

Parker, R. R., and P. A. Larkin. 1959. A concept of growth in fishes. J. Fish. Res. Bd. Can. 16(5):721–745.

Probst, R. T., and E. L. Cooper. 1955. Age, growth, and production of the lake sturgeon (*Acipenser fulvescens*) in the Lake Winnebago Region, Wisconsin. Trans. Amer. Fish. Soc. 84:207–227.

Reimers, N. 1958. Conditions of existence, growth, and longevity of brook trout in a small high-altitude lake of the eastern Sierra Nevada. California Fish Game J. 44(4):319–333.

Richards, F. J. 1959. A flexible growth function for empirical use. J. Exp. Bot. 10(29):290–300.

Ricker, W. E. 1958. Handbook of computations for biological statistics of fish populations. Fish. Res. Bd. Can., Bull. 119. 300 pp.

Ricker, W. E. ed. 1968. Methods for assessment of fish production in fresh waters. Blackwell Scientific Publications, Oxford. 313 pp.

Roelfs, E. W. 1958. Age and growth of whitefish. *Coregonus clupeaformis* (Mitchill), Big Bay de Noc and northern Lake Michigan. Trans. Amer. Fish. Soc. 87:190-199.

Scott, D. M. 1954. A comparative study of the yellowtail flounder from three Atlantic fishing areas. J. Fish. Res. Bd. Can. 11(3):171-197.

Semakula, S. N., and P. A. Larkin. 1968. Age, growth, food and yield of the white sturgeon (*Acipenser transmontanus*) of the Fraser River, British Columbia. J. Fish. Res. Bd. Can. 25(12):2589-2602.

Smith, S. B. 1955. The relation between scale diameter and body length of Kamloops trout, *Salmo gairdneri kamloops*. J. Fish. Res. Bd. Can. 12(5):742-753.

Smith, S. H. 1954. Method of producing plastic impressions of fish scales without using heat. Prog. Fish-Cult. 16(2):75-78.

Sneed, K. E. 1951. A method for calculating the growth of channel catfish, *Ictalurus lacustris punctatus*. Trans. Amer. Fish. Soc. 80:174-183.

Sutro, L. L. 1971. Study of automatic means of determining the age of fish. Massachusetts Inst. Tech., Sea Grant Rep. MITSG-72-2. 41 pp.

Templeman, W., and H. J. Squires. 1956. Relation of otolith lengths and weights in the haddock *Melanogramus aeglefinus* (L.) to the rate of growth of the fish. J. Fish. Res. Bd. Can. 13(4):467-487.

Van Oosten, J. 1923. A study of the scales of whitefishes of known ages. Zoologica II(17):380-412.

_____. 1929. Life history of the lake herring *Leucichthys artedi* LeSueur of Lake Huron as revealed by its scales, with a critique of the scale method. U.S. Bur. Fish. Bull. 44:265-428.

_____. 1944. Factors affecting the growth of fish. Trans. N. Amer. Wildl. Conf. 9:177-183.

_____. 1953. A modification in the technique of computing average lengths from scales of fishes. Prog. Fish-Cult. 15(2):85-86.

Walford, L. A. 1946. A new graphic method of describing the growth of animals. Biol. Bull. 90(2):141-147.

Weber, D. D., and G. S. Ridgeway. 1962. The deposition of tetracycline drugs in bones and scales of fish and its possible use for marking. Prog. Fish-Cult. 24(4):150-155.

Webster, D. A., W. A. Lund, R. W. Wahl, and W. D. Youngs. 1960. Observed and calculated lengths of lake trout (*Salvelinus namaycush*) in Cayuga Lake, New York. Trans. Amer. Fish. Soc. 89(3):274-279.

Went, A. E. and W. E. Frost. 1942. River Liffey Survey. V. Growth of brown trout (*Salmo trutta L.*) in alkaline and acid waters. Proc. Irish Acad., XLVIII, Sect. B, 4:67-84.

Weymouth, F. W., and H. C. McMillin. 1931. Relative growth and mortality of the Pacific razor clam *Siliqua patula* (Dixon) and their bearing on the commercial fishery. U.S. Bur. Fish., Bull. XLVI (1099):543-567.

Whitney, R. R., and K. D. Carlander. 1956. Interpretation of body-scale regression for computing body length of fish. J. Wildl. Manage. 20(1):21-27.

Youngs, W. D. 1958. Effect of the mandible ring tag on growth and condition of fish. New York Fish Game J. 5(2):184-204.

6

Estimating Population Size

Estimation of fish population abundance is necessary for understanding basic changes in population number and composition, for estimating yield, and as a basis for sound management. Occasionally there is an opportunity for direct counting when a population is concentrated and available during some life history stage. More often, indirect methods must be employed either singly or, better, in various combinations to minimize estimation errors. Direct enumeration is the most accurate, and the expense of building a trapping facility and counting may sometimes be as low as the expense of collecting and analyzing data to estimate population size by an indirect method.

Direct Enumeration

Fishes such as the salmon and trout make concentrated spawning migrations and can be guided by weirs either into a collecting box for counting and detailed examination, or past some counting device or observer. Adults can be counted moving upstream, and both young and adults can be counted as they move back downstream.

Biologists in Alaska have taken advantage of the behavior of migrating salmon to erect and utilize counting towers. Salmon follow current patterns along the banks of spawning streams, and in clear streams observers can count the number of fish moving upstream. Observation can be improved by providing a light bottom background, such as painted flooring or metal. Towers need not be manned continuously, but counting periods of ten minutes per hour can be randomized and then expanded for a complete migration period. Species identification is also possible with this method.

Counting towers may be replaced by more sophisticated devices such as a plastic pipe with counting devices in it, so that a fish swimming over one of the apertures in the pipe is automatically recorded on a counter. This device does not enumerate different species and, at present, it counts each fish moving past, whether in an upstream or downstream direction.

Trapping and/or counting facilities should be included in all fishway designs. For many years on the Columbia River dams, alert women sat at stations to identify and enumerate the fish moving across white horizontal counting boards at the exits of the fishways. Now a closed-circuit TV camera, connected to a digital counter with a video-tape machine, is used to record salmon as they move upstream. At the exit of the fishway, water is passed through two stainless-steel tunnels, placed end to end, with the inside walls insulated. Electrodes are imbedded on the inner surfaces of each tunnel, producing a weak electrical field in the water between a pair of electrodes. When a fish penetrates the field in the first tunnel, the circuit is broken, the videotape machine is activated, and the fish is on TV. As the fish proceeds past the second tunnel, the tape the tape machine is automatically shut off. If the fish does not pass through the second tunnel, the machine turns off automatically. Fish passage is periodic at various times of the year and during the day; the night count, for example, is 2 to 8 percent of the day count. By using video tape it is possible to compress the total day's fish passage into about a one-hour reel and, on replay, fish can be accurately identified by stopping the tape when necessary.

Aerial photography to enumerate populations has always tempted biologists and might, under certain ideal conditions, produce reasonably accurate results. Photographing fish migrating over a light background in a river is possible, but only when times of migration and exceptional water clarity coincide.

Small fish ponds should always be designed with facilities for easy draining and with catchment basins for population enumeration. Such devices combine facilities for efficient exploitation with excellent opportunities to study the inter- and intrarelationships of fishes.

Area Density

The area-density method of estimating population consists of counting the number of animals in a series of sample strips or plots distributed systematically or randomly throughout the total area whose population is to be determined. The sample count is then expanded to an estimate of the total population by multiplying the aggregate sample count by the fraction: total area divided by the sum of sample areas. More formally stated, an estimator for an area-density or partial-time coverage (which is the same as sampling a subarea except the subarea is time instead of space) is derived from the following formula:

$$\hat{N} = \frac{A}{a} \sum_{i=1}^{a} N_i$$

where $\hat{N}$ is the estimated population, A is the number of equal units of area (or time) occupied by the total population, a is the number of units sampled, and N_i is the number counted in the ith sample area. The estimated variance for this estimator is:

$$\hat{V}(\hat{N}) = \frac{A^2 - aA}{a} \cdot \frac{a \sum_{i=1}^{a} N_i^2 - \left(\sum_{i=1}^{a} N_i \right)^2}{a(a-1)}$$

The area-density method is especially useful for determining populations in streams sufficiently small to permit sample stream sections to be blocked off with seines and the total population in each section captured. Sample sections should be selected randomly. Sampling should be stratified so that each habitat type will be proportionately represented in the sampling. Sample stratification by habitat categories provides valuable information on the distribution of fish by size, age, and abundance in each habitat type; however, this method of sampling should not be attempted unless enough information is available to do it reliably. When habitat stratification is inadvisable, sampling areas may be chosen uniformly from among sequential stream segments of equal length.

The area-density method may also be applied to any bottom-dwelling forms or to nonschooling fish, especially benthic species, in areas of specified size. This method has been applied to clam populations, using a clamshell bucket that collects all of the top layer to a depth of several centimeters over an area of 0.5 square meters. Another possible method of area-density sampling involves an underwater camera triggered to photograph a defined area when the camera has reached a specified distance above the bottom.

An interesting application of the area-density method has been described for estimating spawning populations of Atlantic herring. With the aid of scuba equipment the investigators quantified the areas of those strata within the spawning ground where herring-egg density was very light, light, intermediate, medium, and heavy. The average number of eggs per square yard in each stratum was determined, to give an estimate of the total number of eggs deposited. Dividing this by the number of eggs per female and multiplying by 2 (assuming one male for each spawning female) gave an estimate of 185,000,000 in the spawning population.

Another time to get information is after chemical reclamation, when scuba-diver counts of dead fish can be made along compass lines across the bottom of a pond to estimate the total population of dead fish by the area-density method, providing the width and length of the strips sampled are known. The estimate is, of course, supplemented by enumeration of the dead fish afloat on the surface and those washed ashore. Often, however, populations from reclamation operations can be more easily and accurately estimated by the Petersen ratio (discussed below), if a fraction of the population can be marked and released before reclamation.

Most indirect methods of population estimation depend on a ratio or on a regression. Many modifications and refinements are available and leave the biologist with the choice of method or combination of methods likely to provide him with the more accurate estimate. Satisfying the basic assumptions inherent in every method is the major problem.

Mark-Recapture: Single Census

One of the simplest and most practical of population estimations is the well-known Petersen ratio using marked fish. A sample of fish is collected, marked, and released, and then at a later time a second "recapture" sample is taken which includes both marked and unmarked fish. The method is based on the general assumption that the proportion of marked fish recovered is to the total catch in the second sample as the total number of marked fish released is to the total population. Specifically, the validity of the Petersen estimate rests on the following assumptions: that marked fish, during the period between release and recapture, suffer no greater mortality nor emigrate further than unmarked fish; that no marks are lost, nor are any recaptured marked fish overlooked; and further that marked fish are caught at the same rate as unmarked; that marked fish are randomly distributed, or if not, the recaptures are; and that there will have been no additions to the population.

The estimator for the Petersen method is $\hat{N} = MC/R$, where $\hat{N}$ is the estimate of population size, M is the number of fish marked and released, C is the recapture sample size which includes both marked and unmarked fish, and R is the number of marked fish which are recaptured. The estimate is that of the population present at the time of the first (marking and release) sample, not at the time of recapture.

The variance of the Petersen estimator depends upon the manner in which the recapture sample is drawn. Most fishery biologists have not been very particular about the way recapture samples were taken and have therefore frequently not used appropriate variance estimates when calculating confidence intervals. There are several ways in which recapture samples may be collected that will provide known distributions and variance estimators. The more common procedure would be to predetermine the recapture sample size, C, and then take a sample of this size, counting the number of marked fish in this sample. The variance of the estimated population is then as follows:

$$\hat{V}(\hat{N}) = \hat{N}^2 \, \frac{(\hat{N} - M) \, (\hat{N} - C)}{M \, C \, (\hat{N} - 1)}$$

The Petersen estimator with the above type of sampling is negatively biased, that is, the estimates of N are less on the average than the true population size. This bias is large with small sample sizes but reduces to acceptable levels when $M \cdot C > 4N$. An example will illustrate the procedure for a Petersen estimate, assuming the following data have been collected; marked fish released, $M = 550$; predetermined recapture sample, $C = 500$; recaptured marked fish, R (included in C) $= 157$; then:

$$\hat{N} = \frac{550 \times 500}{157} = 1751$$

$$\hat{V}(\hat{N}) = 1751^2 \, \frac{(1751 - 550) \, (1751 - 500)}{550 \cdot 500 \, (1750)} = 9571$$

A confidence interval may be placed on the estimate of N, utilizing the standard error, as follows:

$$\hat{N} \pm 1.96 \, \sqrt{\hat{V}(\hat{N})} = 1751 \pm 1.96 \cdot 97.83 = 1751 \pm 192$$

and therefore:

$$P \, (1559 \leqslant N \leqslant 1943) = 0.95$$

This statement means that if sampling were done in such a way that all recapture samples of size C had an equal chance of being taken, and if sampling were repeated an infinite number of times, then 95 percent of the confidence intervals computed in a similar fashion would contain the actual population number N. A check of $M \times C$ shows that it is greater than $4N$ and our estimate of N is therefore not strongly biased.

Charts are available relating the sizes of M and C necessary for a stated level of precision in a population estimate. Values for M and C necessary to insure at the 95 percent confidence level that any population estimate will differ from the actual population number by no more than 25 and 50 percent are shown respectively in the accompanying charts (see Figures 6-1 and 6-2).

It is seen from these figures that some prior knowledge of population

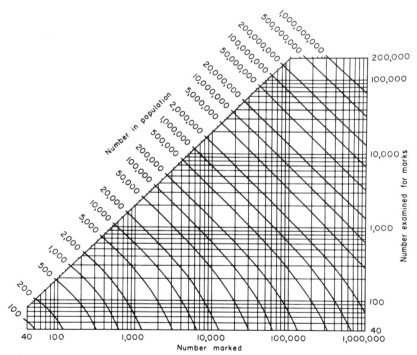

Figure 6-1. Sample size when $1 - \alpha = 0.95$ and $\rho = 0.25$; recommended for management studies. Data based on normal approximation to the hypergeometric distribution. Courtesy of D. S. Robson and H. A. Regier, 1964, Trans. Amer. Fish. Soc. 93(3):221.

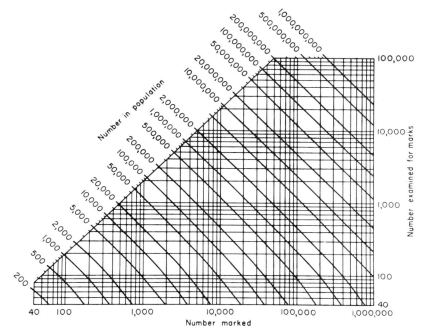

Figure 6-2. Sample size when $1 - \alpha = 0.95$ and $\rho = 0.50$; recommended for preliminary studies and management surveys. Data based on normal approximation to the hypergeometric distribution. Courtesy of D. S. Robson and H. A. Regier, 1964, Trans. Amer. Fish. Soc. 93(3):220.

size is necessary in order to use the charts. A rough estimate may therefore be necessary prior to the actual estimation. The 50 percent level of precision is suggested for surveys while the 25 percent level is recommended for management studies.

In our example above, we had released 550 marked fish (M) and had predetermined a recapture sample of 500 (C). Using the 50 percent level-of-confidence chart, we find $M = 550$ on the Number Marked axis and follow this line up to the intersection with the curved line that represents our estimated population size of approximately 2,000. This intersection shows that approximately 60 fish should be examined for marks. The same level of precision could be obtained in our example from as few as 180 marked fish with 200 in the recapture sample or with 700 marked fish and 40 in the recapture sample. Thus we could take into consideration the cost of obtaining fish for marking and the cost of examining the catch for recaptures and optimally allocate resources to obtain the desired level of precision. Most fishery problems involve nearly the same cost of operation for marking and for examination of recapture sample. Therefore, an equal allocation of resources for both the marking and the recovery operations would be appropriate.

Other types of sampling lead to different estimators of variance. One other type is binomial sampling as, for example, when a fish is taken from the population, examined for a mark, and then returned to the population. Observations of marked and unmarked fish by scuba divers or from towers would be other examples of this type of sampling. The variance estimate for the estimated population for binomial sampling is:

$$\hat{V}(\hat{N}) = \frac{M^2 \, C \, (C - R)}{R^3}$$

With this type of sampling approximate confidence intervals may also be obtained from charts or tables of the binomial or Poisson distributions. In using such tables, R/C is the variable for which limits are determined, and the limits for N are obtained by inverting these limits. This procedure is illustrated in the following example where M (marked fish) = 200; C (recapture sample) = 300; R (marked fish recaptured) = 75:

$$\frac{R}{C} = \frac{75}{300} = 0.25$$

Then:

$$\frac{1}{\hat{N}} = \frac{R}{C} \cdot \frac{1}{M} = \frac{0.25}{200} = 0.00125$$

Since R is binomially (or Poisson) distributed, binomial (or Poisson) charts or tables may be used to determine the limits. The 95 percent confidence limits for 0.25, when C equals 300, are approximately 0.20 and 0.30. Our confidence interval for the reciprocal of the population estimate is:

$$P \, (0.0010 \leqslant \frac{1}{N} \leqslant 0.0015) = 0.95$$

which may be inverted to obtain the confidence interval for N:

$$P \, (667 \leqslant N \leqslant 1000) = 0.95$$

Mark-Recapture Models: Multiple Census

Case 1: No Mortality

All too frequently it is not possible either to mark or to recapture a sufficient number of fish to provide the desired precision in a Petersen-type population estimate. The biologist often is operating in a situation where sampling is continued for some time period with new fish being marked, previously marked fish being recaptured, and all fish released. Assumptions under these models are random sampling, or random mixing of marked and unmarked, recognition of all marks, and no recruitment. Mortality will determine the model used. The first model is applicable when no mortality takes place, the second model when there is known mortality, and the third model when there is unknown mortality.

The model assuming no mortality was proposed by Darroch, and population number is taken as the iterative solution to the following:

$$1 - \frac{u}{N} = \left(1 - \frac{C_1}{N}\right)\left(1 - \frac{C_2}{N}\right) \cdots \left(1 - \frac{C_k}{N}\right)$$

where C_i is the catch of the ith sample and u is the sum of the unmarked fish taken in all k samples. For example assume a population is sampled on four successive days with catches of $C_1 = 100$, $C_2 = 120$, $C_3 = 90$, and $C_4 = 150$. Recaptures are 12, 19, and 42 for the second, third, and fourth samples respectively. Our estimate would be that number which makes the left-hand side equal the right-hand side in the above equation. Substituting numbers in the equation gives the following:

$$1 - \frac{387}{\hat{N}} = \left(1 - \frac{100}{\hat{N}}\right)\left(1 - \frac{120}{\hat{N}}\right)\left(1 - \frac{90}{\hat{N}}\right)\left(1 - \frac{150}{\hat{N}}\right)$$

Choose as an initial value $\hat{N} = C_1 C_2 / R_1 = 1000$ and calculate both sides of the above expression. The left-hand side is equal to 0.613000, the right-hand side gives a value of 0.612612. Robson and Regier provide a method to help in the iterative solution to the above expression. Let the right-hand side of the expression evaluated at the initial guess, called N_o, be P_{N_o}. The next value to try for N, call it N_1, may be found by the following:

$$N_1 = N_o + \frac{N_o P_{N_o} - (N_o - u)}{\Delta_{N_o}}$$

with

$$\Delta_{N_o} = 1 + (N_o - 1) \, P_{N_o-1} - N_o P_{N_o}$$

and P_{N_o-1} is analogous to P_{N_o} being the right-hand side evaluated at $N_o - 1$. This process continues until two consecutive estimates of N are essentially equal. The estimated variance for $\hat{N}$ may then be calculated as:

$$\hat{V}(\hat{N}) = \frac{\hat{N} - u}{\Delta_{\hat{N}}}$$

with $\Delta_{\hat{N}}$ being the Δ expression calculated for the last trial in the estimation of N. Finishing the above example provides the following:

$$\hat{N}_1 = 1000 + \frac{1000 \times 0.612612 - (1000 - 387)}{\Delta_{\hat{N}_o}} = 994$$

$$\Delta_{\hat{N}_o} = 1 + (1000 - 1) \times 0.612291 - 1000 \times 0.612612$$

$$\Delta_{\hat{N}_o} = 1 + 611.679148 - 612.612612 = 0.066536$$

$$\hat{N}_2 = 994 + \frac{(994 \times 0.610681) - (994 - 387)}{0.067732} = 994$$

$$\Delta_{\hat{N}_1} = 1 + (994 - 1) \times 0.610357 - 994 \times 0.610681$$

$$= 0.067732$$

Since the two values, $\hat{N}_1$ and $\hat{N}_2$, are essentially the same, 994 is taken as the estimate for N. The estimated variance of 994 is calculated as:

$$\hat{V}(\hat{N}) = \frac{\hat{N} - u}{\Delta_{\hat{N}}} = \frac{994 - 387}{0.067732}$$

$$= 8961.79$$

The approximate 95 percent confidence interval assuming normal theory is:

$$P \, (805 \leq N \leq 1183) = 0.95$$

Case 2: Known Mortality

Known mortality can be accounted for in the above model by subtracting known deaths from the population size N. For example, suppose that sampling caused 5, 6, 4, and 7 fish to die for the four respective samplings giving catches of 100, 119, 89, and 148. Assume that recaptures of marked fish were 0, 11, 18, and 41 for the respective samples. Unmarked fish are equal to catch less recaptures or in this example $u = 456 - 70 = 386$. The equation as given before is set up as follows using the above data as an example:

$$\frac{\hat{N}\text{-}386}{\hat{N}} = \left(\frac{\hat{N}\text{-}100}{\hat{N}}\right) \left(\frac{\hat{N}\text{-}119\text{-}5}{\hat{N}\text{-}5}\right) \left(\frac{\hat{N}\text{-}89\text{-}5\text{-}6}{\hat{N}\text{-}5\text{-}6}\right) \left(\frac{\hat{N}\text{-}148\text{-}5\text{-}6\text{-}4}{\hat{N}\text{-}5\text{-}6\text{-}4}\right)$$

$$\frac{\hat{N}\text{-}386}{\hat{N}} = \left(\frac{\hat{N}\text{-}100}{\hat{N}}\right) \left(\frac{\hat{N}\text{-}124}{\hat{N}\text{-}5}\right) \left(\frac{\hat{N}\text{-}100}{\hat{N}\text{-}11}\right) \left(\frac{\hat{N}\text{-}163}{\hat{N}\text{-}15}\right)$$

Taking $C_1 C_2 / R_1 = 1082$ as the initial value for $\hat{N}$ gives a value of 0.637540 for the right-hand side. The same iterative procedure given above is used to obtain the next trial value for N.

Calculate $\hat{N}_1$ as:

$$\hat{N}_1 = \hat{N}_o - \frac{\hat{N}_o P_{\hat{N}_o} - (\hat{N}_o - u)}{\Delta_{\hat{N}_o}}$$

$$\hat{N}_1 = 1082 + \frac{(1082 \times 0.637540) - (1082 - 386)}{0.055535} = 971$$

where

$$\Delta_{\hat{N}_o} = 1 + (\hat{N}_o - 1) P_{\hat{N}_o - 1} - \hat{N}_o P_{\hat{N}_o}$$
$$= 1 + (1081)(0.637256) - (1082)(0.637540)$$

Repeating this procedure yields $\hat{N}_2 = 980$, and $\hat{N}_3 = 980$. Since $\hat{N}_2 = \hat{N}_3$ we take 980 as the estimate of N. The $\Delta_{\hat{N}_3}$ is used to calculate the estimated variance of $\hat{N}$ from the expression:

$$\hat{V}(\hat{N}) = \frac{\hat{N} - u}{\Delta_{\hat{N}}}$$

For this example the values are:

$$\hat{V}(\hat{N}) = \frac{980 - 386}{0.066825}$$

$$= 8888.88$$

The approximate 95 percent confidence interval assuming normal theory is:

$$P = (795 \leq N \leq 1165) = 0.95$$

Case 3: Unknown Mortality

The procedure given below for handling the situation with unknown mortality requires that the mark placed on fish be a serially numbered tag. Similar to the Petersen-type estimator a random sample is taken from the population to be estimated, tagged with serially numbered tags, and released. The date of release must be recorded. Samples taken on subsequent dates are treated in like fashion with tag number and date being recorded for recaptures, and for newly tagged fish as well. All fish are released.

An unbiased estimator for population size at time t_i is given by:

$$\tilde{N}_i = \frac{(m_i + 1)(c_i + 1)}{r_i + 1} - 1$$

where

$\tilde{N}_i$ = population size at time t_i
m_i = the sample from N_i at time t_i
c_i = total number of distinct fish captured *after* t_i
r_i = total number of distinct recaptures from the release of m_i tagged fish.

The estimated variance of $\tilde{N}_i$ is given by:

$$\tilde{V}(\tilde{N}_i) = (\tilde{N}_i + 1)(\tilde{N}_i + 2) - \frac{(m_i + 1)(m_i + 2)(c_i + 1)(c_i + 2)}{(r_i + 1)(r_i + 2)}$$

Assumptions for this method are no recruitment, random sampling on all occasions, complete reporting of all tagged fish, no loss of tags, and no effect of tagging on released fish. A tagged fish recaptured on a given date, t_i, becomes part of the m_i tagged fish released on that date. Perhaps an example is the most efficient way of demonstrating the technique. The example is hypothetical and limited in terms of tagged fish in the interest of conserving space.

Suppose that a trapnet is used to sample a population of trout in a small pond and further assume that the pond is not suitable for trout reproduction. The pond is stocked with a group of fall fingerling fish and sampled on three occasions at six-month intervals. The sampling program could lead to the following data set:

Sample	Sample size $= c_i$	Recaptures (r_i samples 2,3)		$m_{i+1} - r_{i+1} = c_i$
1 May 79	100	10	3	(76-7) + (68-0)
2 Oct 79	76		7	(68-0) = 68
3 May 80	68		0	

Of the 13 recaptures of fish tagged during the first sample, one tagged fish was caught on both of the subsequent sampling periods. The r for estimating population size at the first sample period is therefore 12, the number of distinct subsequent recaptures.

The population present at the May 1979 sampling date would be estimated as:

$$\tilde{N}_{May} = \frac{(m + 1)(c + 1)}{r + 1} - 1 = \frac{(100 + 1)(137 + 1)}{12 + 1} - 1$$

$$\tilde{N}_{May} = 1071$$

The October 1979 population would be estimated by:

$$\tilde{N}_{Oct} = \frac{(76 + 1)(68 + 1)}{7 + 1} - 1$$

$$\tilde{N}_{Oct} = 663$$

An estimate of the variance of this estimate would be:

$$\tilde{V}(\tilde{N}) = (\tilde{N} + 1)(\tilde{N} + 2) - \frac{(m + 1)(m + 2)(c + 1)(c + 2)}{(r + 1)(r + 2)}$$

$$= (664)(665) - \frac{(77)(78)(69)(70)}{(8)(9)}$$

$$= 38657.5$$

Using normal theory as a reasonable approximation, 95 percent confidence interval would be:

$$P(270 \leq N \leq 1056) = 0.95$$

Survey Removal

Another type of ratio estimator, also referred to as the change in composition or dichotomy method, is useful where two different classes are recognizable and have different rates of exploitation affecting each class. This method uses the ratio of the two classes before and after harvest, and the number of fish harvested in each of the two categories. The method depends upon the following assumptions: no natural mortality, recruitment, or migration during the time of sampling and harvest; the ratio can be estimated without bias; the two classes or categories are always distinguishable; and total catch or harvest is known exactly. Addition to the population would also provide necessary information, but there is a strong possibility that there would be behavioral differences between the two groups; that is, they might not mix randomly, thus biasing the ratio estimates. Classes might be on the basis of sex, marked and unmarked, juvenile and adult, or two different species.

When the two classes are X and Y, the needed information may be defined as follows; n_1, n_2 = sample size before and after harvest; x_1, x_2 = number of X in n_1 and n_2; y_1, y_2 = number of Y in n_1 and n_2; C_x, C_y = total catch of X and Y respectively during harvest; and further defined as: $C = C_x + C_y$; $p_1 = x_1/n_1$ and $p_2 = x_2/n_2$.

If N_x and N_y are the numbers of X and Y in the population before harvest, estimators are as follows:

$$\hat{N}_x = \frac{p_1 (C_x - p_2 C)}{p_1 - p_2}$$

and:

$$\hat{N} = \hat{N}_x + \hat{N}_y = \frac{C_x - p_2 C}{p_1 - p_2} = \hat{N}_x/p_1$$

The method is illustrated in the following example. Suppose that a lake has brook trout and white sucker present and that trapnet samples provide the following data before and after a special ice-fishing season; $n_1 = 90$ fish, of which 30 are brook trout (x_1) and 60 are white sucker (y_1); and $n_2 = 58$ fish of which 14 are brook trout (x_2) and 44 are white sucker (y_2). Therefore:

$$p_1 = \frac{30}{90} = 0.3333 \text{ and } p_2 = \frac{14}{58} = 0.2414$$

A complete census tabulated a total catch of 160 brook trout (C_x) and 160 white sucker (C_y). We can estimate the number of brook trout before the special season as:

$$\hat{N}_x = \frac{p_1 (C_x - p_2 C)}{p_1 - p_2}$$

$$= \frac{0.3333 (160 - 0.2414 \times 320)}{0.3333 - 0.2414}$$

$$= 300$$

The number of brook trout plus white sucker before the season was:

$$\hat{N} = \hat{N}_x / p_1 = 300 \div \frac{1}{3} = 900$$

and therefore the original number of white sucker was:

$$\hat{N}_y = \hat{N} - \hat{N}_x = 900 - 300 = 600$$

The number of white suckers could have been estimated directly by using the above formula for $\hat{N}_x$ and substituting the y subscript with corresponding values.

Approximate estimates for variances of $\hat{N}_x$ and $\hat{N}$ are:

$$\hat{V}(\hat{N}_x) \doteq \frac{\hat{N}^2 \hat{p}_2{}^2 \hat{V}(\hat{p}_1) + (\hat{N} - C)^2 \hat{p}_1{}^2 \hat{V}(\hat{p}_2)}{(\hat{p}_1 - \hat{p}_2)^2}$$

and:

$$\hat{V}(\hat{N}) \doteq \frac{\hat{N}^2 \hat{V}(\hat{p}_1) + (\hat{N} - C)^2 \hat{V}(\hat{p}_2)}{(\hat{p}_1 - \hat{p}_2)^2}$$

These variances are dependent upon the manner in which the samples are drawn when estimating p_i. Two cases are considered: sampling with replacement and sampling without replacement. Where sampling with replacement, as for example catching one fish at a time and returning it to the water or with direct observation such as by scuba swimmers, the estimated variance for $\hat{p}_i$ is:

$$\hat{V}(\hat{p}_i) = \frac{\hat{p}_i(1 - \hat{p}_i)}{n_i}$$

When sampling without replacement, as for example by observing the trap-net catch, the estimated variance for $\hat{p}_i$ is:

$$\hat{V}(\hat{p}_i) = \frac{\hat{p}_i(1 - \hat{p}_i)}{n_i - 1} \left(1 - \frac{n_i}{\hat{N}_i}\right)$$

The appropriate variance for $\hat{p}_i$ is substituted in the above variance formula for either $\hat{N}_x$ or $\hat{N}$, and the resulting variance can be used for the construction of confidence intervals, providing that the n_i are not small.

A confidence interval for the brook trout-white sucker example given above would be computed as follows. Sampling was without replacement, therefore:

$$\hat{V}(\hat{p}_1) = \frac{(0.3333)\,(0.6667)}{90 - 1}\left(1 - \frac{90}{900}\right) = 0.002247$$

and:

$$\hat{V}(\hat{p}_2) = \frac{(0.2414)\,(0.7586)}{58 - 1}\left(1 - \frac{58}{580}\right) = 0.002891$$

The estimated variance for $\hat{N}_x$ is then:

$$\hat{V}(\hat{N}_x) = \frac{900^2 \times 0.2414^2 \times 0.002247 + 580^2 \times 0.3333^2 \times 0.002891}{(0.3333 - 0.2414)^2}$$

$$= \frac{155.18}{0.00844561}$$

$$= 18373.63$$

If we assume that the n_i are large enough (borderline in this example), the confidence interval for N_x may be computed, assuming the $\hat{N}_x$ are normally distributed, as approximately:

$$P(29 \leq N_x \leq 571) = 0.95$$

These are not very inspiring confidence limits for an estimate, but notice that by taking larger samples, n_1 and n_2, the variance estimate of $\hat{p}_i$ may be reduced, which in turn will reduce the variance estimate of $\hat{N}_x$.

Catch-Effort

Population size may also be estimated from catch and effort data. Three methods are commonly used depending upon catchability and constancy of effort. The Leslie and DeLury methods are appropriate for variable effort. Assumptions necessary for these methods are a closed population; sampling with respect to effort is a Poisson process such that qf is the probability an individual fish is caught with q being the constant catchability coefficient and f being effort; and all individual fish have the same chance of being caught in any given sample. The method of Moran and Zippin is appropriate when effort is constant and assumes that the population is closed, that chance of capture is equal for all individual fish and remains constant from sample to sample. In a practical sense there must be enough fish removed so that the population shows a decline.

In the Leslie method the assumption of catch-per-unit effort to stock abundance is related symbolically as:

$$\frac{C_t}{f_t} = qN_t$$

where subscript t refers, of course, to the time period under consideration. The population at the start of time period t is equal to some original population less what has been caught up to the beginning of period t:

$$N_t = N_o - \sum_{x=1}^{t-1} C_x$$

or, letting

$$\sum_{x=1}^{t-1} C_x = K_t$$

that is, defining K_t as the accumulative catch up to, but not including, time period t, then:

$$N_t = N_o - K_t$$

When a substitution is made for N_t in the catch-per-unit-effort relationship, then:

$$\frac{C_t}{f_t} = qN_t = qN_o - qK_t$$

This is a linear relationship, implying that if catch-per-unit effort is plotted against cumulative catch, up to that time, the result will be a straight line with slope equal to catchability and intercept equal to the original population times catchability. We may estimate N_o by dividing the intercept from regression analysis by the slope or by finding the point where the regression line intersects the abscissa.

The following data are appropriate for a Leslie estimate. The effort is measured in hours of electrofishing spent on a specified area of a river. The catches of brown trout from each of the three electrofishing periods are held in separate live cages.

Sample	Catch (C)	Effort (f)	Catch/Effort (C/f)	Accumulated catch (K_t)
1	150	2	75.0	0
2	203	3	67.6	150
3	86	1.5	57.3	353

Our regression analysis uses catch per effort as the dependent (Y) variable and accumulated catch as the independent variable. Estimates from this analysis are:

Slope = catchability coefficient = $q = 0.05$
Intercept = $qN_o = 75.07$
and therefore:

$$\hat{N}_o = \frac{\text{Intercept}}{\text{Slope}} = \frac{qN_o}{q} = \frac{75.07}{0.05} = 1501$$

The method of DeLury makes the following similar starting assumptions:

$$\frac{C_t}{f_t} = qN_o \left(\frac{N_t}{N_o} \right)$$

If the fraction of the population taken per unit effort is small, $q = 0.02$ or less, then:

$$\frac{N_t}{N_o} \doteq e^{-q} \sum_{i=1}^{t-1} f_i$$

where f_i is fishing effort. Letting

$$G_t = \sum_{i=1}^{t-1} f_i$$

and, combining equations, then:

$$\frac{C_t}{f_t} = qN_o \, e^{-qG_t}$$

Taking natural logarithms of this equation results in the following:

$$ln\left(\frac{C_t}{f_t}\right) = ln \, qN_o - qG_t$$

Again this is a linear relationship, and the plot of $ln \, (C_t/f_t)$ against cumulative effort up to time period t should be a straight line with slope equal to catchability and intercept equal to $ln \, (qN_o)$. Therefore:

$$\hat{N}_o = \frac{e^{\,\text{Intercept value}}}{\text{Slope}}$$

The following is an example of the DeLury method.

Catch (C)	Effort (f)	C/f	G	ln (C/f)
133,500	89	1500.0	0	7.3132
133,110	97	1372.3	89	7.2242
149,449	120	1245.4	186	7.1272
122,608	111	1104.6	306	7.0072

A regression analysis for natural logarithms of catch-per-unit effort and cumulative effort up to but not including the current time period provides estimates for catchability coefficient and intercept value. These values are 0.001 and 7.3132 respectively. From the above equation it was shown that the intercept value is equal to $ln \, (qN_o)$. We therefore estimate N_o as follows:

$$\hat{N}_o = \frac{e^{\,\text{Intercept value}}}{\text{Slope}} = \frac{e^{7.3132}}{0.001}$$

$$\hat{N}_o = 1,499,985$$

Estimators for population size and q using the method of Moran and Zippin, when equal effort is extended for each sample, are given by:

$$\hat{N} = \frac{C}{1 - (1-q)^n} \quad \text{and} \quad \frac{1-q}{q} - \frac{n \, (1-q)^n}{1 - (1-q)^n} = \frac{\displaystyle\sum_{i=1}^{n} (i-1)C_i}{C}$$

where C is total catch and n is the number of samples. For example, suppose that the following trapnet catches were made when fishing a small pond for rainbow trout, and all fish were marked by temporary caudal clip. The catches reported are just for unmarked trout; the marked ones having been "removed." Catches are as follows: $C_1 = 163$; $C_2 = 99$; $C_3 = 53$; and let C equal total catch of 315. We first must iteratively solve for q given in the above equation as follows:

$$\frac{1 - q}{q} - \frac{n (1 - q)^n}{1 - (1 - q)^n} = \frac{\sum_{i=1}^{n} (i - 1)C_i}{C}$$

$$\frac{1 - q}{q} - \frac{3 (1 - q)^3}{1 - (1 - q)^3} = \frac{(1 - 1)163 + (2 - 1)99 + (3 - 1)53}{315}$$

$$\frac{1 - q}{q} - \frac{3 (1 - q)^3}{1 - (1 - q)^3} = \frac{205}{315} = 0.65079$$

$$q = 0.423$$

Therefore:

$$\hat{N} = \frac{315}{1 - (1 - 0.423)^3} = \frac{315}{0.8079} = 390$$

Estimated variances for $\hat{N}$ and $\hat{q}$ when N is large are:

$$\hat{V}(\hat{N}) = \frac{\hat{N} [1 - (1 - \hat{q})^n] (1 - \hat{q})^n}{[1 - (1 - \hat{q})^n]^2 - (n\hat{q})^2 (1 - \hat{q})^{n-1}}$$

and:

$$\hat{V}(\hat{q}) = \frac{[(1 - \hat{q})(\hat{q})]^2 (1 - (1 - \hat{q})^n)}{\hat{N}\{(1 - \hat{q}) [1 - (1 - \hat{q})^n]^2 - (n\hat{q})^2 (1 - \hat{q})^n\}}$$

Confidence limits may be calculated for N using normal theory, assuming normality for $\hat{N}$.

The two sample case of the above method reduces to the following explicit estimators of Seber-LeCren:

$$\hat{N} = \frac{C_1^2}{C_1 - C_2}$$

$$1 - q = C_2/C_1$$

with estimated variance

$$\hat{V}(\hat{N}) = \frac{C_1{}^2 C_2{}^2 (C_1 + C_2)}{(C_1 - C_2)^4}$$

and:

$$\hat{V}(\hat{q}) = \frac{C_2(C_1 + C_2)}{C_1{}^3}$$

References

Abramson, N. J. 1968. A probability sea survey plan for estimating relative abundance of ocean shrimp. California Fish Game J. 54(4):257–269.

Allen, K. R. 1966. Some methods for estimating exploited populations. J. Fish. Res. Bd. Can. 23(10):1553–1574.

Atwood, E. L. 1956. Validity of mail survey on bagged waterfowl. J. Wildl. Manage. 20(1):1–16.

Bailey, N. T. J. 1951. On estimating the size of mobile populations from recapture data. Biometrika 39:293–306.

Barr, L. 1971. Methods of estimating the abundance of juvenile spot shrimp in a shallow nursery area. Trans. Amer. Fish. Soc. 100(4):781–787.

Beuchner, H. K., I. O. Buss, W. M. Longhurst, and A. C. Brooks. 1963. Numbers of migration of elephants in Murchison Falls National Park, Uganda. J. Wildl. Manage. 27(1):36–53.

Beverton, R. J. H., and S. J. Holt. 1957. On the dynamics of exploited fish populations. Min. Agric. Fish and Food (U.K.), Fish. Invest. Ser. II, 19. 533 pp.

Bratten, D. O. 1969. Robustness of the DeLury population estimator. J. Fish. Res. Bd. Can. 26(2):339–355.

Brock, V. E. 1954. A preliminary report on a method of estimating reef fish populations. J. Wildl. Manage. 18(3):297–308.

Buck, H. P., and C. F. Thoits, III. 1965. An evaluation of Petersen estimation procedures employing series in 1-acre ponds. J. Wildl. Manage. 29(3):598–621.

Caddy, J. F. 1970. A method for surveying scallop populations from a submersible. J. Fish. Res. Bd. Can. 27(3):535–549.

Chapman, D. G. 1952. Inverse, multiple and sequential sample censuses. Biometrics 8(4):286–306.

——. 1954. The estimation of biological populations. Ann. Math. Statist. 25:1–15.

——. 1955. Population estimation based on change of composition caused by selective removal. Biometrika 42:279–290.

Chapman, D. G., and G. I. Murphy. 1965. Estimates of mortality and population from survey-removal records. Biometrics 21(4):921–935.

Chapman, D. G., and W. S. Overton. 1966. Estimating and testing differences between population levels by the Schnabel estimation method. J. Wildl. Manage. 30(1):173–180.

Cleary, R. E., and J. Greenbank. 1954. An analysis of techniques used in estimating fish populations in streams, with particular reference to large non-trout streams. J. Wildl. Manage. 18(4):461–477.

Clemens, H. P. 1952. Cove selection and use of gill nets in fish population surveys in large impoundments. J. Wildl. Manage. 16(3):393–396.

Cole, L. C. 1954. The population consequences of life history phenomena. Q. Rev. Biol. 29(2):103–137.

Cooper, G. P. 1952. Estimation of fish populations in Michigan lakes. Trans. Amer. Fish. Soc. 81:4–16.

———. 1953. Population estimates of fish in Sugarloaf Lake, Washtenaw County, Michigan, and their exploitation by anglers. Pap. Michigan Acad. Sci. Arts Lett. 38:1963–1986.

Cooper, G. P., and W. C. Latta. 1954. Further studies on the fish population and exploitation by anglers in Sugarloaf Lake, Washtenaw County, Michigan. Pap. Michigan Acad. Sci. Arts Lett. 39:209–223.

Crowe, W. R. 1954. An analysis of the fish population of Big Bear Lake, Otsego County, Michigan. Pap. Michigan Acad. Sci. Arts Lett. 38:187–206.

———. 1955. Numerical abundance and use of a spawning run of walleyes in the Muskegon River, Michigan. Trans. Amer. Fish. Soc. 84:125–136.

Darroch, J. N. 1958. The multiple-recapture census. I: Estimation of a closed population. Biometrika 45:343–359.

———. 1959. The multiple-recapture census. II: Estimation when there is immigration or death. Biometrika 46:336–351.

Davis, W. S. 1964. Graphic representation of confidence intervals for Petersen population estimates. Trans. Amer. Fish. Soc. 93(3):227–232.

DeLury, D. B. 1947. On the estimation of biological populations. Biometrics 3(4):145–167.

———. 1951. On the planning of experiments for the estimation of fish populations. J. Fish. Res. Bd. Can. 8(4):281–307.

———. 1958. The estimation of population size by a marking and recapture procedure. J. Fish. Res. Bd. Can. 15(1):19–25.

Dowd, R. G., E. Bakken, and O. Nakken. 1970. A comparison between two sonic measuring systems for demersal fish. J. Fish. Res. Bd. Can. 27(4):727–742.

Eberhardt, L. L. 1969. Population estimates for recapture frequencies. J. Wildl. Manage. 33(1):28–39.

Eicher, G. J., Jr. 1953. Aerial methods of assessing red salmon populations in Western Alaska. J. Wildl. Manage. 17(3):521–527.

Embody, D. R. 1940. A method of estimating the number of fish in a given section of a stream. Trans. Amer. Fish. Soc. 69:231–236.

Eschmeyer, R. W. 1938. The significance of fish population studies in lake management. Trans. N. Amer. Wildl. Conf. 3:458–468.

———. 1942. The catch, abundance, and migration of game fishes in Norris Reservoir, J. Tennessee Acad. Sci. 17(1):70–122.

Foerster, R. E. 1954. On relation of adult sockeye salmon (*Oncorhynchus nerka*) returns to known smolt seaward migrations. J. Fish. Res. Bd. Can. 11(4):339–350.

Gerking, S. D. 1953. Vital statistics of the fish population of Gordy Lake, Indiana. Trans. Amer. Fish. Soc. 82:48–67.

Gulland, J. A. 1964. Manual of methods of fish population analysis. FAO, U.N., Fish. Tech. Pap. 40. 62 pp.

Haskell, D. C. 1940. An electrical method of collecting fish. Trans. Amer. Fish. Soc. 69:210–215.

Holton, G. D. 1953. A trout population study on a small creek in Gallatin County, Montana. J. Wildl. Manage. 17(1):62–82.

Holton, G. D., and C. R. Sullivan, Jr. 1954. West Virginia's electrical fish-collecting methods. Prog. Fish-Cult. 16(1):10–18.

Hourston, A. S. 1953. The spawning population of herring in northern British Columbia in 1952. Fish. Res. Bd. Can., Prog. Rep. 97:21–26.

Johnson, M. C. 1965. Estimates of fish populations in warmwater streams by removal method. Trans. Amer. Fish. Soc. 94(4):350–357.

Kennedy, W. A. 1950. The determination of optimum size of mesh for gill nets in Lake Manitoba. Trans. Amer. Fish. Soc. 79:167–179.

Ketchen, K. S. 1953. The use of catch-effort and tagging data in estimating a flat-fish population. J. Fish. Res. Bd. Can. 10(8):459–485.

Latta, W. C. 1959. Significance of trap-net selectivity in estimating fish population statistics. Pap. Michigan Acad. Sci. Arts Lett. XLIV:123–138.

Leslie, P. H. 1952. The estimation of population parameters from data obtained by means of the capture-recapture method. II. The estimation of total numbers. Biometrika 39:363–368.

Leslie, P. H., D. Chitty, and H. Chitty. 1953. The estimation of population parameters from data obtained by means of capture-recapture method. III. An example of practical application of the method. Biometrika 40:137–169.

Loeb, H. A. 1958. Comparison of estimates of fish populations in lakes. New York Fish Game J. 5(1):66–76.

MacLulich, D. A. 1957. The place of chance in population processes. J. Wildl. Manage. 21(3):293–299.

McFadden, J. T., G. R. Alexander, and D. S. Shetter. 1967. Numerical changes and population regulation in brook trout, *Salvelinus fontinalis*. J. Fish. Res. Bd. Can. 24(7):1425–1459.

Midttun, L. 1971. Acoustic methods for estimation of fish abundance. *In* Symposium on remote sensing in marine biology and fishery resources. Texas A. and M. Univ., College Station, Texas, pp. 218–226.

Moran, P. A. P. 1951. A mathematical theory of animal trapping. Biometrika 38:307–311.

Mraz, D., and C. W. Threinen. 1957. Angler's harvest, growth rate and population estimate of the largemouth bass of Brown's Lake, Wisconsin. Trans. Amer. Fish. Soc. 85:241–255.

Muir, B. S. 1964. Vital statistics of *Esox masquinongy* in Nogies Creek, Ontario. Population size, natural mortality, and effect of fishing. J. Fish. Res. Bd. Can. 21(4):727–746.

Murphy, G. I. 1952. An analysis of silver salmon counts at Benbow Dam, South Fork of Eel River, California. California Fish Game J. 38(1):105–112.

Murray, A. R. 1958. A direct-current electrofishing apparatus using separate excitation. Can. Fish. Cult. 23:27–32.

Outram, D. N., and F. H. C. Taylor. 1964. A quantitative estimate of the number of Pacific herring in a spawning population. J. Fish. Res. Bd. Can. 21(5):1317–1320.

Overton, W. S. 1965. A modification of the Schnabel estimator to account for removal of animals from the population. J. Wildl. Manage. 29(2):392-395.

Overton, W. S., and D. E. Davis. 1969. Estimating the numbers of animals in wildlife populations. *In* R. H. Giles, Jr., ed. Wildlife management techniques. 3rd rev. ed. The Wildlife Society, Washington, D.C., pp. 403-455.

Patterson, D. L. 1953. The walleye population of Escanaba Lake, Vilas County, Wisconsin. Trans. Amer. Fish. Soc. 82:34-41.

Paulik, G. J., and D. S. Robson. 1969. Statistical calculations for change-in-ratio estimators of population parameters. J. Wildl. Manage. 33(1):1-27.

Radovich, J., and E. D. Gibbs. 1954. The use of a blanket net in sampling fish populations. California Fish Game J. 40(4):353-365.

Regier, H. A., and D. S. Robson. 1966. Selectivity of gill nets, especially to whitefish. J. Fish. Res. Bd. Can. 23(3):423-454.

_____. 1967. Estimating population number and mortality rates. *In* S. D. Gerking, ed. The biological basis of freshwater fish production. John Wiley and Sons, Inc., New York, pp. 31-66.

Ricker, W. E. 1940. Relation of "catch per unit effort" to abundance and rate of exploitation. J. Fish. Res. Bd. Can. 5(1):43-70.

_____. 1942. Creel census, population estimates and rate of exploitation of game fish in Shoe Lake, Indiana. Indiana Dep. Conserv., Invest. Indiana Lakes Streams 2(12):215-253.

_____. 1945. Abundance, exploitation and mortality of the fishes in two lakes. Indiana Dep. Conserv., Invest. Indiana Lakes Streams 2(17):345-448.

_____. 1946. Some application of statistical methods of fishery problems. Biometrics 1(6):73-79.

_____. 1948. Methods of estimating vital statistics of fish populations. Indiana Univ., Sci. Ser., Publ. 15. 101 pp.

_____. 1958. Handbook of computations for biological statistics of fish populations. Fish. Res. Bd. Can. Bull. 119. 300 pp.

Robson, D. S., and W. A. Flick. 1965. A non-parametric statistical method for culling recruits from a mark-recapture experiment. Biometrics 21(4):936-947.

Robson, D. S., and H. A. Regier. 1964. Sample size in Petersen mark-recapture experiments. Trans. Amer. Fish. Soc. 93(3):215-226.

_____. 1968. Estimation of population number and mortality rates. *In* W. E. Ricker, ed. Methods for assessment of fish production in fresh waters. Blackwell Scientific Publications, Oxford, pp. 124-158.

Robson, D. S., and G. R. Spangler. 1978. Estimation of population abundance and survival. *In* S. Gerking, ed. Ecology of Freshwater Fish Production. Blackwell Scientific Publications, Oxford, pp. 26-51.

Rupp, R. S. 1966. Generalized equation for the ratio method of estimating population abundance. J. Wildl. Manage. 30(3):523-526.

Saunders, J. W., and M. W. Smith. 1955. Standing crops of trout in a small Prince Edward Island stream. Can. Fish Cult. 17:32-39.

Schnabel, Z. E. 1938. The estimation of the total fish population of a lake. Amer. Math. Mon. 45(6):348-352.

Schumacher, F. X., and R. W. Eschmeyer. 1942. The recapture and distribution of tagged bass in Norris Reservoir, Tennessee. J. Tennessee Acad. Sci. 17(3):253-268.

_____. 1943. The estimate of fish populations in lakes or ponds. J. Tennessee Acad. Sci. 18(3):228–249.

Seber, G. A. F. 1973. The estimation of animal abundance and related parameters. Charles Griffin and Co., Ltd., London. 506 pp.

Seber, G. A. F., and E. D. LeCren. 1967. Estimating population parameters from catches large relative to the population. J. Anim. Ecol. 36:631–643.

Seber, G. A. F., and J. F. Whale. 1970. The removal method for two and three samples. Biometrics 26(3):393–400.

Selleck, D. M., and C. M. Hart. 1957. Calculating the percentage of kill from sex and age ratios. California Fish Game J. 43(4):309–316.

Shetter, D. S., and J. W. Leonard. 1943. A population study of a limited area in a Michigan trout stream, September, 1940. Trans. Amer. Fish. Soc. 72:35–51.

Tepper, E. F. 1967. Statistical methods in use in mark-recapture data for population estimation. U.S. Dep. Inter., Bibliogr. 4. 65 pp.

Thorne, R. E. 1971. Investigations into the relation between integrated echo voltage and fish density. J. Fish. Res. Bd. Can. 28(9):1269–1273.

Thorne, R. E., J. Reeves, and A. E. Millikan. 1971. Estimation of the Pacific hake (*Merluccius productus*) population in Fort Susan, Washington, using an echo integrator. J. Fish. Res. Bd. Can. 28(9):1275–1284.

Tibbo, S. N., D. J. Scarratt, and P. W. G. McMullon. 1963. An investigation of herring (*Clupea harengus* L.) spawning using free-diving techniques. J. Fish. Res. Bd. Can. 20(4):1067–1079.

Widrig, T. M. 1954. Method of estimating fish populations with application to Pacific sardine. U.S. Fish Wildl. Serv., Fish. Bull. 56:141–166.

Wohlschlag, D. E. 1952. Estimation of fish populations in a fluctuating reservoir. California Fish Game J. 38(1):63–72.

Wood, G. W. 1963. The capture-recapture technique as a means of estimating populations of climbing cutworms. Can. J. Zool. 4(1):47–50.

Youngs, W. D., and D. S. Robson. 1978. Estimation of population number and mortality rates. *In* T. Bagenal, ed. Methods for assessment of fish production in fresh waters. Blackwell Scientific Publications, Oxford, pp. 137–164.

Zippin, C. 1956. An evaluation of the removal method of estimating animal populations. Biometrics 12(2):163–169.

_____. 1958. The removal method of population estimation. J. Wildl. Manage. 22(1):82–90.

7

Mortality

Knowledge of mortality is essential for management of most fisheries; as a general rule, a fishery manager would like to have control of man-induced mortality and therefore must be able to estimate this component. Death is not a repetitive process, and for every death there is a specific cause. Mortality ascribed to natural causes has actually included death resulting from many causes, as for example from predation, floods, or disease. The development of competing-risks-of-death theory depends upon each cause of death being independent and having associated with it a force of mortality. If all of these forces of mortality when added together turn out to be the same numerical value every year, we can then assume a "constant" natural mortality. Nearly all fishery studies have implicitly made this assumption and thus have dealt with two causes of death: fishing and natural.

Empirical observations over varying periods of time have shown that the number of fish of a particular cohort decline at a rate proportional to the number of fish alive at any particular point in time. This is illustrated in the accompanying graph (see Figure 7-1). This is more apparent when dealing with large numbers of fish and is patently ridiculous when the extreme of one is reached. However, for numbers of fish, at least several hundred, we may approximate this condition by the equation $dN/dt = -ZN$, which expresses the rate of change of cohort or year-class numbers (N) with time as being proportional (Z) to the number present at that time. The negative sign means that numbers are declining. Rearranging and integrating this equation gives the following:

$$N(t) = N(o)e^{-Zt}$$

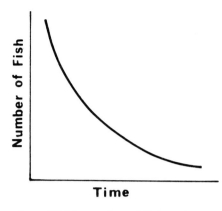

Figure 7-1. Relationship of number of fish with time.

where $N(t)$ is number of fish at time t, $N(o)$ is number of fish at time equal 0, and Z is the force of total mortality.

The definition of survival for any period is the number of fish alive at the end of the period relative to the number alive at the beginning of the period or symbolically (with above notation): S (survival) $= N_t/N_o$. An estimate of the force of total mortality (mortality coefficient and instantaneous mortality rate are equivalent terms) is provided by the natural logarithm of survival and then changing sign, as follows: $Z = -lnS$. The force of total mortality is the sum of the forces for all other mortalities in the population. The consideration here is for just two sources of mortality: fishing and natural. This is expressed as: $Z = F + M$, where F is the force of fishing mortality and M is the force of natural mortality. Therefore:

$$N(t) = N(o)e^{-(F+M)t} = N(o)e^{-Ft}e^{-Mt}$$

It is apparent that survival is equal to the probability of living for a specified time period which is equal to the probability of not being caught by fishing, e^{-Ft}, multiplied by the probability of not dying from natural causes, e^{-Mt}. Mortality is the complement of survival; if A is mortality, then $A = 1 - S$.

An alternate model, to the differential equation one, appropriate for any number of fish is the binomial model with survival being the usual chance of success or of the event happening. The binomial model is

$$P (N_t = x|N_o) = \binom{N_o}{x} S^x (1 - S)^{N_o-x} \quad x = 0, 1, 2, \cdots, N_o$$

where $S = e^{-Zt}$. The expected number of fish at time t is N_oS^t or equivalently N_oe^{-Zt}. This appears to be the same as the differential equation model but the two models are entirely different. The differential equation model is deterministic and has no variance associated with $N(t)$ whereas the binomial model is stochastic and has an expected value and variance. The variance of N_t in the binomial model is $N_o S (1-S)$.

Expectation of death for a specific risk of death is the probability that a fish dies from that risk when all risks are affecting the population. If death from fishing and natural death are the two risks, then the expectation of death from fishing, E, is the chance or probability that a fish dies from fishing during the time period being considered—when both fishing and natural risks are present. The expectation of death from natural causes, D, is the probability that a fish dies of natural causes during the time period—when both natural and fishing risks are present.

Rate of mortality for a specific risk is the probability of death from that risk when that is the only risk of death affecting the population. Thus the annual rate of fishing mortality is the probability of death from fishing when fishing is the only cause of death affecting the population.

The following relationship is very important and will be used repeatedly:

$$\frac{F}{Z} = \frac{E}{(1 - S)}$$

where E is the expectation of death from fishing, and other symbols are as previously defined. This relationship is strictly valid only when F/Z is constant for the time period. The above relationships and definitions are conventionally considered on an annual basis. However, any time period may be used.

The estimation of survival and expectation of death from fishing form the basis of many management programs. Some methods for estimating these parameters will be considered in this section. Natural mortality has almost always been assumed to be constant in the above model. Two situations may exist: survival may be a constant value implying by the model that fishing is likewise constant; or survival may vary, implying nonuniform fishing. Again, these estimates are generally on an annual basis and will be considered as such for the purpose of this discussion.

Constant Survival and Constant Recruitment

A population in a steady state with constant mortality will have the number of recruits exactly equal to the number of fish dying. A nonselective sample from this population would always show the same ratios of ages or year classes. For example, let us assume a constant annual survival of 50 percent. If each year there are 1,000 yearling fish entering the population, then after a period of, say, 11 years the population will have the following age structure:

Age	1	2	3	4	5	6	7	8	9	10	11
Number of fish	1000	500	250	125	62	31	16	8	4	2	1

and the structure would stay the same as long as 1,000 yearling recruits were added each year and the survival was always 50 percent.* A graph of the number of fish plotted against age would look like the illustration (Figure 7-1) at the beginning of this chapter and is of the form of the negative exponential, that is, $N_t = N_o e^{-Zt}$. The same data plotted on semilog paper form a straight line as in the following illustration (see Figure 7-2). The slope of this line is equal to $-Z$, the force of total mortality.

Most fishing gear is selective, that is, it does not catch all age groups in proportion to their abundance. Suppose that the relative efficiency for 1-year-old fish is 0.2 and for 2-year-old fish is 0.6, with the rest of the ages being sampled in proportion to their abundance. Let's assume a 10 percent sample which, with the gear efficiencies, would give the following sample from our hypothetical population:

Age	1	2	3	4	5	6	7	8	9	10	11
Number of fish	20	30	25	12	6	3	2	1	0	0	0

A plot of these data on semilog paper does not result in a straight line (see Figure 7-3). Notice that the curve has an ascending portion or leg, a dome area, and a descending leg. This is typical of catch curves, as this type of graph is called. The descending leg is the portion that can be used to estimate Z. The age group representing the top of the dome may or may not be completely vulnerable to the gear, and therefore the portion of the descending leg used to estimate Z is shifted one age group to the right of the dome. Notice also that the older age groups are represented by very few fish, and any statistical variation in the sample from these age groups can cause a large change in the slope. It is therefore customary not to include in the estimate any age groups beyond the one containing 5 to 10 fish. The shape of the dome also expresses information about recruitment into the

---

*If we allow for statistical variability this would not be true for any given year but would be true for the average of a large number of sample years.

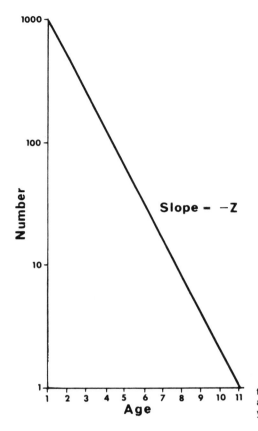

Figure 7-2. Age-frequency distribution resulting from 50 percent survival and constant recruitment of 1,000 yearling fish per year.

population being exploited by that particular gear. A broad, flat dome and slowly ascending leg indicates recruitment over several age groups. A steep ascending leg and a narrow, sharp dome indicates recruitment over few age groups.

Survival may also be estimated from steady-state populations by the method proposed by Jackson. The sum of number at age from first fully recruited age through all but the last age is divided into the sum of all number at age except the first fully recruited age for an estimate of survival. From the previous example, age 3 is the first fully recruited age, so the Jackson estimate would be:

$$\hat{S} = \frac{n_4 + n_5 + n_6 + n_7 + n_8}{n_3 + n_4 + n_5 + n_6 + n_7} = \frac{12 + 6 + 3 + 2 + 1}{25 + 12 + 6 + 3 + 2} = \frac{24}{48} = 0.5$$

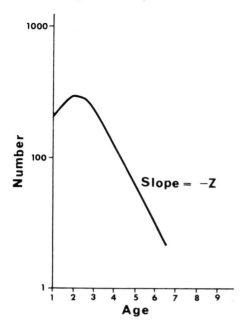

Figure 7-3. A catch curve.

as was used to generate these data. This estimator may in practice take on values greater than one.

The following estimator by Heincke is similar to Jackson's:

$$\hat{S} = \frac{n - n_o}{n} = \frac{49 - 25}{49} = \frac{24}{49} = 0.5$$

where n is the total of all age frequencies in the sample, including the first fully recruited age, and n_o is the frequency of the first fully recruited age. This is an unbiased estimator of S and will never have a value greater than one.

Yet another estimator by Chapman and Robson is available and is the unique minimum-variance unbiased estimator of survival. Using the same example as above, the following demonstrates this estimator:

Age	3	4	5	6	7	8
Coded age	0	1	2	3	4	5
Frequency	25	12	6	3	2	1

The ages are coded so that the youngest age in the sample is set to zero, the next youngest age is 1, and so on to the oldest age. If T equals the sum of the coded age times its frequency, then:

$$T = \sum_{x=o}^{k} xf_x = (0 \times 25) + (1 \times 12) + (2 \times 6)$$
$$+ (3 \times 3) + (4 \times 2) + (5 \times 1) = 46$$

and:

$$n = \sum_{x=o}^{k} f_x = 25 + 12 + 6 + 3 + 2 + 1 = 49$$

Then:

$$\hat{S} = \frac{T}{n + T - 1}$$

and in this example:

$$\hat{S} = \frac{46}{49 + 46 - 1} = \frac{46}{94} = 0.49.$$

The estimated variance of this estimator is:

$$\hat{V}(\hat{S}) = \hat{S}\left(\hat{S} - \frac{T - 1}{n + T - 2}\right)$$

and in this example:

$$\hat{V}(\hat{S}) = .49\left(.49 - \frac{46 - 1}{49 + 46 - 2}\right) = 0.003.$$

Considering the properties of the Chapman-Robson estimator, it would seem that theirs is the best one to use when making estimates with the stated condition of equal recruitment and constant survival existing in the population. It is also obvious that if these conditions exist, a sample taken in a single year will be sufficient for estimation of survival.

Constant Survival and Unequal, Known Recruitment

The next logical step would be the situation of unequal but known recruitment still with constant survival. The following table of age fre-

Table 7-1. Age-frequency distribution for a hypothetical fish population with 50 percent survival and variable number of yearling fish stocked

Year	\multicolumn{11}{c}{Age}
	1	2	3	4	5	6	7	8	9	10	11
1	1000										
2	1500	500									
3	2500	750	250								
4	2000	1250	375	125							
5	1200	1000	625	188	62						
6	1000	600	500	312	94	31					
7	1800	500	300	250	156	47	16				
8	2200	900	250	150	125	78	24	8			
9	2400	1100	450	125	75	62	39	12	4		
10	1600	1200	550	225	62	38	31	20	6	2	
11	1500	800	600	275	112	31	19	16	10	3	1

quencies (see Table 7–1) will help to visualize what is happening in such a population. A survival rate of 50 percent per year will again be used. Assume again that these fish are stocked as yearlings so that the number under age 1 will then refer to the number stocked or recruits.

The diagonals follow a particular year class with time. Survival could be estimated by using the catches of a single year class in successive years. In doing this there is an implicit assumption of equal effort and catchability between sample years.

Alternatively, age frequencies may be adjusted to a common basis by dividing frequencies by the number of one-year-old fish for that particular year class. For example, in the 11th year the following adjustment would be appropriate on a per-thousand basis:

Age	1	2	3	4	5	6	7	8	9	10	11
No. per thousand	$\frac{1500}{1.5}$	$\frac{800}{1.6}$	$\frac{600}{2.4}$	$\frac{275}{2.2}$	$\frac{112}{1.8}$	$\frac{31}{1.0}$	$\frac{19}{1.2}$	$\frac{16}{2.0}$	$\frac{10}{2.5}$	$\frac{3}{1.5}$	$\frac{1}{1.0}$
Adjusted number	1000	500	250	125	62	31	16	8	4	2	1

Adjusted data may then be used in any of the above estimators for survival.

If recruitment is not known, it is possible to generate a known recruitment by distinctly marking a portion of the first fully recruited age each year for a number of years. Year-class adjustment may then be made on the basis of the marked fish, and estimation may be done by the above methods. This procedure of course requires a long period of time in order to place year-class groups of marked fish in the population.

Variable Survival: A Tag-Recapture Model

Survival estimation becomes more complicated when survival is variable between years. A varying degree of fishing success between years causes different fishing mortality, hence different survival. A general procedure is available when recruitment is known or when a group of marked fish can be used as known recruits. This will be demonstrated with an example using tagged fish as the known recruitment.

In this example we will assume fish are sampled from the population and are tagged with numbered tags, so that upon recapture it will be possible to identify the year of tagging. Also assume that the sampling is done in a short period of time and at yearly intervals for a number of years, and that all ages have the same susceptibility to the risks of death. The samples after the first year will contain tagged fish, and these recaptures will form the data base for our estimation. Consider the following example from trap-net samples of a lake trout population. Let R_i be the total number of recaptures from a particular tagging release, and let C_i be the total number of fish recaptured in any particular sample year. Define $T_1 \equiv R_1$ and $T_i = T_{i-1} + R_i - C_{i-1}$. The R's are row totals of recaptures; the C's are column totals of recaptures, and the T's are recapture totals of all recaptures of tags recaptured after the last year of tagging. These are shown below (see Table 7-2) as the sum of all recaptures enclosed by the rectangles formed by extending the crosses.

The fraction of the population taken by sampling may be estimated by the following where M_i is the number of fish tagged in year i:

$$\hat{f}_i = \frac{R_i}{M_i} \cdot \frac{C_i}{T_i}$$

Survival between sample periods may be estimated by the following, in which we let $r_i = R_i/M_i$:

$$\hat{S}_i = \frac{r_i - \hat{f}_i}{r_{i+1}}$$

with estimated variance:

$$\hat{V}(\hat{S}) \doteq \hat{S}^2 \left[\frac{1 - r_i}{r_i M_i} + \frac{1 - r_{i+1}}{r_{i+1} M_{i+1}} + \frac{\hat{f}_i}{T_i(r_i - \hat{f}_i)} \right]$$

From the above example, the fraction of population taken by sampling in

Table 7-2. Tag-recapture distribution by year of tagging and year of recapture

Year tagged	Number tagged	Year recaptured					
		1967	1968	1969	1970	1971	R
1966	700	49 $T_1 = 81$	21	8	2	1	81
1967	800		65 $T_2 = 130$	25	6	2	98
1968	600			56 $T_3 = 117$	13	4	73
1969	900				79 $T_4 = 132$	25	104
1970	500					46 $T_5 = 78$	46
C		49	86	89	100	78	

second year, f_2, would be calculated as follows:

$$\hat{f}_2 = \frac{R_2}{M_2} \cdot \frac{C_2}{T_2} = \frac{98 \times 86}{800 \times 130} = 0.081$$

and S_2 would be estimated as:

$$\hat{S}_2 = \frac{r_2 - \hat{f}_2}{r_3} = \frac{\dfrac{98}{800} - 0.081}{\dfrac{73}{600}} = 0.341$$

The same procedure could be used with a known recruitment, as for example a stocking of fin-clipped fish each year and then samples in subsequent years with recaptures being fin-clipped fish by year of stocking. This technique may also be used for tags returned by anglers, in which case

Table 7-3. Chi-square contingency test of tag recaptures

Year tagged	Year recaptured				Total
	2	3	...	k	
1	R_{12}	R_{13}	...	R_{1k}	$T_2 - R_2$
2	R_{22}	R_{23}	...	R_{2k}	R_2
Total	C_2	R_{23}^*	...	R_{2k}^*	R_2
1 or 2		R_{23}^*	...	R_{2k}^*	$T_3 - R_3$
3		R_{33}	...	R_{3k}	R_3
Total		C_3	...	R_3^*	T_3

Table 7-4. Chi-square contingency test of tag recapture data given in Table 7-2

Year tagged	Year recaptured				Total
	1968	1969	1970	1971	
1966	21	8	2	1	32
1967	65	25	6	2	98
Total	86	33	8	3	130
		$\chi^2 = 0.15$ (3 d.f.)			
1966 or 1967		33	8	3	44
1968		56	13	4	73
Total		89	21	7	117
		$\chi^2 = 0.10$ (2 d.f.)			
1966, 1967 or 1968			21	7	28
1969			79	25	104
Total			100	32	132
		$\chi^2 = 0.01$ (1 d.f.)			

the f's will be estimates of minimum rates of exploitation. Estimates of survival are unaffected.

Any data set should be tested for conformity to the model being proposed. For the tag-recapture model a chi-square contingency test of the model may be constructed by a series of tables as shown in part in Table 7–3. The table continues in similar fashion for the entire set of data. The individual chi-squares may be added for a combined test of the model. The degrees of freedom (d.f.) in the combined test are $(k - 1)(k - 2)/2$. The example given in Table 7–2 is tested by this procedure in Table 7–4.

The overall chi-square of 0.26 would have 6 degrees of freedom. If pooling to avoid less than 5 expected recaptures there would be a corresponding reduction in degrees of freedom. Data in the last two columns in the first chi-square table should be pooled together giving 2 degrees of freedom for that test.

Variable Survival and Unknown Recruitment

The last general case we will consider is unequal survival with unknown recruitment. If all ages have a different susceptibility to fishing and therefore a different expectation of death from fishing for a given amount of effort, the only estimator of survival is the ratio of the number of a given cohort or the ratio of population estimates of the cohort at two time intervals. Such estimates may be poor, due to small numbers of fish involved. In cases where all ages are equally sampled by a particular type of gear, and all ages have the same mortality affecting them, it may be possible to obtain a somewhat better estimate of survival. The estimate is better in the sense that it is based on a larger sample size. The method is introduced by the following example of trawl catches of walleye. The catches are segregated by year of capture and age. The results are shown in Table 7-5.

Survival between two sample years is estimated by the ratio of the sum of fish of ages 2 through 13 in the second sample year to the sum of fish of ages 1 through 12 in the first sample year, assuming fish of all ages are equally vulnerable to capture. Survival between sample year 1969 and 1970, for example, would be estimated by:

$$
\hat{S} = \frac{N_2 + N_3 + \cdots + N_{13} \text{ [for 1970]}}{N_1 + N_2 + \cdots + N_{12} \text{ [for 1969]}}
$$

$$
= \frac{2271 + 748 + 224 + 215 + 110 + 41 + 43 + 8 + 1 + 1 + 1 + 1}{3785 + 1247 + 374 + 358 + 184 + 69 + 72 + 13 + 2 + 2 + 1}
$$

$$
= \frac{3664}{6107} = 0.6
$$

In the general case, the ratio would be the sum of all fishable age frequencies greater than the first fishable age relative to the sum of all age frequencies except the last fishable one. If all ages are fishable, the last age would generally be represented by very few fish and therefore would have little effect on the estimate of survival. In this case, one may simply age the sample in two groups: fish of the first completely fishable age, and older fish. Survival between sample years may then be rapidly calculated. Again using data from the example above, we have the following:

Year of capture	Age 1	Older than 1
1967	3120	5851
1968	4157	3587
1969	3785	2322
1970	3089	3664
1971	2906	4728
1972	3527	2290
1973	4006	2913

Survival between 1967 and 1968 is estimated by:

$$
\hat{S} \text{ [between 1967 and 1968]} = \frac{\text{Older than 1 in 1968}}{\text{Age 1 + older than 1 in 1967}}
$$

$$
= \frac{3587}{8971} = 0.40
$$

Table 7-5. Age-frequency distribution for a hypothetical fish population with unequal survival and unequal recruitment (continued)

Year of capture	Age												
	1	2	3	4	5	6	7	8	9	10	11	12	13
1967	3120	2986	1531	575	602	104	21	18	10	3	1	0	0
1968	4157	1248	1194	612	230	241	42	8	7	4	1	0	0
1969	3785	1247	374	358	184	69	72	13	2	2	1	0	0
1970	3089	2271	748	224	215	110	41	43	8	1	1	1	1
1971	2906	2162	1590	524	157	151	77	29	30	6	1	1	0
1972	3527	872	649	477	157	47	45	23	9	9	2	0	0
1973	4006	1764	436	325	239	79	24	23	12	5	5	1	0

and between 1968 and 1969 by:

$$\hat{S}[\text{between 1968 and 1969}] = \frac{\text{Older than 1 in 1969}}{\text{Age 1 + older than 1 in 1968}}$$
$$= \frac{2322}{7744} = 0.30$$

This estimator of survival is biased, but the bias reduces with increasing recruitment (unknown) and increasing exploitation. The variance of this estimator is not exact and is complicated with the additional problem of being dependent upon the sampling procedure.

Any of the techniques using age frequencies may also be used with groups of tagged or marked fish in the population. The assumption in doing this is that the marked fish are in every sense representative of the unmarked portion of the population. In this manner the estimates derived from the marked population are applied directly to the whole population.

Fishing Mortality

The discussion thus far has been concerned primarily with estimation of survival. While knowledge of survival is sufficient for some applications, as a rule it is desirable, and may be necessary, to estimate the fishing and natural components of total mortality. Mortality from fishing is normally estimated, and then natural mortality is assigned to the remaining mortality.

There are two general approaches to estimating fishing mortality. One procedure considers the fraction of the population harvested as a measure of the amount of exploitation, while the other procedure considers some specified amount of gear effort as being proportional to the force of fishing mortality.

The rate of exploitation or expectation of death from fishing is defined as the probability that a fish would die from fishing during some specified time period when all causes of death are working on the population. If a population contains 1,000 fish at the beginning of a fishing year, for example, and 350 are taken by fishing, then the rate of exploitation, or expectation of death from fishing, would be 350/1,000 or 0.35.

The definition of rate of exploitation immediately suggests methods for estimating this parameter. The total catch divided by the initial population provides the value for the parameter. If both catch and initial population are known, the ratio is the parameter value and not an estimate. Rarely will this occur, so other estimators have been developed.

The release of a group of marked fish immediately at the start of an angling season provides a population of known number. Total recoveries of marked fish throughout the season relative to the number released provides an estimate of exploitation. The assumption of equal representation of tagged to untagged is made, and the estimate is applied to the entire population. With this method, incompleteness of tag or mark recovery may lead to underestimation of the actual rate of exploitation, while mortality associated with the marking process may lead to overestimation (if fishing coverage is complete).

Exploitation may also be estimated from a creel census of total catch relative to the estimated original size of the fishable population. Creel censuses are seldom complete but are expanded to a complete coverage on the basis of some sampling scheme. The resulting rate-of-exploitation estimate does not have any well-defined properties; it is just an estimate. It cannot be stated that the estimate is a minimum, but it can be stated with generality that this type of estimator is biased.

The rate of exploitation estimated from virtual-population information provides a maximum estimate. The virtual population is the sum of the catches for a particular cohort throughout its lifetime. Therefore, the catch in a particular year relative to the catch in that year plus all succeeding years is the estimate. It is maximum because the virtual-population estimate is a minimum estimate; it does not include fish which die from causes other than reported fishing.

One other estimator of rate of exploitation was given previously in conjunction with a group of tag releases and recaptures over a period of years. The estimator was given as:

$$\hat{f}_i = \frac{R_i}{M_i} \cdot \frac{C_i}{T_i}$$

with symbols as previously defined. The variance for this estimator is approximately equal to:

$$\hat{V}\,(\hat{f_i}) \doteq \frac{\hat{f_i}}{R_i}\left[\hat{f_i}\,(1\,-\,r_i)\,+\,\frac{R_i\,(r_i\,-\,\hat{f_i})}{T_i}\right]$$

Estimates of rate of exploitation are not sufficient by themselves to allow managers to make decisions. However, by transforming rate of exploitation to force of fishing mortality, all the independent information in the model may be estimated, and from these estimates one can derive any dependent functions of the model. In doing this kind of transformation we use the following from the general model description:

$$F = \frac{E\,Z}{1\,-\,S}$$

in which F = the force of fishing mortality, E = the rate of exploitation, Z = the force of total mortality, and S = the survival rate. It is apparent that S and E are sufficient to define the situation, since Z is $-lnS$ and M is $Z\,-\,F$.

The consequences of altering the harvest are of interest to the manager; he would like to predict beforehand the effect on mortality of, for example, liberalizing a season, changing mesh size, or changing bag or size limits. If he wants to know what the survival rate would be if fishing were allowed to harvest an additional 10 percent when the present rate of exploitation is 0.3 with a survival rate of 0.3, then, since survival is 0.3, Z must be 1.204, and therefore:

$$F = \frac{E\,Z}{1\,-\,S} = \frac{0.3\,\times\,1.204}{1\,-\,0.3} = 0.516$$

and then:

$$M = Z\,-\,F = 1.204\,-\,0.516 = 0.688$$

By increasing exploitation 10 percent, the new exploitation rate would be 0.4, as in the following:

$$\frac{F}{E} = \frac{Z}{1\,-\,S}$$

$$\frac{F}{0.4} = \frac{F\,+\,0.688}{1\,-\,e^{-F}e^{-0.688}}$$

$$F - Fe^{-F} e^{-0.688} = 0.4F + 0.688 \times 0.4$$

$$\frac{0.6F}{e^{-0.688}} - Fe^{-F} = \frac{0.688 \times 0.4}{e^{-0.688}}$$

$$F = 0.756*$$

The new Z then equals:

$$Z = F + M = 0.756 + 0.688 = 1.444$$

and therefore the new survival rate is:

$$S = e^{-Z} = e^{-1.444} = 0.236$$

If, for example, one wished to determine the new rate of exploitation when in a given fishery the survival rate changes from 0.5 to 0.4 and the previous exploitation was 0.4, then from this information the force of fishing can be determined as:

$$F = \frac{E Z}{1 - S} = \frac{0.4 \times 0.693}{0.5} = 0.555$$

and:

$$M = Z - F = 0.693 - 0.555 = 0.138$$

Since the new force of total mortality is 0.916, the new force of fishing must be $F = Z - M = 0.916 - 0.138 = 0.778$. Again using the proportional relationship, the new rate of exploitation is:

$$E = \frac{F(1 - S)}{Z} = \frac{0.778 \times 0.6}{0.916} = 0.51$$

One point to keep in mind when using these models is the assumed constancy of force of natural mortality and of the F/Z ratio.

The second general approach to exploitation normally associated with commercial fisheries, assumes that force of fishing mortality is propor-

---

*This value is calculated iteratively; choose an F until both sides of the equation differ by some preselected small amount.

tional to the amount of gear or the effort expended on the fishery, symbolically represented by: $F = qf$. Effort or gear are generally known or estimateable for commercial fisheries, since licenses and logs are required by law. Harvest is also generally known.

The problem of partitioning mortality into the fishing and natural components is approached as a regression problem on the basis of the above assumption of force of fishing mortality being proportional to effort or gear. In this procedure force of total mortality is regressed upon effort or gear with the regression model being $Z = M + qf$. The intercept value from the regression is an estimate of force of natural mortality; the slope is referred to as the coefficient of catchability.

The force of fishing mortality is used for computing yield. Estimates of force of fishing mortality and force of natural mortality are sufficient to describe the relationships in the above model of competing risks of death. In combination with parameters from the growth models there is sufficient information to compute yield.

References

Baranoff, T. I. 1918. On the question of the biological basis of fisheries. (Trans. by W. E. Ricker, 1945, 53 pp., mimeogr.) Original title and citation: K voprosu o biologischeskiĭ osnovanñakh rybnovo khoziaĭstva. Nauchnyĭ issledovatelskiĭ iktiologisheskiĭ Institut, Izvestiia 1(1):81–128.

Beverton, R. J. H., and S. J. Holt. 1957. On the dynamics of exploited fish populations. Min. Agric. Fish and Food, (U.K.). Fish. Invest. Ser. II, 19. 533 pp.

Brownie, C., D. R. Anderson, K. P. Burnham, and D. S. Robson. 1978. Statistical inference from band recovery data—a handbook. U.S. Fish. Wildl. Resource Publ. 131. 212 pp.

Chapman, D. G. 1961. Statistical problems in the dynamics of exploited fish populations. Proc. 4th Berkeley Symp. 1960, 4:153–168.

Chapman, D. G., and G. I. Murphy. 1965. Estimates of mortality and population from survey-removal records. Biometrics 21(4):921–935.

Chapman, D. G., and D. S. Robson. 1960. The analysis of a catch curve. Biometrics 16(3):354–368.

Chiang, C. L. 1968. Introduction to stochastic processes in biostatistics. John Wiley and Sons, New York. 313 pp.

Eberhardt, L. L. 1969. Population analysis. *In* R. H. Giles, Jr., ed. Wildlife management techniques. 3rd rev. ed. The Wildlife Society, Washington, D.C., pp. 457–495.

Edser, T. 1908. Note on the number of plaice at each length in certain samples from the southern part of the North Sea, 1906. J. R. Stat. Soc. 71:686–690.

Graham, M. 1935. Modern theory of exploiting a fishery, and application to North Sea trawling. J. Conseil Int. Explor. Mer 10:264–274.

_____. 1939. Rates of fishing and natural mortality from the data of marking experiments. J. Conseil Int. Explor. Mer 13:76–90.

Gulland, J. A. 1964. Manual of methods of fish population analysis. FAO, U.N., Fish. Tech. Pap. 41. 62 pp.

Hazen, W. E. 1964. Readings in population and community ecology. W. B. Saunders Company, Philadelphia. 388 pp.

Heincke, F. 1913. Investigations of the plaice. General report. 1. The plaice fishery and protective measures. Preliminary brief summary of the most important points of the report. Conseil Int. Explor. Mer., Rapp. 16. 67 pp.

Holt, S. J., J. A. Gulland, C. Taylor, and S. Kurita. 1959. A standard terminology and notation for fishery dynamics. J. Conseil Int. Explor. Mer 24(2):239–242.

Jackson, C. H. N. 1939. The analysis of an animal population. J. Anim. Ecol. 8:238–246.

Kennedy, W. A. 1950. The determination of optimum size of mesh for gill nets in Lake Manitoba. Trans. Amer. Fish. Soc. 79:167–179.

_____. 1954. Growth, maturity and mortality in the relatively unexploited lake trout, *Cristivomer namaycush,* of Great Slave Lake. J. Fish. Res. Bd. Can. 11(6):827–852.

Leslie, P. H. 1952. The estimation of population parameters from data obtained by means of the capture-recapture method. II. The estimation of total numbers. Biometrika 39:363–388.

Leslie, P. H., and D. Chitty. 1951. The estimation of population parameters from data obtained by means of the capture-recapture method. I. The maximum likelihood equations for estimating the death-rate. Biometrika 38:269–292.

Leslie, P. H., D. Chitty, and H. Chitty. 1953. The estimation of population parameters from data obtained by means of capture-recapture method. III. An example of practical application of the method. Biometrika 40:137–169.

Moran, P. A. P. 1953. The estimation of death-rates from capture-mark-recapture sampling. Biometrika 39:181–188.

Mraz, D., and C. W. Threinen. 1957. Angler's harvest, growth rate and population estimate of the largemouth bass of Brown's Lake, Wisconsin. Trans. Amer. Fish. Soc. 85:241–255.

Muir, B. S. 1964. Vital statistics of *Esox masquinongy* in Nogies Creek, Ontario. II. Population size, natural mortality, and effect of fishing. J. Fish. Res. Bd. Can. 21(4):727–746.

Murphy, G. I. 1952. An analysis of silver salmon counts at Benbow Dam, South Fork of Eel River, California. California Fish Game J. 38(1):105–112.

Needham, P. R., and H. J. Rayner. 1939. The experimental stream, a method for study of trout planting problems. Copeia 1939(1):31–38.

Ness, J. C., W. T. Helm, and C. W. Threinen. 1957. Some vital statistics on a heavily exploited population of carp. J. Wildl. Manage. 21(3):279–292.

Olson, D. E. 1958. Statistics of a walleye sport fishery in a Minnesota lake. Trans. Amer. Fish. Soc. 87:52–72.

Paloheimo, J. E. 1961. Studies on estimation of mortalities. 1. Comparison of a method described by Beverton and Holt and a new linear formula. J. Fish. Res. Bd. Can. 18(5):645–662.

Regier, H. A., and D. S. Robson. 1967. Estimating population number and mortality rates. In S. D. Gerking, ed. The biological basis of freshwater fish production. John Wiley and Sons, New York, pp. 31–66.

Ricker, W. E. 1940. Relation of "catch per unit effort" to abundance and rate of exploitation. J. Fish. Res. Bd. Can. 5(1):43–70.

———. 1942. Creel census, population estimates and rate of exploitation of game fish in Shoe Lake, Indiana. Indiana Dep. Conserv., Invest. Indiana Lakes Streams 2(12):215–253.

———. 1944. Further notes on fishing mortality and effort. Copeia 1944(1):23–44.

———. 1945a. Natural mortality among Indiana bluegill sunfish. Ecology 26(2):111–121.

———. 1945b. Fish catches in three Indiana lakes. Indiana Dep. Conserv., Invest. Indiana Lakes Streams 2(16):325–344.

———. 1945c. Abundance, exploitation and mortality of the fishes in two lakes. Indiana Dept. Conserv., Invest. Indiana Lakes Streams 2(17):345–448.

———. 1946. Some application of statistical methods of fishery problems. Biometrics 1(6):73–79.

———. 1948. Methods of estimating vital statistics of fish populations. Indiana Univ., Sci. Ser., Publ. 15. 101 pp.

———. 1949. Mortality rates in some little-exploited populations of freshwater fishes. Trans. Amer. Fish. Soc. 77:114–128.

———. 1950. Cycle dominance among the Fraser sockeye. Ecology 31(1):6–26.

———. 1958. Handbook of computations for biological statistics of fish populations. Fish. Res. Bd. Can., Bull. 119. 300 pp.

———. 1975. Computation and interpretation of biological statistics of fish populations. Fish. Res. Bd. Can., Bull. 191. 382 pp.

Robson, D. S. 1963. Maximum likelihood estimation of a sequence of annual survival rates from a capture-recapture series. Conseil Int. Explor. Mer, Rapp. proc.-verb. 370:330–335.

Robson, D. S., and D. G. Chapman. 1961. Catch curves and mortality rates. Trans. Amer. Fish. Soc. 90(2):181–189.

Robson, D. S., and H. A. Regier. 1968. Estimation of population number and mortality rates. In W. E. Ricker, ed. Methods for assessment of fish production in fresh waters. Blackwell Scientific Publications, Oxford, pp. 124–158.

Robson, D. S., and G. R. Spangler. 1978. Estimation of population abundance and survival. In S. Gerking, ed. Ecology of Freshwater Fish Production. Blackwell Scientific Publications, Oxford, pp. 26–51.

Robson, D. S., and W. D. Youngs. 1971. Statistical analysis of reported tag recaptures in the harvest from an exploited population. Cornell Univ., Biom. Unit, Ser. BU-369-M. 15 pp. Mimeogr.

Schaeffer, M. B. 1943. The theoretical relationship between fishing effort and mortality. Copeia 1943(2):79–82.

Seber, G. A. F. 1970. Estimating time-specific survival and reporting rates for adult birds from band returns. Biometrika 57(2):313–318.

———. 1973. The estimation of animal abundance and related parameters. Charles Griffin and Co., Ltd., London. 506 pp.

Selleck, D. M., and C. M. Hart. 1957. Calculating the percentage of kill from sex and age ratios. California Fish Game J. 43(4):309–316.

Silliman, R. P. 1950. Studies on the Pacific pilchard or sardine (*Sardinops caerulea*): a method of computing mortalities and replacements. U.S. Fish Wildl. Serv., Fish. Bull. 58(133):215–252.

Tester, A. L. 1955. Estimation of recruitment and natural mortality rate from age-composition and catch data in British Columbia herring populations. J. Fish. Res. Bd. Can. 12(15):649–681.

Youngs, W. D. 1972. Estimation of natural and fishing mortality from tag recaptures. Trans. Amer. Fish. Soc. 101(3):542–545.

———. 1974. Estimation of the fraction of anglers returning tags. Trans. Amer. Fish. Soc. 103(3):616–618.

———. 1976. An analysis of the effect of seasonal variability of harvest on the estimate of exploitation rate. Trans. Amer. Fish. Soc. 105(1):45–47.

Youngs, W. D., and D. S. Robson. 1975. Estimating survival rate from tag returns: Model tests and sample size determination. J. Fish. Res. Bd. Can. 32(12):2365–2371.

———. 1978. Estimation of population number and mortality rates. *In* T. Bagenal, ed. Methods for assessment of fish production in fresh waters. Blackwell Scientific Publications, Oxford, pp. 137–164.

8

Recruitment and Yield

Recruitment can be defined as the addition of new members to the aggregate under consideration. In a fishery recruitment is the supply of fish that becomes available at some particular stage in their life history, generally that stage at which the fish first become vulnerable to the gear used in the fishery. Although simple in concept, recruitment has been one of the more difficult population attributes to express in a quantitative manner.

Procedures to provide predictive estimates for recruitment have not been successful. The reasons for this become apparent when the process of recruitment is examined. An adult reproductive stock is obviously necessary to produce recruits. But what numbers of adults are necessary? With a very few adults, recruit production may be low, and likewise with large numbers of adults, the number of recruits may be low. Size and growth characteristics also influence the production of young since fecundity is greater at one stage of life than at another, measured, that is, by the number of eggs per unit weight of adult female. There are feedback mechanisms between numbers and growth rate such that the number of adults may not be an appropriate unit.

Nearly all discussions of the problem of relating adult stock to recruits have assumed two basic types of mortality to be operative in a population: density-independent and density-dependent. Floods, droughts, extreme temperatures during certain life stages, excessive wind, and pollution are examples of density-independent types of mortality. These occur at more or less unpredictable times and are not a result of any action of the fish population. Density-dependent factors on the other hand are closely associated with actions of the fish population such as, for example, cannibalism, disease, some forms of predation, and exhaustion of food. Density-dependent factors are normally compensatory, that is, the occurrences tend to regulate the population in the direction of the long-term average.

Reproduction curves or stock recruitment relationships have been developed using a major underlying assumption that density-dependent mortality is operating in the population. These types of mortality generally affect the early life stages, as for example the larval stage. A dense population would be associated with a higher mortality, whereas a sparse population would favor survival. If such an assumption is true, the reproduction curve would be expected to have a dome shape with the left side of the dome showing recruitment greater than the stock producing the recruitment. There would be a point to the right of the dome where recruitment and the stock producing that recruitment would be equal, and then further to the right a region where recruitment would be less than the stock producing that recruitment. A graph of the stock recruitment relationship would therefore have the general appearance shown in Figure 8-1.

A reproduction curve of this type can be expressed by a simple mathematical relation:

$$\frac{R}{R_\epsilon} = \frac{S}{S_\epsilon} \; e^{[(S_\epsilon - S)/S_m]}$$

where R is filial generation, i.e., recruitment; S is parent generation, i.e., stock; S_ϵ is the equilibrium stock (size of stock which exactly replaces itself); R_ϵ is the filial generation produced by S_ϵ; and S_m is the stock which produces maximum recruitment. By placing R and S in the same

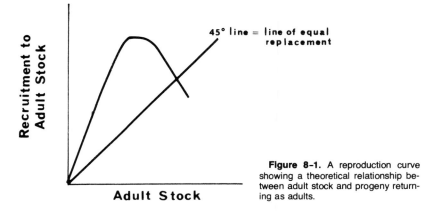

Figure 8-1. A reproduction curve showing a theoretical relationship between adult stock and progeny returning as adults.

units, that is, eggs or spawning adults, the denominators are equal and thus reduce the equation to:

$$R = S \, e^{[(S_\epsilon - S)/S_m]}$$

with the two parameters S_ϵ and S_m determining a particular curve from this family.

Cycles of a population are inherent in a reproduction curve. The simplest case occurs when only one age group is reproducing on a yearly basis. If for the moment we ignore density-independent causes of mortality, there are certain characteristics worthy of note concerning the reproduction curve and cycles. The slope of the descending leg determines the type of population cycles that would occur. If the population is in the region of the descending leg, then a slope of -1 will result in undampened oscillations of equal magnitude about the equilibrium point. Slopes between 0 and -1 will cause dampened oscillations about the equilibrium point. Slopes between -1 and $-_\infty$ will cause oscillations in the population along the descending leg which result in increases up to the apex of the dome, and then the series of oscillations would be repeated.

Multiple-age spawning stocks generally result in smaller deflections from equilibrium. Populations with reproduction curves having descending legs with slopes between 1 and -1 all eventually become stable at the equilibrium level. Multiple-age spawning populations having a reproduction curve with slope between -1 and $-\infty$ will have permanent cycles of abundance. The period of cycles is double the average egg-to-egg interval.

The effect that exploitation of the mature stock will have on the population depends upon the shape of the reproduction curve and the number of ages spawning. For single-age spawning stocks, the population may be held at equilibrium by removal of a fraction of the mature stock before spawning. The rate of exploitation is calculated as the complement of the reciprocal of the slope of a line joining the origin to the desired point on the reproduction curve. In the following figure (see Figure 8–2), suppose that point A on the reproduction curve represents the desired equilibrium level. The exploitation rate would be calculated as:

$$E = 1 - \frac{OC}{AC} = 1 - \frac{BC}{AC}$$

In the figure shown, if point B is 0.9 units of stock, and point A is 1.2 units, then the calculation would be $E = 1 - (0.9/1.2) = 0.25$ or 25

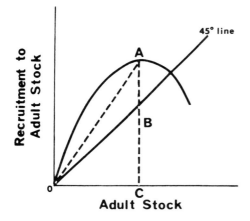

Figure 8-2. Reproduction curve showing points used to calculate exploitation rate for a desired equilibrium level.

percent of the stock. The application refers to the mature stock just prior to spawning as would occur, for example, on a salmon fishery when the harvest was taken as the adult stock moved upstream to spawn.

The exploitation of multiple-age spawning stocks will result in stability if the population was stable, and will also result in a change in the abundance of the stock. This change may be an increase or a decrease depending upon occurrences that shape the reproduction curve and the extent of exploitation. Exploitation of multiple-age spawning stocks that are in stable oscillation results in an increase in the stock until exploitation becomes excessive, which then causes a decline, and the amplitude and period of oscillation decreases. The inclusion of larger immature fish in the fishery does not change the population-control mechanisms as outlined above but does of course influence the level of equilibrium.

Thus far the discussion has centered on the ideal situation, that of a population under control of density-dependent factors of mortality. Such situations are seldom observed in actuality outside of some experimental populations. Generally, the density-dependent and density-independent factors of mortality operate concurrently. The density-independent causes seem to be overriding and thus mask the density-dependent control.

Computer simulation studies of the interactions of compensatory and noncompensatory mortality provide insight to the general problem, but little development has resulted from data analysis and estimation. It is clear how such data might be generated, but thus far good fits of data collected in actual fisheries have not been obtained.

Yield is, of course, dependent upon the number of recruits coming into the fishery. However, many of the yield models have been developed on a per recruit basis which does not take the stock-recruitment relationship into account. Other models have attempted to lump all factors, such as stock recruitment and mortality, into a single parameter. Each method has shortcomings which are discussed at more length below.

Yield

Yield is the portion of a fish population taken by man. The units of measurement are generally in terms of weight per unit time per unit area. There are numerous ways of estimating yield and potential yield, ranging from what have been referred to as macromodels to explicit dynamic pool models. Before going into detail for some of these models, it will be worthwhile to review basic definitions for some of the terms that will be used.

The standing crop is the concentration of the population for a given area at a given point in time. Measurement of standing crop may be in any of a number of different units, such as number of animals, weight, or energy content. We will be most interested in the weight aspect which is generally referred to as biomass. Biomass is therefore a particular measure of standing crop: specifically, the total weight of the referred organism present in a specified area (volume) at a specified time.

Production is the total elaboration of new biomass at the trophic level under consideration, again generally with reference to some spatial and temporal units. Thus lake trout production would be the total amount of new lake trout flesh, say, for example, in Lake Superior for a one-year period. Production may be broken down into gross and net. Gross production is the total increase of biomass regardless of whether or not that biomass lives through the period under consideration. Gross production would then include biomass used up by metabolism as well as biomass which was lost through death. Net production is the difference between gross production and losses due to metabolism and death. These ideas will be dealt with more explicitly in the trophic pool models. But first we will examine a group of quite general macromodels of value to fishery scientists.

Production Indices

In a general way there are certain gross characteristics that determine within some bounds the potential yield from various types of waters. Many characteristics have been examined both from a limnological attempt to classify lakes and from a fishery-management viewpoint of harvesting the most pounds of fish. Nearly all these characteristics have been externalities of the biological components. Again, nearly all these attempts have considered the characteristics in a static situation. Thus we have such physical characteristics as area, mean depth, maximum depth, shore development, and flushing time. Physicochemical characters have included mean temperature, highest mean temperature, oxygen, and total dissolved solids. Some studies have considered biological factors such as plankton, bottom fauna, and fish catch in the assessment of productivity of waters.

Ryder's morphoedaphic index is one such simple model, relating fish yield (Y) to total dissolved solids (T) and mean depth ($\bar{D}$). Slightly over 70 percent of the variability in fish yield could be accounted for by a logarithmic transformation of the ratio of total dissolved solids to mean depth in a group of northern temperate lakes under similar fishing intensities. Subsequent workers have shown that the logarithmic transformation may not be linear over the entire range. However, this simple index is potentially of great value and should be considered in management planning. This index is approximately as follows:

$$Y = 2 \sqrt{\frac{T}{\bar{D}}}$$

Other more elaborate models, in the sense of including additional variables, have been developed and may be of value in more specific situations. These models have all been multiple regression analyses of various combinations of the factors listed above. In general, the factors having the most apparent influence are area, mean depth, and total alkalinity. It would appear that Ryder's is the more useful index for quick estimation of yield.

These types of models do not readily provide insight into fish population or community dynamics; they do provide some broad guidelines within which more refined models might find utility. There are several types of more refined models: one group considers the population response in terms of production and is called the surplus yield model (such as Graham, Schaefer); another group considers rather detailed aspects of growth and mortality (Ricker, Beverton and Holt) and is called the dynamic pool model; and yet another group considers broad aspects but within the context of the community association as a dynamic entity, including some economic and social factors, and is called the trophic dynamic model. The choice of a model will be dependent upon the context or purpose. Ryder's index would be of little value in managing a specific fishery in any specific lake but would be of value for predicting potential yield if land-use practice were altered in such a manner as to reduce total dissolved solids in a region.

Surplus Production Models

A fish population in a given situation will have some long-term average abundance, as was shown in the section on recruitment. As a simplistic explanation, a population which is subject to increased mortality will respond by attempting an increase in biomass, that is, by increasing growth, reproduction, and survival; whereas a population at its highest levels of biomass will react by increasing mortality, decreasing growth and reproduction. This idea forms the basis for a group of management models best exemplified by Schaefer's work.

The above concept of population regulation may be expressed mathematically as follows:

$$\frac{dP}{dt} = Pg(P)$$

where P is the population biomass and $g(P)$ is a single-valued function of the population biomass, P, which gets smaller as P gets larger. This function, $g(P)$, reaches zero when P reaches the limiting population biomass. The simplest assumption would be to approximate $g(P)$ as the linear function:

$$g(P) = k(P_\infty - P)$$

where k is a constant and P_∞ is the limiting population biomass. The relationship then becomes:

$$\frac{dP}{dt} = kP(P_\infty - P)$$

The catch from a fishery is obviously a function of the size of the population present and of the fishing effort extended, and therefore the population size is also a function of effort. In the chapter on mortality an expression was developed which assumed force of fishing mortality to be proportional to fishing effort. Rate of catch at any point in time would then be:

$$\frac{dC}{dt} = qfP$$

where q (as defined in Chapter 7) is a catchability coefficient and f is fishing effort.

The above formulae state that a biomass at any point will increase to P_∞, where it would supposedly remain. Any exploitation would cause the biomass to be smaller than P_∞, and the response would be to affect an increase in P. To hold a population in equilibrium at some level less than P_∞ would mean that rate of catch would have to equal what the population would replace from that level of removal. Thus we can equate the following:

$$\frac{dC}{dt} = \frac{dP}{dt}$$

or:

$$qfP = kP(P_\infty - P)$$

hence:

$$P = P_\infty - \frac{q}{k} f$$

This expression states that population size is a linear function of fishing effort. Thus a plot of population size against fishing effort should be a straight line with negative slope. If catches for unit time intervals of equal effort are compared with unit time interval catches for different but constant effort, the following relationship results:

$$C = qf(P_\infty - \frac{q}{k} f)$$

which is the form appearing in Figure 8-3. The highest catch appears at an intermediate fishing effort. The dome of this curve, or more precisely, the point of zero slope, corresponds to a population size one-half the size of the original unexploited stock.

Early surplus yield models have some obvious shortcomings. Major

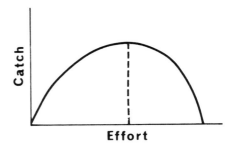

Figure 8-3. Relationship of catch and effort, assuming a surplus yield model.

among these is the implicit assumption that time lag between spawning and recruitment has no effect on population. Another condition imposed by the Schaefer model is the symmetry of the yield-effort relationship. The latter condition has been corrected by models of Fox and Pella and Tomlinson, while the former and latter conditions are corrected in a modification of these models by Walter. These models are referred to as time-lag or delay-differential models. The basic idea is that the population may be controlled by the population present at some earlier time or times as well as by the immediate population.

An example of such a model with one time lag could be:

$$\frac{dP}{dt} = (a_o - a_1 \log P(t) + a_2 \log P(t - \delta) - F(t))\, P$$

where the a's are empirically determined constants and δ is the time lag. These constants should be estimated from a series of catch and effort data by multiple regression. Solution to such differential equations may not always be explicit, but this poses no problem since numerical solution is possible using computer techniques. The general response may also be visualized by computer simulation or solution with a graphic or oscilloscope output. This will be discussed further when multiple species associations are considered below.

Dynamic Pool Models

The dynamic pool models for yield determination consider explicitly the factors of growth and mortality in affecting biomass. This type of analysis stems from a mathematical treatment of the definition of yield. It was stated that yield was the harvest taken by man, which is simply the sum over time of the product of force of fishing mortality times the standing crop of any given time. The standing crop in weight at any given time is the product of the average individual weight times the number of individuals present at that time. This concept may be expressed in a general way by:

$$Y = \int_T F(t)W(t)N(t)dt$$

where $F(t)$ is a time function of force of fishing mortality, $W(t)$ is a time function for weight, and $N(t)$ is a time function for number of fish in the population. The product of these functions on an instant-by-instant basis is summed over time period T. Two specific applications of this concept will be considered next; the first formulation is exemplified by Ricker, the second by Beverton and Holt. The major difference between the two procedures is the function used to express growth.

Ricker's method breaks the time period into intervals and life stages so that the rates of growth and mortality may be considered constant within the time interval without any appreciable error being introduced. The stock change and yield for each interval and age group are summed over intervals to provide an estimate of total yield. Growth is assumed to be expressed by:

$$W(t) = W(o)e^{gt}$$

and numbers of fish alive by:

$$N(t) = N(o)e^{-Zt}$$

Initial biomass is $W(o)N(o)$. Biomass for the next unit interval (assuming constant growth and mortality rates) is then the initial biomass times the growth function times the mortality function:

$$\text{Biomass} = N(o)W(o)e^{g}e^{-Z}$$

Given the initial biomass for any interval, the factor by which this changes is e^{g-Z} which may be easily evaluated from a table of exponential functions. The force of total mortality is partitioned into the fishing and natural components, and by an iterative procedure, the force of fishing giving maximum yield may be determined. Effects on catch-per-unit effort and average size of fish may also be estimated by such tabular procedures.

To compute yield by this method it is necessary to have frequent measurement of size by age as well as knowledge of natural mortality. The yield is given on a per recruit or per assumed initial weight of stock basis, as is true of the dynamic pool models in general. Population estimates for each year class would provide an actual basis for numbers present.

The following abbreviated example will demonstrate the above procedure for determining yield. Largemouth bass from a small lake were sampled at periodic intervals with fishing and natural mortality determined from tagging studies, creel census, and population estimates. Information obtained is presented in Table 8-1. The same kind of calculation would be made for all ages in the population; we have presented only two ages to show computations. Different values for F may be substituted to determine the one giving the largest yield.

The yield model of Beverton and Holt starts from the familiar expression for yield:

$$Y = \int_T F(t)W(t)N(t)dt$$

with von Bertalanffy's growth equation for $W(t)$. The expression for $N(t)$ is common to all models in fishery literature, but is broken down into time periods corresponding to age at recruitment and age at capture. If recruitment, R, to the area of a fishery occurs at age t_r, then:

$$N(t) = \mathrm{Re}^{-M(t - t_r)}$$

Table 8-1. Calculation of yield by Ricker's method

Age	Weight (g)	g	F	M	Z	g − Z	Weight change factor exp(g − Z)	Initial weight (kg)	Average weight	Yield
II	86							1000		
		0.51	0	0.2	0.2	0.31	1.36		1181	0
II¼	143							1361		
		0.36	0.02	0.2	0.22	0.14	1.15		1463	29
II½	205							1565		
		0.18	0.17	0.2	0.37	−0.19	0.83		1432	243
II¾	246							1299		
		0	0	0.2	0.2	−0.4	0.67		1085	0
III	246							870		
		0.31	0.2	0.2	0.4	−0.09	0.91		831	166
III¼	335							792		
		0.14	0.2	0.2	0.4	−0.26	0.77		701	140
III½	385							610		
		0.10	0.2	0.2	0.4	−0.30	0.74		531	106
III¾	427							451		
									Total	684

Further, if age when fish become vulnerable to the fishing gear, t_c, is greater than t_r, and both are less than age t, we have the following:

$$N(t) = Re^{-M(t_c - t_r)}e^{-(F + M)(t - t_c)}$$

The first exponential considers the time period when only natural mortality affects recruits after reaching the fishing grounds but before reaching a size vulnerable to the capture gear. The second exponential considers the time period after reaching capture size when subject to both fishing and natural mortality. This introduces another parameter, age at capture, which can be controlled by a regulatory agency for management of a fishery. Making the substitution:

$$R' = Re^{-M(t_c - t_r)}$$

the final expression to be used is:

$$N(t) = R'e^{-(F + M)(t - t_c)}$$

Force of fishing mortality is treated as constant:

$$F(t) = F$$

The von Bertalanffy expression for growth was developed to express growth in lengths; we wish to have an expression at this point relating to weight. If we assume isometric growth and make the substitution:

$$W = \text{Constant} \times L^3,$$

the growth equation in terms of weight becomes:

$$W(t) = W_\infty (1 - e^{-K(t - t_o)})^3$$

where W_∞ is the ultimate or limiting weight and other terms are as defined in Chapter 5. Expanding this expression results in the following:

$$W(t) = W_\infty [1 - 3e^{-K(t - t_o)} + 3e^{-2K(t - t_o)} - e^{-3K(t - t_o)}]$$

which may be more simply written as:

$$W(t) = W_\infty \sum_{n=0}^{3} U_n e^{-nK(t - t_o)}$$

where $U_o = 1$, $U_1 = -3$, $U_2 = 3$, and $U_3 = -1$.
The yield model now becomes:

$$Y = \int_{t_c}^{t_l} FRW_\infty \; e^{-F(t - t_c) - M(t - t_r)} \sum_{n=0}^{3} U_n e^{-nK(t - t_o)} \, dt$$

Upon substitution with R' and then integrating, the yield equation is:

$$Y = FR'W_\infty \sum_{n=0}^{3} \frac{U_n}{F + M + nK} \; e^{-nK(t_c - t_o)} [1 - e^{-(F + M + nK)(t_l - t_c)}]$$

Where the upper limit of integration, t_l (corresponding to oldest age in fishery), is large, the yield equation simplifies to:

$$Y = FR'W_\infty \sum_{n=0}^{3} \frac{U_n}{F + M + nK} \; e^{-nK(t_c - t_o)}$$

Three parameters in this yield equation may be controlled by management agencies: the force of fishing mortality (F), age of entry into the fishery (t_c), and in some situations upper age in the fishery (t_l). With t_c, K, M, and R' held constant, there is generally a value of F which produces a maximum yield. The shape of this relationship is demonstrated in Figure 8–4 which also shows the effects of different levels of natural mortality.

Age at entry into the fishery has a similar type of relationship as shown in Figure 8–5, again showing effects of different levels of natural mortality. It is apparent that different combinations of F and t_c may produce the same yields. Plotting t_c against F provides what is called a yield isopleth diagram as in Figure 8–6.

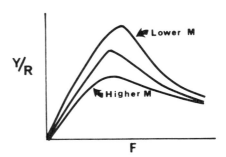

Figure 8–4. Relationship of yield per recruit (Y/R) to force of fishing mortality (F).

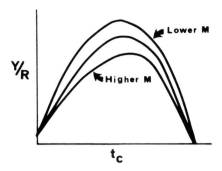

$\dfrac{Y}{R}$

t_c

Figure 8-5. Relationship of yield per recruit (Y/R) and age at entry into fishery (t_c).

Yield values are easily calculated on electronic calculators or computers. However, extensive tables have been compiled for a modified version of the above yield function and for a yield function using allometric growth parameters, that is, where $w = al^b$. These will not be discussed, but the interested reader is referred to Jones (1957), Wilimovsky and Wicklund (1963), and Beverton and Holt (1966).

The lake trout data in the Beverton-Holt model result in the following when we assume a large oldest age in fishery (reasonable for lake trout):

$$Y/R = W_\infty F \sum_{n=0}^{3} \frac{U_n}{F + M + nK} \; e^{-nK(t_c - t_o)}$$

$$= 4710 \; F \sum_{n=0}^{3} \frac{U_n}{F + 0.10 + 0.196n} \; e^{-0.196n(4 - 0.057)}$$

Various values of F are used in the formula which gives the graph shown in Figure 8-7 when yield is plotted against F.

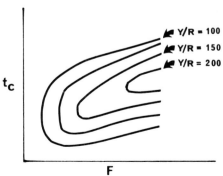

t_c

Y/R = 100
Y/R = 150
Y/R = 200

F

Figure 8-6. Relationship of age at entry into fishery (t_c) and force of fishing mortality (F).

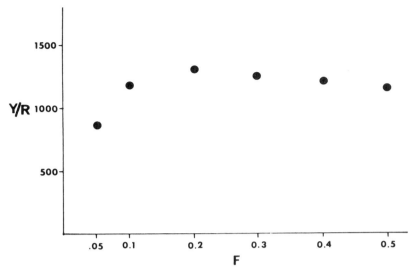

Figure 8-7. Relationship of yield per recruit (Y/R) and force of fishing mortality (F) for lake trout example.

A major criticism of the surplus yield and dynamic pool models is the treatment of the species under consideration as if it were living in a non-dynamic environment. The trophic dynamic models have attempted to overcome this objection. This type of model considers the system as an entity with various components of the system being represented by a differential equation. The simultaneous solution of this set of differential equations results in a description of how the system responds. The more realistic the equations, the greater the number of parameters there are to estimate, resulting in rather bulky analysis and necessitating a large data base. It is virtually impossible to handle this type of model without use of a computer, and therefore most of the work on this class of models has been in the last ten years. However, the simple case of two interacting species, that is, predator-prey, host-parasite, has been studied for over fifty years. With these simple systems explicit solutions are possible, and a great deal of work has been done on the analysis of such systems.

The trophic dynamic models may be introduced in principle by looking at an example from Plant et al. (1973). This example will use mass transfer through the system; other transfers might be in terms of energy or numbers. The coefficients will be dependent upon the type of transfer under consideration. The flow will also be dependent upon the units and hence upon the model itself. We will use a portion of the Cayuga Lake system to first construct a block model with arrows showing the direction of mass transfer. The biological components of this system will be the phytoplankton, zooplankton, alewife, and lake trout. Studies on Cayuga have shown that the alewife feeds primarily on zooplankton, and the lake trout feeds on alewife for a major portion of the time. A mass flow diagram is shown in Figure 8-8. In this simple system there is a mass flow from nutrient, N, to phytoplankton, P, to zooplankton, Z, to alewives, A, and finally to lake trout, L. Each of the biological compartments has some mass that is lost through death and which is eventually recycled. This is shown by the arrows going back to N. The whole system is driven by solar energy, with the processes being temperature dependent. The nutrient in this particular case is phosphorus, since studies have shown this to be the limiting nutrient. Equations for this model are as follows:

$$\frac{dN}{dt} = K\,[I_N - P_N \cdot G_P]$$

$$\frac{dP}{dt} = K\,[G_P - R_P \cdot P - G_Z]$$

$$\frac{dZ}{dt} = K\,[G_Z - R_Z \cdot Z - G_A]$$

$$\frac{dA}{dt} = K\,[G_A - R_A \cdot A - G_L]$$

$$\frac{dL}{dt} = K\,[G_L - (R_L + F_L) \cdot L]$$

where K is a temperature coefficient equal to $(T - 4)/20$, G_i is a growth rate term, and R_i is a respiration term. I_N is nutrient input to the lake, P_N

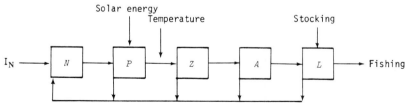

Figure 8-8. Mass flow diagram for a Cayuga Lake food chain. Courtesy of R. Plant, R. Lance, and W. D. Youngs, 1973, Cornell Univ. Water Res. Mar. Sci. Center, Tech. Rep. 69.

is the nutrient pool, and F_L is the force of fishing mortality for lake trout. The growth rate terms have a mass transfer coefficient, X_i, as shown in the following:

$$G_P = X_P \cdot E \cdot N \cdot P \cdot \exp - (T - 30)^2/82.8$$
$$G_Z = X_Z \cdot P \cdot Z \cdot \exp - (T - 25)^2/23.4$$
$$G_A = X_A \cdot Z \cdot A$$
$$G_L = X_L \cdot A \cdot L$$

Solar energy is expressed by:

$$E = 0.6 + 0.6 \cos 2 \pi t$$

and temperature is expressed by:

$$T = 4 + 16 \exp - (t - 236)^2/9590$$

The constants in the above expression are chosen to fit the data.

The set of differential equations may be solved by numerical integration using a computer. A simulation for one year is shown in Figure 8-9.

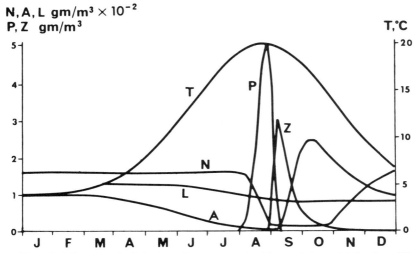

Figure 8-9. Computer simulation for various components of Cayuga Lake model. Courtesy of R. Plant, R. Lance, and W. D. Youngs, 1973, Cornell Univ. Water Res. Mar. Sci. Center, Tech. Rep. 69.

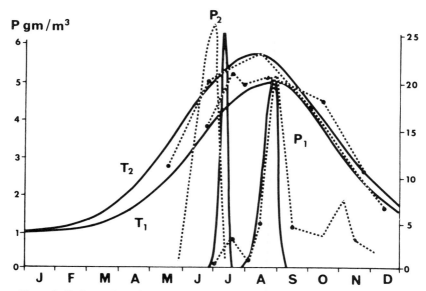

Figure 8-10. Comparison of computer simulations (solid line) of temperature, *T*, and phytoplankton, *P*, with actual observations. Courtesy of R. Plant, R. Lance, and W. D. Youngs, 1973, Cornell Univ. Water Res. Mar. Sci. Center, Tech. Rep. 69.

A comparison between simulation (solid line) and observed data (broken line) for temperature and phytoplankton for two different years is shown in Figure 8–10.

The effect of various fishing rates, F_L, on the whole system may be examined by varying the values for F_L. As with all models, the model is assumed to represent the system under consideration. The only means of testing this assumption is by the degree of approximation provided by the model when compared with observed information. This generally leads to further adjustment of the model and comparison until one is satisfied with the results. The optimum yield of lake trout, in view of other aspects of the system, may be obtained iteratively. Such a model may also be expanded to include land use effects on nutrient input, I_N, for example, or how human population might influence land use. Area or regional models of this type are being developed.

References

Beverton, R. J. H., and S. J. Holt. 1957. On the dynamics of exploited fish populations. Min. Agric. Fish and Food (U.K.), Fish. Invest. Ser. II, 19. 533 pp.

_____. 1966. Manual of methods for fish stock assessment. Part II. Tables of yield functions. FAO, U.N. Fish. Tech. Pap. 38 (Rev. 1). 97 pp.

Cole, L. C. 1951. Population cycles and random oscillations. J. Wildl. Manage. 15(3):233–252.

_____. 1954a. Some features of random population cycles. J. Wildl. Manage. 18(1):2–24.

_____. 1954b. The population consequences of life history phenomena. Q. Rev. Biol. 29(2):103–127.

Fox, W. W., Jr. 1970. An exponential surplus-yield model for optimizing exploited fish populations. Trans. Amer. Fish. Soc. 99(1):80–88.

Gulland, J. A. 1969. Manual of methods for fish stock assessment. Part I. Fish population analysis. FAO, U.N., Man. Fish. Sci. 4. 154 pp.

Gulland, J. A., and L. K. Boerema. 1973. Scientific advice on catch levels. Natl. Oceanogr. Atmos. Adm., Fish. Bull. 71(2):325–335.

Graham, M. 1935. Modern theory of exploiting a fishery, and application to North Sea trawling. J. Conseil Int. Explor. Mer 10:264–274.

Jones, R. 1957. A much simplified version of the fish yield equation. Document P. 21, presented at the Lisbon joint meeting of ICNAF/ICES/FAO. 8 pp.

Larkin, P. A., and H. S. Hourston. 1964. A model for simulation of the population biology of Pacific salmon. J. Fish. Res. Bd. Can. 21(5):1245–1265.

Larkin, P. A., and W. E. Ricker. 1964. Further information on sustained yields from fluctuating environments. J. Fish. Res. Bd. Can. 21(1):1–7.

Lindeman, R. L. 1942. The trophic-dynamic aspect of ecology. Ecology 23:399–418.

Lotka, A. J. 1925. Elements of physical biology. Williams and Wilkins Co., Baltimore. 460 pp.

_____. 1956. Elements of mathematical biology. Dover Publications, Inc., New York. 465 pp.

Nicholson, A. J. 1933. The balance of animal populations. J. Anim. Ecol. 2:132–178.

Nicholson, A. J., and V. A. Bailey. 1935. The balance of animal populations. Part I. Proc. Zool. Soc. London 1935:551–598.

Oglesby, R. T. 1977. Relationships of fish yield to lake phytoplankton standing crop, production, and morphoedaphic factors. J. Fish. Res. Bd. Can. 34(12):2271–2279.

Parrish, B. B., ed. 1973. Fish stocks and recruitment. Rapp. P.-v. Réun. Cons. int. Explor. Mer. 164. 372 pp.

Patten, B. C. 1969. Ecological systems analysis and fisheries science. Trans. Amer. Fish. Soc. 98(3):570–581.

Patten, B. C., ed. 1971. Systems analysis and simulation in ecology. Vol. 1. Academic Press, New York. 607 pp.

_____. 1972. Systems analysis and simulation in ecology. Vol. 2. Academic Press, New York. 592 pp.

Paulik, G. J. 1969. Computer simulation models for fisheries research, management, and teaching. Trans. Amer. Fish. Soc. 98(3):551–559.

Pella, J. J., and P. K. Tomlinson. 1969. A generalized stock production model. Inter-Amer. Trop. Tuna Comm., Bull. 14:421–496.

Plant, R., R. Lance, and W. D. Youngs. 1973. Computer simulation of trophic level interrelationships in Cayuga Lake. Cornell Univ. Water Res. Mar. Sci. Center, Tech. Rep. 69. 26 pp.

Rawson, D. S. 1952. Mean depth and the fish production of large lakes. Ecology 33(4):513–521.

———. 1955. Morphometry as a dominant factor in the productivity of large lakes. Verh. int. Ver. Limnol. 12:164–175.

———. 1961. A critical analysis of the limnological variables in assessing the productivity of northern Saskatchewan lakes. Verh. int. Ver. Limnol. 15:160–166.

Regier, H. A., and H. F. Henderson. 1973. Towards a broad ecological model of fish communities and fisheries. Trans. Amer. Fish. Soc. 102(1):56–72.

Ricker, W. E. 1946. Production and utilization of fish populations. Ecol. Monogr. 16:373–391.

———. 1954. Stock and recruitment. J. Fish. Res. Bd. Can. 11(5):559–623.

———. 1958a. Maximum sustained yields from fluctuating environments and mixed stocks. J. Fish. Res. Bd. Can. 15(5):991–1006.

———. 1958b. Handbook of computations for biological statistics of fish populations. Fish. Res. Bd. Can., Bull. 119. 300 pp.

Ricker, W. E., and R. E. Foerster. 1948. Computation of fish production. Bingham Oceanogr. Collect., Bull. 11(4):173–211.

Roedel, P. M., ed. 1975. Optimum sustainable yield as a concept in fisheries management. Amer. Fish. Soc. Spec. Publ. 9. 89 pp.

Ryder, R. A. 1965. A method for estimating the potential fish production of north-temperate lakes. Trans. Amer. Fish. Soc. 94(3):214–218.

Schaefer, M. B. 1959. Biological and economic aspects of the management of commercial marine fisheries. Trans. Amer. Fish. Soc. 88(2):100–104.

———. 1968. Methods of estimating effects of fishing on fish populations. Trans. Amer. Fish. Soc. 97(3):231–241.

Schaefer, M. B., and R. J. H. Beverton. 1963. Fishery dynamics. Their analysis and dynamics. In M. N. Hill, ed. The sea. Vol. 2, John Wiley and Sons, New York, pp. 464–483.

Schaefer, M. B., and R. Revelle. 1959. Marine resources. In M. R. Huberty and W. L. Flock, eds. Natural Resources. McGraw-Hill, New York, pp. 73–109.

Scott, A. D., ed. 1970. Economics of fisheries management. Univ. British Columbia, Vancouver, H. R. MacMillan Lecture, Fish., 1969. 115 pp.

Silliman, R. P. 1969. Analog computer simulation and catch forecasting in commercially fished populations. Trans. Amer. Fish. Soc. 98(3):560–569.

———. 1971. Advantages and limitations of "simple" fishery models in light of laboratory experiments. J. Fish. Res. Bd. Can. 28(8):1211–1214.

Thompson, W. F., and F. H. Bell. 1934. Biological statistics of the Pacific halibut fishery. 2. Effect of changes in intensity upon total yield and yield per unit of gear. Rep. Int. Fish. (Pacific Halibut) Comm. 8. 49 pp.

Walter, G. G. 1973. Delay-differential equation models for fisheries. J. Fish. Res. Bd. Can. 30(7):939–945.

Walter, G. G., and W. J. Hogman. 1971. Mathematical models for estimating changes in fish populations with applications to Green Bay. Internat. Assoc. Great Lakes Res. Proc., 14th Conf. Great Lakes Res.:170–184.

Watt, K. E. F. 1968. Ecology and resource management. A quantitative approach. McGraw-Hill, New York. 450 pp.

Watt, K. E. F., ed. 1966. Systems analysis in ecology. Academic Press, New York. 276 pp.

Wilimovsky, N. J., and E. C. Wicklund. 1963. Tables of the incomplete beta function for the calculation of fish population yield. Univ. British Columbia. Vancouver, Inst. Fish. 291 pp.

9

Fish Marking

Fishery scientists have been searching for the ideal mark since 1873 when Atlantic salmon, tagged in the Penobscot River, Maine, were later recovered in fair numbers. The ability to account for the presence of a particular fish or group of fish in time and space furnishes a basic tool for fishery resource management. Among questions that can be answered by recognizing fish are degree of homing, racial studies, age and growth, mortality rates, speed and migration routes, survival of hatchery fish, and survival and growth of transplants.

Spectacular tag returns emphasize the migrations and growth that may be determined for species from marking programs. The longest migration of a tagged halibut was from the Aleutian Islands to northern California, a distance of 3,680 km. The fish had been tagged for 6 years. Actually, only a relatively few halibut have been tagged and a very few are reported as recaptured, but a recent tag return reported by the International Pacific Halibut Commission records a fish tagged in Russia and recovered south of the Alaska Peninsula near the Shumagin Islands. During the two years the fish had been at large it had grown 12.5 cm and gained 7.2 kg, and likely traveled a distance of 1,600 km. California Fish and Game biologists report a 13.5 kg striped bass caught in the Sacramento River that had carried its tag for 18 years, 7 months. The previous California record belongs to the striper that had been at large for 17 years, 5 months. A skipjack tuna tagged by the Inter-American Tropical Tuna Commission off the southern coast of Baja California in 1973 was taken about 4.8 km south of Honolulu. The fish, during its 14 months of freedom, had moved at least 4,400 km, increased in length from 45 to 75 cm, and gained 9.5 kg.

Fish marks can be divided generally into those where the fish is mutilated in some manner, as by removing a fin or part of the maxillary bone, or by tagging, which involves the insertion, attachment, or injection of a foreign object or substance. Natural tags such as parasites or bacterial fauna and chemoprints are useful. Color marks and radioactive tags may be injected or the fish may be immersed in the solutions.

The problems in developing and choosing the proper mark are easily illustrated by the following list of characteristics for the ideal mark, which should:

1. Remain unaltered during the lifetime of the fish.
2. Have no effect on fish behavior or make the fish more available to predators.
3. Not tangle with weeds or nets.
4. Be inexpensive and easily obtained.
5. Fit any size fish with little alteration.
6. Be easy to apply without anesthetic and with little or no stress to the fish.
7. Permit enough variation to at least separate groups.
8. Create no health hazard.
9. Cause no harm to fish as food or to aesthetics.
10. Be easy to detect in the field by untrained individuals.
11. Cause no confusion in reporting.
12. Remain unaffected by preservation.

Choice of mark is of vital importance to the success of any large-scale program because many of the parameters, such as mortality rates, cannot be properly evaluated if the mark is changed during the course of the program. Among the factors the biologist should consider in deciding the best mark are:

1. Length of time the mark is to remain on the fish. Whether the mark is intended to last from the smolt stage to the return of the adult salmon requires a very different selection than choosing a mark to check time lag in fish passing a questionable fish ladder.
2. Personnel available for applying the mark. Some tags require skilled taggers, and recoveries may be suspect if there is a lack of uniformity in technique. Fin-clipping requires careful attention if fin regeneration is not to occur and cloud results. Rough handling and improper techniques should be avoided.
3. Species of fish. Strap tags may give excellent results attached to a halibut, but poor results on cod, haddock, and pollock. No adequate external tags for mackerel-like fishes have been developed. Internal

tags are possible with most fishes, but the problem of recovery is always present.

4. Methods of capture and handling. Species that are handled individually and cleaned at the time of catching offer a larger choice of tags. A fair-sized brightly colored celluloid body-cavity tag can usually be found when the fish is cleaned. That it will remain with the fish for a long period often outweighs the possibility of its being overlooked in cleaning. Body-cavity tags may prove worthless if fish are shipped in the round and may not be discovered until the fish is thousands of miles from the coast or lake. A study of handling techniques in commercial operations is needed before deciding which mark to use. If sport fishermen and commercial fishermen are to identify and report a fin mark, then considerable care should be taken to avoid any confusion as to what fin to tag or whether on the left or right side of the fish.

5. Interstate or international programs for fish that cross state and national boundaries put additional restraints on the choice of marks. Coordination of fin-clipping programs is necessary if fish from different states or nations are recovered in other's territories. A good example is the careful coordination between the Maine Atlantic Salmon Commission and the Fisheries Agencies of Canada to insure that smolts and adults apt to be taken in Canadian waters are fin-clipped according to a schedule designed to avoid confusion as to the origin of fish. This is a particularly difficult proposition with fin-clipping, as only a limited number of combinations of fin removal are available.

As much as possible, coordination among fisheries as to the kinds of tags should be planned in any large-scale tagging program, as with the present Atlantic salmon program and because of the concern over fishing off the coast of Greenland. Tags of several nations are coordinated through the Department of Fisheries in Halifax, Nova Scotia.

Every proposal for marking fish should be accompanied with a detailed plan for recovering marks. A basic problem is to insure equal effort on mark recovery during the duration of the program. As often as possible, trained personnel should examine the fin clip or tag to be as certain as possible that information reported is accurate. Programs should be advertised on radio, T.V., newspapers, trade journals, leaflets, and public meetings. One of the problems, of course, is to keep the fishermen interested. Many times they are enthusiastic in the first and second years but may not continue to cooperate in later years.

Personal contact by the biologists and other fishery personnel for tag pick-up and fish examination stimulates interest. The tag itself is helpful if it has clear instructions for return of information. Prompt reply to tag returns with a little story of the project, where the fish was originally tagged, how much it weighed, how long it was, and something about its activities during the period at large is long-term insurance for additional recoveries. Furnishing commercial vessels or guides and cooperative sport fishermen with log books helps to obtain complete information reported in a standard way.

Rewards for the return of information on marked fish take many forms from lotteries to straight payment. Fish and game clubs and other interested service groups will frequently sponsor a drawing for prize money or for an outboard motor or some other pieces of outdoor equipment. Sporting-goods stores may provide a headquarters for information on marked fish.

Monetary rewards need to be high enough to motivate the fishermen. A tag return poster from Greenland advertised a \$25.00 payment for full information on tagged Atlantic salmon. A 1978 announcement from the International Pacific Halibut Commission increases rewards for halibut tags from \$2.00 to \$5.00 with special rewards of \$100 for preselected tags. Once a program of paying rewards is begun it should be continued for the duration of the program. Further, payment of rewards for one project and not another can lead fishermen to assume that information from the nonreward mark is not as important. The temptation would be to pay rewards for tags, but not pay for thousands of fin-clipped fish.

Estimates of the time required to mark fish vary widely. Tags are so varied in attachment that rates would have to be considered for all the variations. As many as 8,000 fish may be marked per hour with fluorescent particles, and thousands may be marked by immersion in dyes. Experienced women marking various fin combinations on smolt-sized Atlantic salmon averaged 335 fish per hour on a total of 199,401 smolts; in this program the women were given a 10-minute break per hour, and the fish were brought to them and returned to raceways by hatchery workers. Lake trout fin-clipping rates at two National Fish Hatcheries are reported as an average of 549 per hour for one fin and 453 per hour for two fins. The workers should not be rushed nor should competition be encouraged; speed may mean more fish marked but not as carefully.

The classic work of Rounsefell and Kask of 1945 defined and described 18 general types of tags. Some of those tags are obsolete or have never been used extensively and have not been included here.

Archer tag

This tag consisted originally of a single plate attached by two wires, one at either end, that pierce the tissues, and are then twisted or clinched. Variations include the use of pins through holes in the plate to replace the wire, and in the use of two plates, one usually being on each side of the part pierced. The original tag was made of silver and was only 10 mm long, with two 8-mm points, and 4 mm wide. The latest modification (used on Canadian salmon) is two plastic plates, each 31 mm by 8 mm by 0.6 mm thick. One plate is placed on each side of the dorsal fin, and the plates are held in place by two nickel pins 32 mm in length of no. 20 B. & S. gauge (0.9 mm) with a head 2 mm in diameter. This tag has been used principally on adult salmon, attached to the dorsal or adipose fin. It has not been adequately tested and, at present, is not highly regarded.

Atkins Tag

This tag is extremely simple, consisting of a bead or flat plate attached by a thread or wire that pierces the tissues, forming a loop. In its simplest form it closely resembles a luggage tag. It was used in 1873 by Charles G. Atkins on Atlantic salmon on the Penobscot River. Various sizes, shapes, and materials have been tried, as well as different points of attachment. Good results have been obtained using an oblong silver plate 24 mm by 9 mm attached with soft silver wire through the back of the fish at the front of the dorsal fin. It is highly approved by Europeans for marking salmon.

Ph. Wolf of the Salmon and Trout Association, Malmo, Sweden, has used an Atkins tag consisting of a piece of white Celluloid printed with a serial number and mailing directions. This is sealed between two pieces of transparent Celluloid of the same size. It is attached by a very fine wire threaded on a needle in order to leave a very small wound in attaching it. One size is 25 mm by 8 mm, and a smaller size is only 14 mm by 5 mm.

Bachelor-Button Tag

This tag consists of two plates or disks held together rigidly by a shaft that pierces the tissues. The first employment was in 1908 by Charles W. Greene on Pacific salmon in the Colombia River. It was given an extensive trial on sockeye salmon in Puget Sound and on Pacific halibut. The silver type with concave edges cut off the circulation, and the tissues decayed so that the tag fell off, leaving a hole in the operculum. The flat aluminum

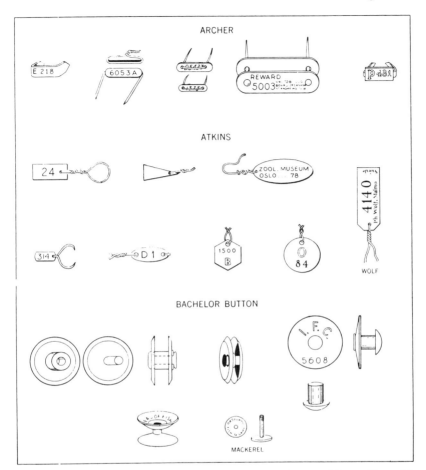

Figure 9-1. Types of Archer, Atkins, and bachelor-button tags. Courtesy of George Rounsefell, 1945, Trans. Amer. Fish. Soc. 73:320-363.

type (disks 19 mm by 1 mm, the solid shaft 4 mm by 9 mm, the hollow shaft 7 mm by 7 mm, weight 2.6 g), used on the opercle of sockeye salmon (*Oncorhynchus nerka*), gave as good returns as the strap tag on the caudal peduncle. Neither the concave silver type nor the flat rivet type of Monel metal were efficient on halibut, as many tags came off. The cupped bachelor-button was devised in 1932 by William C. Herrington for use on haddock. The outer cupped disk of aluminum is 15.7 mm in diameter, 0.5 mm in thickness, and is cupped 3 mm, outside measurement. It has a hollow shaft tapering from 5.0 to 4.5 mm to a shoulder. Beyond the shoulder, the shaft is 3.3 mm in diameter to fit through a hole in the plastic disk. This narrow portion is only 1.5 mm in length and is crimped onto the plastic disk. The shaft from the disk to the shoulder is 3.5 mm in length for use on larger fish and 2.3 mm for smaller fish. The second disk (to use inside the opercle), is very slightly concave, 15.5 mm in diameter and 0.7 mm in thickness with a central hole 3.5 mm in diameter. Because of the cupped shape this tag did not stop the circulation of blood in the tissues. The cupping was supposed to prevent overgrowth by opercular tissue, but within one year these tags are sometimes completely hidden by opercular tissue. The rigid shaft may contribute to this overgrowth. This tag is as good as or better than the modern Petersen tag on haddock, but it requires special pliers that are difficult to keep in adjustment, and it cannot be attached as quickly as the Petersen tag.

A very small bachelor-button tag with the disks only 9 mm in diameter has been tried experimentally on mackerel with indifferent success. The flat disk of aluminum has a hollow shaft that pierces a hole in the opercle and is then clinched through the other disk of red plastic. The shaft is about 3 mm in outside diameter, and the disks are 3 mm apart when the tag is clinched. This tag has definite possibilities and should be given further tests on other species.

Barb Tag

This is any tag consisting of a straight shaft, with or without an attached plate, that is pushed into the tissues and that depends for holding wholly on one or more barbs. It was first tried by Heldt to mark tuna and swordfish in Tunis, but it was not adequately tested. Templeman used a plastic barb to mark lobsters in Newfoundland by pushing it between the segments in the abdominal side of the tail segments. The same tag has also been used by the Marine Laboratory of the University of Miami to mark spiny lobsters. One of these tags consists of a piece of stiff white plastic over ½ mm thick, measuring 40 mm by 6 mm. One end is sharpened to a point 6 mm long, with three triangular-shaped notches (or barbs) on each side. Each notch is 1 mm deep and 2½ mm long. The front edge of each notch is perpendicular to the long axis of the tag. One side of the tag is printed with the address, and the reverse gives directions, "Tell where, when, and by whom caught." The two surfaces are covered (over the printing) with a very thin layer of transparent plastic.

Several types of plastic barb tags are available, ranging from small sizes to be applied on trout to larger sizes (up to six inches in length) with a shaft large enough to hold printed instructions for return of the tag. The plastic barb tags are available in different colors, and instructions or identification can be printed directly on the shaft.

Body-Cavity Tag

This tag can be defined as any material inserted loosely into the body cavity. It was invented by Robert A. Nesbit for marking squeteague in Chesapeake Bay in 1931. His original tag consisted of a strip of colored Celluloid about 0.7 mm thick that was usually about 4 by ½ cm. Later he changed the shape to make the tags wider. This type of tag is attached by making a small vertical incision in the body wall, usually with a scalpel, and inserting the tag.

Various sizes of body-cavity tags have been used. The very large wedge-shaped tag was devised in 1935 by John L. Kask. It is made of red Celluloid, 1.5 mm in thickness, for use on halibut. Sizes as small as 12 mm by 3 mm by 1 mm have been used on various other species including trout, salmon, cod, haddock, and mackerel. This tag can be applied to very small fish and will remain to be returned after growth to adult size.

The magnetic body-cavity tag was devised by George A. Rounsefell and Edwin H. Dahlgren in 1932. The first tags were used on Alaska herring. They were very small (13 mm by 3 mm by 0.7 mm) and made of pure nickel. The chief advantage of the magnetic tag lay in the ability to recover tags by electromagnets from the reduction plants as the fish are processed into meal. These tags were modified by Rounsefell and Dahlgren in 1934 to a larger size (19 mm by 4 mm by 1.0 mm), and they were made of steel, since the nickel was not sufficiently magnetic to be readily recoverable by the magnets except under ideal conditions. Although most of the steel tags were nickel-plated, it was soon discovered that the bare steel did not corrode in the body cavity. However, plating renders the tags much brighter and easier to detect when searching for the magnets.

Attempts have been made to improve the shape of these tags to prevent shedding of tags. Dumbbell-shaped tag and tags with square ends were tried, but in neither case was there any significant improvement, either in the prevention of shedding or in the magnetic recovery. As soon as the magnetic body-cavity tags proved their value on the Alaska herring, they were widely adopted along the Pacific Coast for use on herring, pilchards (sardines) (*Sardinops caerulea*), and mackerel (*Pneumatophorus diego*).

An electronic tag detector was developed in 1935 by Edwin H. Dahlgren with the aid of the electrical-engineering faculty of the University of Washington. With this instrument, fish bearing the metal tags could be detected as they were unloaded from the fishing vessel instead of after being processed into metal. It was now possible for the first time to gain very accurate data on the locality of recapture of small schooling fish.

Collar Tag

This is a ring of any material attached wholly by encirclement without piercing any tissues. It may bear an attached plate. Early attempts by

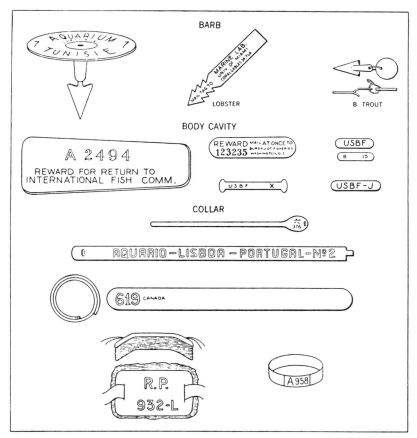

Figure 9-2. Types of barb, body-cavity, and collar tags. Courtesy of George Rounsefell, 1945, Trans. Amer. Fish. Soc. 73:320-363.

Atkins in 1872 using rubber bands and by Sella in 1911 using copper chains yielded no returns. Rubber bands on yellowtail (*Seriola quinqueradiata*) in Japan gave a few returns, and later the rubber bands were tried on mackerel in the Black Sea with only slightly better results. The first successful collar tag was a thin flat strip of silver on the caudal peduncle of Japanese yellowtail (9.0 percent recovery) and Japanese mackerel (0.7 percent recovery).

Commencing in 1925 several thousand mackerel were tagged on the Atlantic Coast of the United States and Canada with Celluloid poultry leg bands around the caudal peduncle. Recaptures were usually less than 2 percent. A few remained at liberty as long as three years, but usually the fish were very emaciated, and the caudal peduncle showed severe chafing.

Sette experimented with several types of collar tags on mackerel confined in a large outdoor pool. The Celluloid band used in these experiments was a strip 0.635 mm thick, 8 mm wide, and 50 mm long, molded to form a circle 11.1 mm in inside diameter and overlapping about one third of the circumference. A ring was also tried, made of rods of Celluloid 2.5 mm in diameter and 38 mm in length, cut obliquely at the ends to fit together when molded to a circle of 9.5 mm inside diameter. A third collar tag tested was a rubber band made of drainage tubing 9.5 mm in diameter with walls 0.33 thick cut into sections 9.5 mm wide.

None of the three tags was fully successful, but the Celluloid rings gave the best results, the rubber bands the poorest. These experiments did show that if a collar tag is at all loose on the caudal peduncle it may slip off over the tail. This limits the use of these tags, since sufficient allowance for future growth cannot be made in tagging young mackerel.

Hook Tag

This is a shaft piercing the tissues and held by the curve of the shaft and usually by one or more barbs. Not actually a tag, this is more a method of marking commercial fishing hooks, so that when a fish breaks the line and escapes the fisherman can report the incident, and if the fish is recaptured bearing the hook its movement can be computed. Tuna hooks are marked in Portugal and France with the locality and a number for the year.

A tag developed by R. A. McKenzie of the Fisheries Research Board of Canada in 1942 should probably be classified as a hook tag, since it depends on the bend of the shaft for attachment. It was used on several thousand smelt (*Osmerus mordax*) in the Miramichi River, New Brunswick. The tag consists of a very thin strip of red plastic (less than ½ mm) bent back on itself. It is thrust through a small incision in the opercle. The inner portion is 15 mm long by 2.5 mm wide and pointed. The outer half is of the same width for 2 mm and then swells in width to form a tear-shaped disk 12 mm long by 8 mm wide. This disk is variously notched on the edges to indicate different marking experiments.

Hydrostatic Tag

This tag is attached by a wire piercing the tissues; the tag itself is hollowed so that its specific gravity is very slightly less than that of water. Developed by Einar Lea of Norway, it has given remarkably high recoveries compared with the Petersen tag. It is made of a piece of hollow plastic tubing. Each end is stopped with a short rod. The rear end is tapered. The forward end is flattened and contains a small hole for attaching the wire. Inside of the hollow tag complete directions are printed in two languages on a roll of thin tissue paper 14.5 cm long. A sample tag on hand measures 28 mm over-all and nearly 4 mm in diameter. The tag is very conspicuous, as the center transparent section is of a yellowish tinge, while the ends have been dipped in a bright blue.

Internal Anchor Tag

This tag consists of a flexible chain or thread (with or without any attached material on the outside) that pierces the body wall and is held in place by being attached to material inside the body cavity.

This tag was developed in 1936 by George A. Rounsefell, to meet the need for an externally visible tag in which the wound could heal quite completely and the fish could undergo a very large increase in size without losing the tag. The idea was furnished by John L. Kask, who in 1930 tried a metal tag in the body cavity, with a protruding chain, on a few flounders in a live car, and found that the wound healed completely. The small-sized internal anchor was tested in 1937 on young sockeye salmon 5 to 15 cm in length, that were being held in tanks. The wound completely healed within a few days, but the experiment had to be terminated at the end of 9 months. Haddock were tagged in Maine with two large sizes of internal anchors; returns were comparable to those from the Herrington bachelor button or the improved Petersen, from haddock tagged in the same experiment.

The internal anchor tag has one great advantage over the body-cavity tag. It has long been known that a significant proportion of the fish marked with body-cavity tags lose them within a short time after tagging. Janssen

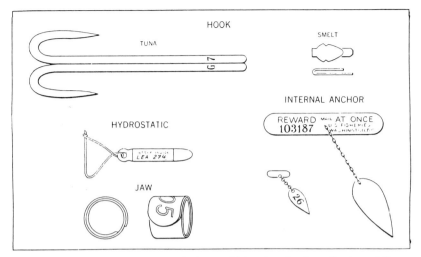

Figure 9-3. Types of hook, hydrostatic, jaw, and internal anchor tags. Courtesy of George Rounsefell, 1945, Trans. Amer. Fish. Soc. 73:320–363.

and Aplin discovered in holding experiments with sardines that 8 to 58 percent of the body-cavity tags were shed within 5 months. The internal anchor tag is not shed, as the chain prevents it from moving either forward or backward in the body cavity, and thus it cannot escape by the insertion wound. Being held in one position against the body wall it is quickly encysted firmly by tissue. Body-cavity tags move freely about at first and often move against the heart. In all of the fish examined that had been tagged with internal anchor tags, the tag was encysted against the body wall where it had been inserted.

Jaw Tag

This is any tag that is attached by encirclement of any of the bones of the jaw. It was developed by David S. Shetter, for use on freshwater fishes, by deforming ordinary strap tags so that they are more or less circular and can be placed around a jaw bone. Later he used a C-shaped tag and special pliers. Recently this practice has been criticized on the grounds that there is evidence that fish so tagged do not feed properly. It is believed that they lose weight and become significantly thinner than unmarked fish. There is also some question whether or not they compare with unmarked fish in their catchability.

John L. Hart of the Fisheries Research Board of Canada has designed a jaw tag for use on the maxillary of the cultus or ling cod (*Ophidon elongatus*). It is a red plastic strip that forms a ring, overlapping itself almost twice. The strip is 13 mm wide and about 0.7 mm thick with rounded and tapered ends. All edges are carefully smoothed, a point which is very important to prevent chafing. He reports that it has not affected growth. This appears to be an ideal tag for this particular species and may be adaptable to many others.

Petersen Tag

This popular tag consists of two plates or disks (sometimes double) attached loosely together by a wire or a pin that pierces the tissues. Invented by Petersen in 1894 for use on the European plaice (*Pleuronectes platessa*), this was widely used and generally successful. It first consisted of two bone disks connected by a silver wire twisted at either end. It is attached to the opercle, the caudal peduncle, the nape, or the back under the dorsal fin. In spite of rapid deterioration of bone in sea water, early investigators favored its use, since they feared that placing metal plates in

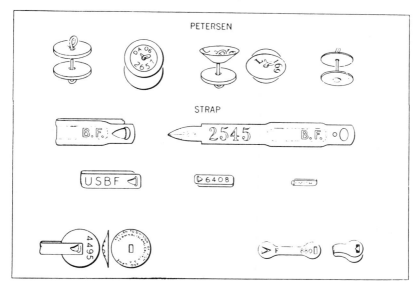

Figure 9-4. Types of Petersen and strap tags. Courtesy of George Rounsefell, 1945, Trans. Amer. Fish. Soc. 73:320-363.

contact with the surface of the fish would cause necrosis of the underlying tissues. The identifying numbers burned into the surface of the bone soon became illegible as the organic material rotted in sea water. To remedy this fault one or more brass disks were attached outside of the bone disks to carry the serial numbers. As many as 4 disks were used, two bone disks next to the tissues, and a brass disk over each bone disk. As early as 1902 Garstang, in marking plaice, used a white bone disk on the underside and an oval, concave disk of brass on the upper side. The concavity of the brass disk was supposed to prevent the tissues from growing over the tag as often happens after tags have been out for some time.

A Petersen tag of two silver disks fastened together with a silver pin, head inside, was used by Hjort for tagging large cod in the opercle in 1913. Commencing about 1905, ebonite disks often were substituted for bone. Soon it became customary to use two ebonite disks, especially for marking cod and plaice. This tag became identified by the name of the Scottish plaice label. Later, plastic was often substituted for ebonite. Various sizes of tags were used varying from about 6.5 to 16.5 mm.

A modification of the Petersen tag, made in 1930 by Robert A. Nesbit, consisted of a pure nickel pin, instead of a silver wire, and thin Celluloid disks with printed instructions on them covered by a thin coat of transparent Celluloid. The Nesbit disks were usually 12.5 mm in diameter and 0.6 mm thick. They were used on striped bass, flounders, scup, shad, cod, and haddock, among others. Usually one red and one white disk were used, with either the red or white disk outside according to the color of the fish. In another experiment a white disk of 13.5 mm with a red center 7.5 mm in diameter was used by Kask to mark sockeye salmon in British Columbia so they could be easily spotted while on the spawning beds. The disks for use on shrimp along the South Atlantic and Gulf Coasts are 10 mm in diameter for large shrimp and 8 mm for small shrimp. Disks of 10 mm have been used for tagging lobsters.

Strap Tag

This is a flat metal strip (which occasionally carries an attached plate or disk), in which one pointed end pierces the tissues and is clinched through a hole in the other end. The original strap tag was adapted from the cattle ear tag by Charles H. Gilbert for tagging salmon in Alaska. It was made of pure aluminum, 69 mm when stretched out, 9.5 mm at the widest point, and about 1.4 mm in thickness. The strap, clinched with special pliers onto the upper lobe of the caudal fin, is still widely used for tagging salmon on the Pacific Coast. Out of over 83,000 Pacific salmon tagged with the strap tags from 1922 to 1936, recoveries averaged 26.9 percent.

Strap tags were used by William C. Schroeder from 1923 until 1933 to tag cod, haddock, and pollock off New England. His strap tag measured 57.5 mm in length, 6.4 mm in width, and 0.7 mm in thickness. He tried tags made of silver, aluminum, copper, silver-plated copper, and Monel metal, but found no obvious differences in recoveries between the different metals, and so after 1923 all of his tags were of Monel metal. From 1923 until 1933 nearly 52,000 cod, over 18,000 haddock, and nearly 5,000 pollock were tagged by the United States and Canada in the Atlantic with strap tags on the upper lobe of the caudal fin. Schroeder states that some were tagged on the jaw in 1927, but although it seemed impossible for the tag to become dislodged, the recoveries from the jaw-tagged fish were not sufficiently better to justify the discontinuance of the tail-marking method. Since the numbers so tagged are not given, it is impossible to verify the soundness of this conclusion. Schroeder mentions the capture by tagging vessels of 42 cod that had been at liberty for at least one year, and at the same time the vessels captured 63 cod with scars on their tails where tags had fallen off. The use of a strap tag on the tail is apparently useful only for short-term experiments.

Commencing in 1925 the International Fisheries Commission used strap tags very successfully on the opercle of halibut. High percentages of recoveries have been made, and fish have been retaken after over a decade at sea. As the opercle grows, the tag maintains about the same position relative to the edge of the opercle, leaving behind a small wound and later a long healed-over scar to show where the tag had been in past years when the opercle was smaller. Their large strap tag was 69 mm by 8 mm by 1.0 mm and weighed 2.6 g. The strap used for smaller halibut was 58 mm by 6.5 mm by 0.6 mm and weighed 1.6 g. Such tags are always attached to the upper side of the halibut.

A smaller strap tag, 35 mm by 3.5 mm by 0.6 mm, has been used unsuccessfully on the opercle of mullet, herring, steelhead trout, and Pacific mackerel.

A very small strap tag, or "fingerling" tag was introduced by Carl L. Hubbs in 1930 for young trout and other small freshwater species. It measures 21 mm (9.4 mm clinched) by 2 mm by 0.3 mm and weighs only 0.0675 g. It has been used with considerable success in freshwater species, attached to fins, opercle, and the upper or lower jaw. These tags of pure nickel were tried on Alaska herring, but recoveries were only 0.5 percent against 4.0 percent for the nickel body-cavity tag, and so it was discontinued.

Four thousand tuna were marked from 1934 to 1938 in California by H. C. Godsil with a sterling silver strap tag bearing a red Celluloid disk. None were recaptured. The tuna were tagged on the preopercle.

Strap tags, because of the uniformity and speed with which they can be applied, have been very popular.

Because individual fish can be recognized by tagging and because reasonably accurate returns can be obtained from sport and commercial fishermen, biologists continue to experiment and develop new kinds of tags. Various plastics have been tried for disk, tubing, and pennant tags. However, evaluation of the various kinds of plastics is difficult because of the variety of products and the many trade names. Some newer types of tags are discussed below.

Modified Carlin Tag

The original Carlin tag consisted of a plastic pennant attached by a wire link to a double wire through the fish's back. The tag was developed in Sweden for salmon and trout smolts. Researchers at the St. Andrews laboratory of the Fisheries Research Board modified the tag, replacing the wire with surgical suture. Observations of returning adult salmon with this tag confirmed its utility, and it was adopted by the Maine Atlantic Salmon Commission. Biologists with the Commission modified the attachment to include only three punctures for attaching the tag to the back of the Atlantic salmon smolts. Many thousand salmon have been tagged, with reasonable recoveries, including some migrants from Maine rivers to the salmon fishery off Greenland.

Spaghetti Tag

Named from the appearance of the plastic tubing, this tag has enjoyed considerable popularity. A loop of vinyl tubing, available in different sizes and colors, is attached to the back of the fish. The major problem seems to be the difficulty of reading the printing on the tag once the fish has been at large for some time.

Coded, Magnetized Wire Tag

Small (1 mm by 0.25 mm) coded, magnetized wire tags have been used successfully in marine fisheries research with chinook salmon and steelhead trout and in freshwater work in cutthroat and rainbow trout. The small tags are injected into the nose cartilage of the fish (3-inch fingerlings). Tags are coded with a series of binary notches providing many combinations. Earlier problems with the injection and shedding of tags appear to have been corrected with more sophisticated application and more skillful taggers. A very sensitive metal detector is used to identify fish with tags. The tag must be dissected from the fish and the identity determined with magnification. A research program at Utah State University in cooperation with the Utah Fish and Wildlife Department is using the coded, magnetized wire tags in an evaluation of six strains of rainbow trout. The Oregon State Department of Fish and Wildlife has developed the ultimate in comfort and efficiency by outfitting a motor home where the wire tags are applied and the tagged fish then returned to the hatchery pond by way of the van's outflow pipe. The van is easily moved and can be set up for operation in about two hours.

Biotelemetry

Two general types of telemetry systems are available for monitoring fish behavior. Sonar or ultrasonic transmitters and receivers have been in use for several years and are available from commercial firms. The second type, still in the developmental stages, employs radio transmitters and receivers. Future developments for both types will involve further miniaturization, longer tag life, and more economical costs.

Initial studies were with single-transistor oscillators, 2.3 cm in diameter and 6 cm long, with an average operating life of 7 hours and a range of 300 feet. At this writing, tags are available for $25 each and receivers for $350. Sonic tags can be detected in fresh water for up to 3 miles and have a battery life of 12 weeks for salmon-sized tags and 4 weeks for the shad-sized tags. Wisconsin workers report the use of a tag for white bass "smaller" than the tip of a pencil, which can be detected up to a half mile and lasts for 15 hours.

The discovery and application of transistors and small, efficient batteries have encouraged development of systems using radio transmitters and receivers. Work at the University of Wyoming and Colorado State University has resulted in the development of tags weighing 38 g in air and 17 g in water. These tags displace nearly 21 cubic centimeters. Of the various methods of attachment studied, a modified Petersen tag appears best. Regular radios will pick up the signals, and while speakers may be used, ear phones help to cut out external noises. A 23-m range is possible in water 1½ to 2.5 m deep; a 30-m range in water ½ to 1 m deep; and ice up to 5 cm has no effect.

Major considerations with the radio transmitter design are that the equipment:

1. Must be able to withstand temperature and pressure change.
2. Must remain waterproof throughout duration of study.
3. Cause as little inconvenience to the fish as possible.

Color Marking and Other Materials

Dyes have been investigated by fishery biologists because of the potential ease in marking large numbers without mutilating or attaching any foreign material in the body. Many dyes have been tried, applied either in the food, by immersion, or by injection for local marks.

A general review of the literature seems to indicate that Bismarck Brown Y is practical for the immersion of small fish for short-term studies—Canadian workers at the Big Qualicum project on Vancouver Island immersed young salmon in Bismarck Brown for short-term population estimates. New York State biologists have used Sudan Black B, by feeding one meal, to produce blue trout, carp, and goldfish. The blue color remained in diminishing intensity for two months.

Although dispersal dyes are preferable because the fish is most easily recognized, local dyes have also been used. An example of localized color is the subcutaneous injection of larval lampreys with India ink which lasted for six weeks. Shrimp, stained locally with Trypan blue stain, can be recognized for nearly four months by the color in the gills on both sides of the head.

Generally, the problems with dyes involve the length of time for which the dye is effective, and the possibility of physiological and behavioral changes. Thousands of dyes remain to be tested.

Radioactive Marking

Radioactive isotopes can be used to mark fish, but the obvious public relations problems will likely restrict their use. The following information, largely from articles in the Journal of the Fisheries Research Board of Canada, sets forth the three basic characteristics of an eligible radioactive isotope.

1. The element must have a slow turnover rate in the fish, that is, a long biological half-life in relation to the fish's longevity.
2. A radioisotope of the element must exist whose physical half-life is reasonably long in relation to the fish's longevity.

3. The radioisotope must emit radiation which is reasonably detectable with field equipment but at the same time has a low-enough energy transfer to the fish so that little or no biological damage is done.

Danger to humans would depend on many factors, but one report indicates that in a tagging operation using Iron-59 in brook trout, 20 radioactive fish would equal the maximum permissible body burden, and 2 fish could provide the maximum permissible body burden for the general public. The iron would concentrate in the blood, and the observation is made that most of this would be removed during the cleaning.

Calcium-45 is reported as having a useful life of approximately one year with the CA-45 concentrated almost entirely in the bony structures. Calcium-45 emits a single relatively soft beta ray with no associated gamma emission. Physical half-life is 164 days. Phosphorus-32 was fed to oligochaetes which were in turn fed to 106,000 young sturgeon. Phosphorus-32 had only a 14.3-day half-life, and the mark remained for 2½ to 3 months. Radioactive caesium has been used to mark lamprey larvae by simply placing them in solutions. Caesium-137 is considered less damaging to the fish because it is distributed generally throughout the body. This, however, makes it less likely that it would be washed out or removed during cleaning. The caesium is highly soluble in water and similar chemically to potassium. Confining the use of caesium to fishes not ordinarily utilized by humans or using the radioactive materials in areas not usually fished to prevent contamination does not appear realistic to us.

Tattooing

Tattooing has been used to some extent. Titanium oxide was used by us in applying a system of dots to fish used in a behavior experiment. The fish retained the dots for several months during the experiment. It was the opinion of psychologists on a reviewing board that the small tattooed dots would have less effect on the behavior of the fish than other marks.

Branding

Both hot and cold branding have been used to mark fish and have the advantage of providing distinctive marks to recognize individual fish. Problems do arise when, as the fish gets older, brands may simply fade into the natural coloration. With either hot or cold branding, care must be taken to insure that the brand is either very hot or very cold. Work with a hot wire and an electric current or battery demonstrated that unpleasant-looking necrotic areas resulted when the wire was not hot enough.

Pennsylvania Fish Commission biologists report and show pictures of 4-year-old brook trout with recognizable brands applied with a modified wood-burning pencil. Trout branded as 10-cm fingerlings also had recognizable brands at 35 cm. A branding iron for sharks has been developed by Canadian workers with letters 4 cm high and 1 cm deep. Branding irons are heated in a gasoline stove and used to mark the fish with various combinations of letters. The sharks are simply brought alongside the boat and branded without removing them from the water.

Fluorescent Pigment

Colorado biologists have been very successful in marking rainbow, kokanee salmon, and fathead minnows with fluorescent pigments. Fish at large for nearly four years retained their marks.

The florescent grit has been used to mark 675,000 sockeye smolts with three single colors and 5 two-color combinations. As many as 32,000 smolts per hour were marked by a 3-man crew. Grit was also sprayed on very small pink and chum salmon fry. Mortality of the sprayed fry amounted to 11 percent over a 20-hour holding period. For best results grit should be sprayed at 100 psi at a distance of 36 to 46 cm. One disadvantage of the method is the need for black light to see the fluorescent particles.

Tetracycline Antibiotics

Tetracycline antibiotics are deposited in the bones and can be detected as a yellowish fluorescence under ultraviolet light. Colorado Game, Fish and Park biologists have successfully marked trout and salmon by introducing the tetracycline through food. Examination of these fish showed that the ribs seemed to be the easiest to dissect and the easiest bones in which to see the mark. The marking has remained visible for several years.

Mutilation

Fin-clipping has been a popular method of recognition for specific groups of fish. A proper fin clip will remain on the fish throughout a long period of very rapid growth, as with salmon smolts. Where several experiments are going on simultaneously this method of marking is limited, since there are only 10 combinations of two-fin marks (excluding the pectorals) available. Because of the occasional regeneration of fins that are not clipped off closely at the base, and the occasional natural occurrence of a fin missing or crippled, it is not safe to mark only one fin.

Partial fin clips, or holes punched in fins, can be used for short-term marking. Partial clips can with care be recognized because a line does exist between the original fin and the regenerated portion. Caution is advised in the use of half-fin marks.

The end of the maxillary bone on some species, such as the smallmouth black bass, may be clipped off for a mark. This is not a conspicuous mark and may easily be overlooked by a commercial or sport fisherman.

References

Al-Hamid, M. I. 1954. The use of dyes for marking fish. Prog. Fish-Cult. 16(1):25–29.

Allen, G. H. 1965. Estimating error associated with ocean recoveries of fin-marked coho salmon. Trans. Amer. Fish. Soc. 94(4):314–326.

Anonymous. 1953. A guide to fish marks used by members of the International Council for the Exploration of the Sea and by some non-participant countries. 2nd ed. J. Conseil Int. Explor. Mer 19(2):241–289.

Arnold, D. E. 1966. Use of the jaw-injection techniques for marking warmwater fish. Trans. Amer. Fish. Soc. 95(4):432–433.

_____. 1966. Marking fish with dyes and other chemicals. U.S. Fish. Wildl. Serv., Tech. Pap. 10. 44 pp.

Bailey, M. M. 1965. Lake trout fin-clipping rates at two national fish hatcheries. Prog. Fish-Cult. 27(3):169–170.

Bellrose, F. C. 1955. A comparison of recoveries from reward and standard bands. J. Wildl. Manage. 19(1):71–75.

Blair, A. A. 1956. Atlantic salmon tagged in east coast Newfoundland waters at Bonavista. J. Fish. Res. Bd. Can. 13(2):219–232.

Blunt, C. E., Jr., and J. D. Messersmith. 1960. Tuna tagging in the eastern tropical Pacific, 1952–1959. California Fish Game J. 46(3):301–369.

Butler, R. L. 1957. The development of a vinyl plastic subcutaneous tag for trout. California Fish Game J. 43(3):201–212.

_____. 1962. Recognition and return of trout tags by California anglers. California Fish Game J. 48(1):5–18.

Butler, T. H. 1957. The tagging of the commercial crab in the Queen Charlotte Islands Region. Fish. Res. Bd. Can., Prog. Rep. 109:16–19.

Calaprice, J. R., and F. P. Calaprice. 1970. Marking animals with micro-tags of chemical elements for identifications by X-ray spectroscopy. J. Fish. Res. Bd. Can. 27(2):317–330.

Calhoun, A. J. 1953. Aquarium tests of tags on striped bass. California Fish Game J. 39(2):209–218.

Chadwick, H. K. 1963. An evaluation of five tag types used in a striped bass mortality rate and migration study. California Fish Game J. 49(2):64–83.

_____. 1966. Fish marking. *In* A. Calhoun. Inland Fisheries Management. California Dep. Fish Game, pp. 18–40.

Chapman, D. W. 1957a. Use of latex injections to mark juvenile steelhead. Prog. Fish-Cult. 19(2):95–96.

_____. 1957b. An improved portable tattooing device. Prog. Fish-Cult. 19(4):182–184.

Clancy, D. W. 1963. The effect of tagging with Petersen disc tags on the swimming ability of fingerling steelhead trout (*Salmo gairdneri*). J. Fish. Res. Bd. Can. 20(4):969–981.

Clemens, H. P., and K. Sneed. 1959. Tattooing as a method of marking channel catfish. Prog. Fish-Cult. 21(1):29.

Coble, D. W. 1967. Effects of fin-clipping on mortality and growth of yellow perch with a review of similar investigations. J. Wildl. Manage. 31(1):173–180.

Cochran, W. W., and R. D. Lord, Jr. 1963. A radio-tracking system for wild animals. J. Wildl. Manage. 27(1):9–24.

Collyer, R. D. 1954. Tagging experiments on the yellowtail, *Seriola dorsalis* (Gill). California Fish Game J. 40(3):295–312.

Collyer, R. D., and P. H. Young. 1953. Progress report on a study of the kelp bass, *Paralabrax clathratus*. California Fish Game J. 39(2):191–208.

Davis, W. S. 1959. Field tests of Petersen, streamer and spaghetti tags on striped bass, *Roccus saxatilis* (Walbaum). Trans. Amer. Fish. Soc. 88(4):319–329.

Dell, M. B. 1968. A new fish tag and rapid cartridge-fed applicator. Trans. Amer. Fish. Soc. 97(1):57–59.

Duncan, R. N., and I. J. Donaldson. 1968. Tattoo-marking of fingerling salmonids with fluorescent pigments. J. Fish. Res. Bd. Can. 25(10):2233–2236.

Ebel, W. J. 1974. Marking fishes and invertebrates. III. Coded wire tags useful in automatic recovery of chinook salmon and steelhead trout. Mar. Fish. Rev., 36(7):10–13.

Eipper, A. W., and J. L. Forney. 1965. Evaluation of partial fin clips for marking largemouth bass, walleyes and rainbow trout. New York Fish Game J. 12(2):233–240.

Eschmeyer, P. H. 1953. The effect of ether anesthesia on fin-clipping rate. Prog. Fish-Cult. 15(2):80–82.

_____. 1955. The movement and recovery of tagged walleyes in Michigan, 1929–1953. Michigan Dep. Conserv., Misc. Publ. 8. 32 pp.

_____. 1959. Survival and retention of tags, and growth of tagged lake trout in rearing pond. Prog. Fish-Cult. 21(1):17–21.

Eschmeyer, P. H., R. Daly, and L. F. Erkkila. 1953. The movement of tagged lake trout in Lake Superior, 1950–1952. Trans. Amer. Fish. Soc. 82:68–77.

Everest, F. H., and E. H. Edmundson. 1967. Cold branding for field use in marking juvenile salmonids. Prog. Fish-Cult. 29(3):175–176.

Everhart, W. H., and R. S. Rupp. 1960. Barb-type plastic fish tag. Trans. Amer. Fish. Soc. 89(2):241–242.

Fleener, G. C. 1958. A method for marking fish with an electrical burning device. Prog. Fish-Cult. 20(3):140–142.

Fletcher, F. T. 1968. A subcutaneous dart tag for fish. J. Fish. Res. Bd. Can. 25(10):2237–2240.

Forrester, C. R., and K. S. Ketchen. 1955. The resistance to salt water corrosion of various types of metal wire used in tagging flatfish. J. Fish. Res. Bd. Can. 12(1):134–142.

Fujihara, M. P., and R. E. Nakatani. 1967. Cold and mild heat marking of fish. Prog. Fish-Cult. 29(3):172–174.

Gerking, S. D. 1958. The survival of fin-clipped and latex-injected redear sunfish. Trans. Amer. Fish. Soc. 87:220–228.

Groves, A. B., and A. J. Novotny. 1965. A thermal-marking technique for juvenile salmonids. Trans. Amer. Fish. Soc. 94(4):386–389.

Henderson, H. F., A. D. Hasler, and G. G. Chipman. 1966. An ultrasonic transmitter for use in studies of movements of fishes. Trans. Amer. Fish. Soc. 95(4):350–356.

Hoss, D. E. 1967. Marking post-larval paralichthid flounders with radioactive elements. Trans. Amer. Fish. Soc. 96(2):151–156.

Imler, R. L. 1969. Design and evaluation of a radiotelemetry system for fish. M.S. thesis, Colorado State Univ. 50 pp.

Jahn, L. C. 1966. Open-water movements of the cutthroat trout (*Salmo clarki*) in Yellowstone Lake after displacement from spawning streams. J. Fish. Res. Bd. Can. 23(10):1475–1485.

Jefferts, K. B., P. K. Bergman, and H. F. Fiscus. 1963. A coded wire identification system for macro-organisms. Nature(Lond.) 197:460–462.

Jensen, A. C. 1967. Effects of tagging on the growth of cod. Trans. Amer. Fish. Soc. 96(1):37–41.

Jensen, A. C., and K. B. Cummin. 1967. Use of lead compounds and tetracycline to mark scales and otoliths of marine fishes. Prog. Fish-Cult. 29(3):166–167.

Joeris, L. S. 1953. Technique for the application of a streamer-type fish tag. Trans. Amer. Fish. Soc. 82:42–47.

Johnson, H. E., and J. M. Shelton. 1958. Marking chinook salmon fry. Prog. Fish-Cult. 20(4):183–185.

Johnson, J. H. 1960. Sonic tracking of adult salmon at Bonneville Dam, 1957. U.S. Fish Wildl. Serv., Fish. Bull. 176:471–485.

Jones, B. F. 1966. Two new tools for applying disk tags to fish. Trans. Amer. Fish. Soc. 95(3):323–325.

Jordan, F. P., and H. D. Smith. 1968. An aluminum stable tag for population estimates of salmon smolts. Prog. Fish-Cult. 30(4):230–234.

Kelly, W. H. 1967a. Marking freshwater and marine fish by injected dyes. Trans. Amer. Fish. Soc. 96(2):163–175.

———. 1967b. Relation of fish growth to the durability of two dyes in jaw-injected trout. New York Fish Game J. 14(2):199–205.

Kennedy, W. A. 1954. Tagging returns, age studies, and fluctuations in abundance of Lake Winnipeg whitefish, 1931–1951. J. Fish. Res. Bd. Can. 11(3):284–309.

Kimsey, J. B. 1956. Largemouth bass tagging. California Fish Game J. 42(4):337–345.

LaFaunce, D. A., J. B. Kimsey, and H. K. Chadwick. 1964. The fishery at Sutherland Reservoir, San Diego County, California. California Fish Game J. 50(4):271–291.

Lawler, G. H., and M. Fitz-Earle. 1968. Marking small fish with stains for estimating populations in Henning Lakes, Manitoba. J. Fish. Res. Bd. Can. 25(2):255–266.

LeMunyan, C. D. 1959. Design of a miniature radio transmitter for use in animal studies. J. Wildl. Manage. 23(1):107–110.

Linhart, S. B., and J. A. Kennedy. 1967. Fluorescent bone labeling of coyotes with demethylchlortetracycline. J. Wildl. Manage. 31(21):217–321.

Loeb, H. A. 1962. Effect of the dye, Plasto Pink, on carp and trout. New York Fish and Game J. 9(2):142–143.

Loeb, H. A., W. H. Kelley, and K. F. Stafford. 1961. Feeding dyes to carp. New York Fish Game J. 8(2):151–153.

Lonsdale, E. M., and G. T. Baxter. 1968. Design and field tests of a radio-wave transmitter for fish tagging. Prog. Fish-Cult. 30(1):47–52.

Marborough, D. 1963. The unsuitability of monel metal opercular strap tags for tagging carp. Prog. Fish-Cult. 25(3):155–158.

McGamon, G. W. 1956. A tagging experiment with channel catfish (*Ictalurus punctatus*) in the lower Colorado River. California Fish Game J. 42(4):323–335.

Mears, H. C., and R. W. Hatch. 1976. Over winter survival of fingerling brook trout with single and multiple fin clips. Trans. Amer. Fish. Soc. 105(6):669–674.

Meister, A. L., and C. Ritzi. 1958. Effects of chloretone and M.S. 222 on eastern brook trout. Prog. Fish-Cult. 20(3):104–110.

Monan, G. E., J. H. Johnson, and G. F. Esterberg. 1975. Electronic tags and related tracking techniques in study of migrating salmon and steelhead trout in the Columbia River Basin. Marine Fish. Rev., Nat. Marine Fish. Serv. 37(2):9–15.

Newell, A. E. 1957. Effects of jaw tags and fin-clipping on returns of stocked trout. Prog. Fish-Cult. 19(4):184.

Nicola, S. J., and A. J. Cordone. 1969. Comparisons of disk-dangler, trailer, and plastic jaw tags. California Fish Game J. 55(4):273–384.

Pelgen, D. E. 1954. Progress report on the tagging of white catfish (*Ictalurus catus*) in the Sacramento-San Joaquin Delta. California Fish Game J. 40(3):313–321.

———. 1955. Second progress report on the tagging of white catfish (*Ictalurus catus*) in the Sacramento-San Joaquin Delta. California Fish Game J. 41(4):261–269.

Phinney, D. B., D. M. Miller, and M. L. Dahlberg. 1967. Mass-marking young salmonids with fluorescent pigment. Trans. Amer. Fish. Soc. 96(2):157–162.

Randall, J. E. 1956. A new method of attaching Petersen disk tags with monofilament nylon. California Fish Game J. 42(1):63–67.

Rawstron, R. R. 1971. Nonreporting of tagged white catfish, largemouth bass, and bluegills by anglers at Folsom Lake, California. California Fish Game J. 57(4):246–252.

Ricker, W. E. 1956. The marking of fish. Ecology 37(4):665–670.

Ripley, W. E. 1949. Tagging salmon with blowgun darts. Copeia 1949(2):97–100.

Rounsefell, G. A., and J. L. Kask. 1945. How to mark fish. Trans. Amer. Fish. Soc. 73:320–363.

Schneider, J. C., P. H. Eschmeyer, and W. R. Crowe. 1977. Longevity, survival, and harvest of tagged walleye in Lake Gogebic, Michigan. Trans. Amer. Fish. Soc. 106(6):566–568.

Scott, D. P. 1961. Radioactive iron as a fish mark. J. Fish. Res. Bd. Can. 18(3):383–391.

_____. 1962. Radioactive caesium as a fish and lamprey mark. J. Fish. Res. Bd. Can. 19(1):49–157.

Sinderman, C. J. 1961. Parasite tags for marine fish. J. Wildl. Manage. 25(1):41–47.

Skinner, J. E., and A. J. Calhoun. 1954. Field tests of stainless steel and tantalum wire with disk tags on striped bass. California Fish Game J. 40(3):323–328.

Slack, K. V. 1955. An injection method for marking crayfish. Prog. Fish-Cult. 17(1):36–38.

Smith, M. W. 1957a. Comparative survival and growth of tagged and untagged brook trout. Can. Fish Cult. 20:1–6.

_____. 1957b. Lea's hydrostatic tag on brook trout and Atlantic salmon smolts. Can. Fish Cult. 20:39–44.

Stasko, A. B., and D. G. Pincock. 1977. Review of underwater biotelemetry, with emphasis on ultrasonic techniques. J. Fish. Res. Bd. Canada. 34(9):1261–1285.

Stauffer, T. M., and M. J. Hansen. 1969. Mark retention, survival, and growth of jaw-tagged and fin-clipped rainbow trout. Trans. Amer. Fish. Soc. 98(2):225–229.

Tracy, H. B. 1968. Development of a shallow-water tagging gun. Prog. Fish-Cult. 30(1):53–56.

Verme, L. J. 1962. An automatic tagging device for deer. J. Wildl. Manage. 26(4):387–392.

Vladykov, V. D. 1957. Fish tags and tagging in Quebec waters. Trans. Amer. Fish. Soc. 86:345–349.

Vrooman, A. M., P. A. Palomo, and R. Jordan. 1966. Experimental tagging of the northern anchovy, *Engraulis mordax*. California Fish Game J. 52(4):228–239.

Warner, K. 1971. Effects of jaw tagging on growth and scale characteristics of landlocked Atlantic salmon, *Salmo salar*. J. Fish. Res. Bd. Can. 28(4):537–542.

Weber, D., and R. J. Wahle. 1969. Effect of fin clipping on survival of sockeye salmon (*Oncorhynchus nerka*). J. Fish. Res. Bd. Can. 26(5):1263–1271.

Webster, D. A. 1956. Rate of fin-clipping. Prog. Fish-Cult. 18(4):185–187.

Wilson, R. C. 1953. Tuna marking, a progress report. California Fish Game J. 39(4):429–442.

Wood, R., and R. A. Collins. 1969. First report of anchovy tagging in California. California Fish Game J. 55(2):141–148.

Yamashita, D. T., and K. D. Waldron. 1958. An all-plastic dart-type fish tag. California Fish Game J. 44(4):311–317.

Young, P. H., J. W. Schott, and R. D. Collyer. 1953. The use of monofilament nylon for attaching Petersen disk fish tags. California Fish Game J. 39(4):445–462.

Youngs, W. D. 1958. Effect of the mandible ring tag on growth and condition of fish. New York Fish Game J. 5(2):184–204.

10

Factors Limiting Abundance

The state of a population of fishes depends on a complicated system of physical and biological checks and balances. Fishes have an awesome reproductive potential with some species capable of egg production in the millions per female, but checks operate to maintain a balance. Usually fish that broadcast the eggs, but provide no parental care, have the greater number of eggs, but mortality may exceed 99 percent. A nest-building fish that guards its young will have fewer eggs, but with an increased survival. Keeping a population exactly in balance would require two replacements for each adult pair. Fish populations, however, are dynamic, not static.

A harsher environment generally has fewer species, but the surviving species increase in number. An environment with less demanding physical conditions, but with more biological competition, will have many different species but fewer numbers of each. Polluted-water areas are frequently identified by their fauna, which is characterized by a paucity of species but tremendous numbers of those present. Harsh conditions of pollution have eliminated many species and much of the biological competition. Unfortunately, polluted areas rarely produce a benefit to mankind.

Fecundity

Fecundity, the ability of fish to produce young, may affect abundance. If each species is to maintain the same relative abundance, then the same relative numbers must survive from generation to generation. The average survival to maturity from fish spawning two million eggs can be no higher than the average survival from the spawning of a species with only two thousand eggs. Species with a low fecundity survive because each individual egg has a greater chance of survival. Generally, species with a low fecundity exhibit smaller fluctuations in abundance because of greater protection from natural phenomena, and because the limited number of eggs does not permit so sudden an increase when conditions are exceptionally favorable.

Accurate determination of the fecundity of a fish is time consuming, and, in fishes producing a large number of small eggs in various stages of development, the problem is especially complicated by the necessity to classify eggs on the basis of maturity. The range of fecundity of largemouth bass (*Micropterus salmoides*) has been reported as between 5,000 and 82,000 eggs. Variation in the fecundity of fish of comparable size poses a problem, and some variation is to be expected from genetic individuality and from differences in physiological condition.

Reproductive potential, even though a gross estimate, is still a usable index and can be estimated from:

1. Relative abundance of the adult population. (This will usually be in pounds of fish caught by some standard amount of fishing effort.)
2. Relative abundance of eggs or larvae. (This is usually a summation of the density of eggs [in the case of pelagic eggs] per cubic meter over the water area inhabited by the particular population under consideration. Estimates of abundance of species spawning in the littoral zone may be based on miles of shoreline utilized for spawning.)
3. Actual numbers of mature adults.

These measures of reproductive potential are each based on one or more assumptions:

1. That the number of eggs spawned is in direct proportion to the number of mature adults and their mean weight (or length). For this to be true, the relation between size of fish and fecundity must be linear. Moreover, if the size composition of the adult population varies from year to year, then the theorem is true only if the regression of eggs on size passes through the origin. The regression formula must be of the form $y = bx$.
2. That the annual sex ratio remains constant.
3. That the regression of number of eggs on size of fish does not vary between years.
4. That the size and/or age at maturity does not vary between years.
5. That the number of eggs is a function of fish size independent of age.
6. That there is no annual variation in the proportion of eggs retained by the females in spawning.

Thus, it is easy to recognize the opportunity for error between numbers of adults and actual reproductive potential. Total annual egg deposition, an estimate in itself, is usually a better estimation of reproductive potential.

Fish of different localities or populations may vary in fecundity, and consideration should be given to such factors as the size of the fish in relation to number of eggs, the age of the fish at maturity, the size of the eggs, the seasonal trends in fecundity in the same locality, and the annual variation in fecundity. The surplus, checks, and balances of fish populations operate to prevent a straight-line relationship between survival and number of eggs. Competition for food, decrease in growth rates, delayed maturity, and decreased fecundity all act to prevent explosion or excess of a particular species.

Critical Stages of Life History

Success or failure of a species in any particular environment may not depend on the suitability of the environment as a whole, but rather on whether the species is able to survive during some particular phase of its life history. Many fish, for example, require special spawning conditions. The spawning run to fresh water is vital to the anadromous fish. Oceans may continue to provide adequate conditions for adult fish, but if spawning tributaries are obstructed, water levels are below minimum, or pollution has destroyed the stream potential, then the species fails because a critical stage in the life history fails.

Fry of marine species floating at the surface of the ocean are subject to weather and currents, and so may be carried to areas where food is unavailable at the critical time when the yolk sac is absorbed. Winds may carry herring (*Clupea harengus*) fry into deep water, and currents along the outer edge of the continental shelf affect the haddock (*Melanogrammus aeglefinus*).

Recognition of critical stages in the life history of fish can be the beginning of effective research and management programs to alleviate the critical stage or to protect the fish during it.

Periphery of Range

Animals are most abundant in what is considered the ideal portion of their environment or range. Conversely, they will be less abundant in those parts of their range where the habitat is marginal. Environmental conditions may change range and abundance. Thus, it may be unrealistic to attempt an increase in species numbers where the environment is generally marginal. Some species of fish may be simply unadapted to conditions presently occurring and, from an evolutionary point of view, may be on their way to extinction. Biologists should consider carefully the question of whether mankind can afford to protect or to support such vanishing species.

Critical Mixing Zone

Work at Woods Hole Oceanographic has demonstrated the importance of the critical mixing zones on Georges Bank in the production of organic material. Agreement between mixing depths, standing crops of plankton, and fish production on the Bank is reported.

Salinity

Salinity is a limiting factor for many species, both directly and indirectly. All species tolerate a range of salinity, for even fresh waters contain dissolved salts. Fresh water will have an average salinity of 0.065 parts per thousand for soft water and up to 0.30 parts per thousand for hard water. By comparison, the general open seas average 34 or 35 parts per thousand.

Salinity change is one of the causes of mass mortalities in the sea. In general, large changes in salinity are confined to coastal waters, usually in bays or estuaries. In the Laguna Madre, off the Texas coast, there are occasional heavy mortalities of fish caused by high salinities because of insufficient exchange of water between the lagoon and the Gulf of Mexico. Evaporation raises the salinity to as high as 75 to 100 parts per thousand, and fish die by the ton.

In the Gulf of St. Lawrence area large amounts of fresh water are discharged during the relatively short period of spring thaw. Mixing of runoff and sea water is slow, and surface freshwater layers over one meter deep are normal in many estuaries at this season. A sharp halocline is present, with the underlying waters close to normal or average sea-water salinity. Animals permanently inhabiting the zone shallower than the halocline are exposed annually to fresh water and are adapted to the change. The soft-shell clam, *Mya arenaria,* and the northern rough periwinkle, *Littorina saxatilis,* are examples. Other animals of the shallow-water fauna, for example the mud snail, *Nassarius obsoletus,* migrate out of the zone, thus avoiding the fresh water. Animals in the deeper zone normally escape any encounter with drastic salinity reductions.

During the spring of 1967 an up-estuary gale caused melt water to build up an extraordinarily deep freshwater layer in the Bideford River, Prince Edward Island. This was followed by extensive mortalities, as salinities fell to less than one part per thousand from top to bottom. The American eel, *Anguilla rostrata,* was reported as a virtually complete mortality. Starfish suffered complete mortality where the fresh water reached the bottom. Pelecypod molluscs, blue mussels, the moon snail, and some specimens of lobster were among the mortalities. Rapid change from 30 to 1 parts per thousand salinity did not allow for needed acclimation even in species that might have been expected to survive.

Oxygen

Fish need oxygen, and one would expect much definitive information would be available about this basic requirement. However, the Environmental Protection Agency has run into problems associated with establishing water-quality criteria that would indicate otherwise. Although 5 ppm. of oxygen has become well established as satisfying the oxygen requirement of most fish, many biologists, supported by experimental results, feel, in fact, that this is a minimum level, particularly for coldwater fish. Most usually accept 6 ppm., but some still feel this is too low for all situations and for thriving populations of fish. The words "enough for survival" or "enough for thriving populations" are important for the right decision here.

Oxygen can become a limiting factor when it is reduced by substances with a high-oxygen demand. Decaying organic matter on lake bottoms tends to exhaust oxygen below the metalimnion during summer months. In some lakes the oxygen level becomes too low to support fish, and fish production is limited to the water area above the oxygen deficiency. A second likely oxygen deficiency may occur in shallow lakes in cold climates with ice cover and heavy snows. Light is shut off by the ice and snow, and the plants, including planktonic algae, consume more oxygen than they produce. The formation, by the decay of organic matter on the bottom, of methane, hydrogen sulfide, carbon dioxide, ammonia, nitrogen, and other gases may consume oxygen and contribute to oxygen depletion and winter kill.

The National Marine Fishery Service reported in 1977 that lack of oxygen had caused the mortality of surf clams off the coast of New Jersey with an estimated loss of clam meat of 59,000 metric tons. The area of the Hudson Shelf Valley out 70 miles from New York along with a tongue extending eastward to Fire Island was also affected. Bottom-dwelling fish and lobsters suffered early mortality but were able to migrate from the affected areas. Low oxygen in the thermocline likely related to extensive blooms of the dino-flagellate, *Ceratium tripos,* along with the contributing causes of higher than normal air temperatures, persistent south winds, an early Hudson River runoff in February, and an early thermocline formation in April. Combinations of natural and man-made limiting factors require extreme care in diagnosing causes for declines or collapses of aquatic populations.

Exceptions to the general oxygen requirements have been reported, including instances of fish present despite very low oxygen levels and of mortalities blamed on supersaturation of water with oxygen.

Temperature

Great concern arises when the temperature of a human being changes a few degrees; by contrast, the ability of aquatic poikilotherms to accommodate to environmental temperatures from almost freezing to high summer temperatures is remarkable. Fish temperature is constantly brought into equilibrium with the water temperature at the exposed gill surfaces. Unable to control external temperatures, a fish survives only in a water area where the external environmental temperature is compatible with the internal tissues. Fish, as gill-breathing vertebrates adapted for aquatic life, are chained to the variations of environmental temperature. The aquatic climate is protected by the high specific heat of water from the extreme fluctuations of atmospheric conditions.

Temperature becomes a limiting factor because it sets lethal limits to life. Acclimatization, rate of development, limits of metabolic rate, congregation within given thermal ranges, and movements to new environmental conditions are controlled by temperature.

Extremes of tolerable thermal environment are set by upper and lower lethal temperatures. Experimental results show the upper lethal temperature to be within a $\pm$ 0.2°C for many species. This precision has been utilized to compare species tolerances, and suggestions have appeared indicating that selective breeding can affect temperature tolerance. Lethal temperature can be defined as that temperature at which 50 percent of a population can exist for infinite time. Studies of temperature tolerance of fish should pay close attention to acclimation of the experimental fish.

Temperature increases during the period of acclimation can result in lethal temperature increases.

Fishery biologists and sport fishermen are aware that fish can withstand higher temperatures than normally expected for short periods. Rivers, during exceptionally warm days, may rise above survival temperatures for coldwater fishes, and yet the fish survive. Or, biologists may net coldwater fishes in the upper warmwater levels of lakes. Resistance time—the period of tolerance prior to death—protects the fish during these critically high-temperature periods.

So many variables affect the ideal temperature that only general comments can be made. Probably 21 C would be the level at which most coldwater and warmwater fishes can be separated. Trout in otherwise acceptable natural conditions do not flourish at temperatures much above 21 C, and warmwater fishes do best at temperatures over 21 C, even into the 30's for channel catfish, for example.

Fish vary considerably in their tolerance both of range of temperature and of the abruptness of temperature changes. Fish that ordinarily live in an environment with a stable temperature are apt to be quite sensitive to change. Tropical fish succumb readily to a few degrees' drop in temperature. Many marine fishes approach the shores in spring as the water warms up, but desert the shoals in late summer when water temperature approaches its peak. On shallow coasts a sudden cold spell will occasionally kill large quantities of marine fish trapped by sudden cooling of the shallows. Warmwater freshwater fishes, such as the black bass, become extremely sluggish as the water cools.

All fishes are affected by temperature. Differences between reactions of different species are demonstrated in the width of the band of temperatures tolerated, in the position of the upper and lower limits, and in the optimum temperature. For short periods fish will tolerate temperatures that might otherwise be lethal. Thus Atlantic salmon (*Salmo salar*) parr will survive temperatures over 27 C and are known to withstand temperatures up to 32 C during the day if the total period is but a few days and if there is a lowering of temperature at night. Even closely related species exhibit differences in their tolerance of high temperatures. Thus, on Atlantic salmon rivers the brook trout (*Salvelinus fontinalis*) cannot endure such high temperatures as the salmon parr. In midsummer, while the salmon parr are thriving in the main rivers, the brook trout desert these warm waters to gather in cool spring holes or enter cooler spring-fed tributaries. High temperatures are thus a definite limiting factor for many species.

The growth rate of fish depends to a large extent on temperature. Digestion proceeds very slowly at low temperatures. This is often shown by the differences in growth rate between populations of the same species inhabiting waters of different temperature. Alewives in the Taunton River which enters the Atlantic Ocean south of Cape Cod mature very predominantly at three years of age, while alewives running up the Damariscotta and Orland Rivers in Maine, which enter the cooler waters of the Gulf of Maine, mature chiefly at four years of age. This decrease in rate of growth with lowered temperatures is also demonstrated by the Pacific salmons. Most of the sockeye salmon (*Oncorhynchus nerka*) of the Fraser River mature at four years, while those in the Karluk River on Kodiak Island mature at five years. In the rivers of the Alaska Peninsula and Bristol Bay large numbers mature at six years.

Obviously a species inhabiting a temperature zone unfavorable to it is at a disadvantage in competing with species better adapted to the conditions. This fact should not be neglected in stocking hatchery-reared fish. If a lake has only a very restricted area of the hypolimnion suitable for coldwater species, it is very unlikely they can be made abundant by heavy stocking. Conversely, the stocking of deep, cold lakes with a limited littoral zone with warmwater fishes may be equally bad, resulting in sparse populations of slow-growing fish yielding little fishing and preying on the young of the coldwater fish. The amount of warm and cold water available in a lake, especially at critical periods, is thus a definite limiting factor.

Temperature affects reproduction. Most species do not spawn unless the water temperature is within certain limits. As the temperature rises later in the year in northern latitudes, spawning is correspondingly delayed. Pacific herring (*Clupea harengus pallasi*) may spawn as early as December at San Diego, but they spawn later and later northward along the coast—as late as June at St. Michaels in Alaska. For many species temperature raises a barrier to their reproduction in northern waters. The oyster (*Crassostrea virginica*) fails to spawn with success in northern New England or on the Pacific Coast, even though young oysters (spat) brought from the south and planted in the same areas may thrive. The low maximum summer-ocean temperatures in northern New England have precluded the successful raising of oysters in these waters except in shallow bays which warm in summer. Experiments with hardier species of oysters may ultimately lead to success.

In addition to the direct effect of limiting or precluding spawning itself, temperature limits fish populations by its effect on the survival of young. This may influence the abundance or lack of sufficient food of the proper kind at a critical period or result in a greatly reduced growth rate which leaves the young susceptible to certain predators over a longer period. Mortality of young sockeye salmon in a lake is extremely high during their earlier stages, falling off steadily as the fish advance in size. If the younger fish grow slowly, their high mortality rate is prolonged, thus causing a greater reduction in numbers.

Space

Species are frequently limited in abundance by the amount of space available during some critical stage of their life history. The spawning area is an obvious example of one stage where space can become a problem. Too many fish in a limited spawning area can result in the superimposition of redds. Behavior and choice of spawning areas may result in fish from different time periods choosing the same area. A limited spawning area may also result in hybridization, particularly if the area is submarginal for one of the kinds.

Although a spawning area may seem too limited, the reproductive potential of the fish is frequently so great that a real surplus can occur, so even though some of the eggs are lost there are still plenty of young to populate the area. This is frequently so with members of the Salmonidae, and particularly with those fish that require an extended nursery period in fresh water, such as the Atlantic salmon. Nursery areas of suitable rubble, flow, and depth of water may then become the primary space problem.

The idea that marine species are able to spawn anywhere is erroneous. Because of strong ocean currents, fish with buoyant eggs must often spawn in rather circumscribed areas if the young are not to be swept out to sea and lost. For example, the Pacific halibut (*Hippoglossus stenolepis*) of the Gulf of Alaska spawn chiefly on the grounds off Yakutat, and many of the eggs and young are carried hundreds of miles westward along the Alaska Peninsula by the northern branch of the Japan Current. The young are spread along the shore when they reach the proper stage for settling down. When they mature, halibut inhabiting this stretch of coast must migrate to the eastward to spawn.

Space may be a critical factor at later stages of the life history. The alewife (*Alosa pseudoharengus*) makes most of its growth in the sea, has a very high fecundity, and is able to spawn under a variety of conditions since its adhesive eggs hatch in a matter of two or three days. However, the young are plankton feeders and spend their first summer in fresh water.

The abundance of each run of alewives depends, therefore, in large measure on lake area available during their first summer. Nursery space for young fish can be a limiting factor for many species.

While most of us can appreciate the problems of space in a small freshwater lake, it is harder to comprehend a density-dependent problem in the oceans. But, unless you imagine that unoccupied habitats are waiting in the hydrosphere for Man to introduce some suitable species, then introductions of new species must be considered a serious problem, even in the ocean.

Total Productivity

The total productivity of any water area is limited. We may find it practical in small ponds to increase natural productivity by fertilization, but such practice on a large scale is impractical. Thus, a very positive limit to maximum abundance is set for every population.

Utilization of total productivity is complicated in fishery management because we are more often interested in certain species or in suitable sizes, usually larger fish. Unfortunately, the species most usually attractive and the most popular sizes are not compatible with extracting total productivity from the water. A fishery manager might suggest a management technique for producing the largest possible standing crop of fish and yet fail because the fish are an unwanted species or because there are no "trophy" fish. What purpose does it serve to provide a large standing crop utilizing as much of the total productivity as possible, if the yield is small? The maximum yield of fish flesh will likely be attained when the largest quantity of food, energy, is being converted into flesh. But, the fisherman may prefer an optimum yield based on preference or size.

Competition

Competition, as a problem in fishery management, can be understood if the definition is confined to situations where more than one organism demands, typically at the same time, the same resources of the environment, and this demand is in excess of immediate supply. There may be competition for habitat as a result of one species, the carp (*Cyprinus carpio*) for example, destroying the aquatic vegetation and roiling the water. Or it may be competition for spawning grounds when one species destroys either the habitat or digs up great quantities of another fish's eggs, as happens with the five species of Pacific salmon.

Overpopulation may emphasize intraspecific competition—which may actually be the most important type of competition. Size of a population is limited by intraspecific competition, as with the Fraser sockeye salmon where the dominance of one out of four of the cycles in the sockeye salmon life history is caused by intraspecific competition. Sockeye of the Fraser mature predominantly at four years of age, and all die after spawning, so that each cycle is really a different population of the same species. Populations of the three cycles following a dominant cycle do not achieve the size permitted by the available food, so that the actual limiting factor is the abundance of the preceding dominant cycle.

Predation

Predation is a powerful force in limiting abundance. Dramatic examples of the effects of predation on fish are available. For example, lake trout were nearly exterminated in Lake Michigan by lamprey attacks; only eight lake trout were caught in 1.7 million linear m of gill net in 1955. Estimated annual production of lake trout fell from 2.7 million kg to less than 45 kg in just slightly more than a decade.

In many bodies of water the removal of a large share of the larger predator fish by intensive fishing has not resulted in better fishing. On the contrary, in waters with favorable conditions for reproduction, the removal of predators often permits a survival of young fish too great for the available food supply. The result is overcrowding of the waters by an abundance of slow-growing fish unable to attain normal size.

Bird predation, too, has received much attention in both commercial and sport fisheries. Pacific herring, during the spring months, deposit adhesive eggs on vegetation growing in shallow water above and below the zero tide mark. A heavy mortality, ranging from 56 to 99 percent, is the estimated egg loss during a 15-day incubation period from all mortality factors. The decrease in egg numbers attributed to bird predation alone ranged from 30 to 55 percent, averaging 39 percent. Research studies indicate that the number of eggs in the stomach of the herring gull (*Larus argentatus*) was 8,500, and the average number of eggs in the stomach of the glaucous-winged gull (*Larus glaucescens*) was 13,800.

Information has been published describing the feeding habits of sea birds (the gannet, the Cape cormorant, and the penguin) off the west coast of the Union of South Africa. The Cape cormorant population consumes 12,000 tons of fish per year, and the gannets consume 30,000 tons of fish. The total number of pilchards consumed by the birds, 43,300 tons, amounted to nearly half the total catch of pilchards in one year.

Researchers at the St. Andrews Biological Station of the Fisheries Research Board of Canada have conducted intensive studies of bird predation on fish with their studies of Atlantic salmon smolt production correlated with American merganser (*Mergus merganser*) and belted kingfisher (*Megaceryle alcyon*) control. Control of these birds made it possible to average a smolt production of not less than five times as great as without control. Control of the type employed to increase the Pollett River smolt output by five times or more must limit merganser activity to a level of only 1 bird on 20 ha of water, or about 24 km of a stream 10 m wide, for the entire open-water season. Control of kingfishers may be impractical unless the incidence of kingfishers should increase much beyond a rate equivalent to about 1 bird per kilometer of stream 10 m wide, or roughly 1 bird for 1 ha of water.

Washington Game Department studies report that mergansers eat from 0.2 to 0.5 kg of fish per day. Some of their other estimates indicate that mergansers consumed as many as 160,000 trout (5 per kg), which was 36 percent of the total trout planted in one lake. The value of the fish loss from 14 lakes was estimated at $18,564.

Diseases and Parasites

Diseases and parasites also affect and to some extent limit fish populations, but here destruction of the host is essentially biological suicide, and elimination of diseases and parasites could well remove a necessary selective process and balancing factor.

References

Adams, E. S. 1975. Effects of lead and hydrocarbons from snowmobile exhaust on brook trout (*Salvelinus fontinalis*) Trans. Amer. Fish. Soc. 104(2):363–373.

Balon, E. K. 1975. Reproductive guilds of fishes: A proposal and definition. J. Fish. Res. Board Can. 32(6):821–864.

Beamish, R. J., and H. H. Harvey. 1972. Acidification of the LaCloche Mountain lakes, Ontario, and resulting fish mortalities. J. Fish. Res. Bd. Can. 29(8):1131–1143.

Beiningen, K. T., and W. J. Ebel. 1970. Effect of John Day Dam on dissolved nitrogen concentrations and salmon in the Columbia River, 1968. Trans. Amer. Fish. Soc. 99(4):664–671.

Brett, J. F. 1956. Some principles in the thermal requirements of fishes. Q. Rev. Biol. 31(2):75–87.

Burns, J. W. 1970. Spawning bed sedimentation studies in northern California streams. California Fish Game J. 56(4):253–270.

Calhoun, A. 1966. Inland Fisheries Management. California Dep. Fish Game, Sacramento. 546 pp.

Carlander, K. D. 1958. Disturbance of the predator-prey balance as a management technique. Trans. Amer. Fish. Soc. 87:34–38.

Clady, M., and B. Hutchinson. 1975. Effect of high winds on eggs of yellow perch, *Perca flavescens,* in Oneida Lake, New York. Trans. Amer. Fish. Soc. 140(3):524–525.

Davies, D. H. 1958. The predation of sea-birds in the commercial fishery. Union of South Africa, Dep. Commer. Ind., Div. of Fish., Invest. Rep. 31. 16 pp.

Davis, J. C. 1975. Minimal dissolved oxygen requirement of aquatic life with emphasis on Canadian species: a review. J. Fish. Res. Board Can. 32(12):2296–2332.

Dillon, P. J., et al. 1978. Acid precipitation in south-central Ontario: Recent observations. J. Fish. Res. Bd. Can. 35(6):809–815.

Elson, P. F. 1962. Predator-prey relationships between fish-eating birds and Atlantic salmon. Fish. Res. Bd. Can., Bull. 133. 87 pp.

Eschmeyer, P. H. 1957. The near extinction of lake trout in Lake Michigan. Trans. Amer. Fish. Soc. 85:102–119.

Everhart, W. H., and C. A. Waters. 1965. Life history of the blueback trout (Arctic char, *Salvelinus alpinus* [linnaeus]), in Maine. Trans. Amer. Fish. Soc. 94(4):393–397.

Foerster, R. E. 1938. Mortality trend among young sockeye salmon (*Oncorhynchus nerka*) during various stages of lake residence. J. Fish. Res. Bd. Can. 4(3):184–191.

Forney, J. L. 1976. Year-class formation in the walleye (*Stizostedion vitreum vitreum*) population of Oneida Lake New York. 1966–73. J. Fish. Res. Board Can. 33(4):783–792.

Gangmark, H. A., and R. D. Broad. 1956. Further observations on stream survival of king salmon spawn. California Fish Game J. 42(1):37–49.

Hayes, F. R. 1957. On the variation in bottom fauna and fish yield in relation to trophic level and lake dimensions. J. Fish. Res. Bd. Can. 14(1):1–32.

Larkin, P. A. 1956. Interspecific competition and population control in freshwater fish. J. Fish. Res. Bd. Can. 13(3):327–342.

Likens, G. E., and F. H. Bormann. 1974. Acid rain: A serious regional environmental problem. Science 184:1176–1179.

MacGregor, J. S. 1957. Fecundity of the Pacific sardine. U.S. Fish Wildl. Serv., Fish. Bull. 121(57):427–449.

Margolis, L. 1966. Parasites as an auxiliary source of information about the biology of Pacific salmons (*Genus Oncorhynchus*). J. Fish. Res. Bd. Can. 22(6):1387–1395.

Mauck, W. L., and D. W. Coble. 1971. Vulnerability of some fishes to northern pike (*Esox lucius*) predation. J. Fish. Res. Bd. Can. 28(7):957–969.

McNeil, W. J. 1964. Redd superimposition and egg capacity of pink salmon spawning beds. J. Fish. Res. Bd. Can. 21(6):1385–1396.

Miller, R. B. 1950. The Square Lake experiment: an attempt to control *Triaenophorus crassus* by poisoning pike. Can. Fish Cult. 7:3–18.

Outram, D. N. 1958. The magnitude of herring spawn losses due to bird predation on the west coast of Vancouver Island. Fish. Res. Bd. Can., Prog. Rep. III:9–13.

Richey, J. E., M. A. Perkins, and C. R. Goldman. 1975. Effects of kokanee salmon (*Oncorhynchus nerka*) decomposition on the ecology of a subalpine stream. J. Fish. Res. Board Can. 32(6):817–820.

Rounsefell, G. A. 1957. Fecundity of North American Salmonidae. U.S. Fish Wildl. Serv., Fish. Bull 122:451–468.

———. 1958. Factors causing decline of sockeye salmon in Karluk River, Alaska. U.S. Fish Wildl. Serv., Fish. Bull. 130:83–169.

Schofield, C. L. 1976. Acid precipitation: effects on fish. Ambio. 5:228–230.

Scott, G. H. 1955. Economic factors in catch fluctuations. J. Fish. Res. Bd. Can. 12(1):85–92.

Schaefer, M. B. 1970. Men, birds and anchovies in the Peru Current—dynamic interactions. Trans. Amer. Fish. Soc. 99(3):461–467.

Shetter, D. S., and G. R. Alexander. 1970. Results of predator reduction on brook trout and brown trout in 4.2 miles (6.76 km) of the North Branch of the Au Sable River. Trans. Amer. Fish. Soc. 99(2):312–319.

Stevenson, W. H., and E. J. Pastula, Jr. 1971. Observations on remote sensing in fisheries. Commer. Fish. Rev. 38(9):9–21.

Thomas, M. L. H., and G. N. White. 1969. Mass mortality of estuaries fauna at Bideford, P.E.I., associated with abnormally low salinities. J. Fish. Res. Bd. Can. 26(3):701–704.

Tsuyuki, H., and E. Roberts. 1960. Inter-species relationships within the Genus *Oncorhynchus* based on biochemical systematics. J. Fish. Res. Bd. Can. 23(1):101–107.

Ware, D. M. 1972. Predation by rainbow trout (*Salmo gairdneri*): the influence of hunger, prey density, and prey size. J. Fish. Res. Bd. Can. 29(8):1193–1201.

Wells, L. 1970. Effects of alewife predation on zooplankton populations in Lake Michigan. Limnol. Oceanogr. 15(4):556–565.

Yeo, R. R., and M. J. Risk. 1979. Intertidal catastrophes: Effect of storms and hurricanes on intertidal benthos of the Minas Basin, Bay of Fundy. J. Fish. Res. Bd. Can. 36(6):667–669.

Yoder, W. G. 1972. The spread of *Myxosoma cerabralis* into native trout populations in Michigan. Prog. Fish-Cult. 34(2):103–106.

11

Habitat Improvement

Many of the tools and management techniques available to the fishery biologist are designed for habitat improvement. Fishways and fish-guiding devices, spawning channels, pollution abatement, removal of obstructions to fish migration, pond reclamation, and proper overall watershed management are kinds of habitat management. Biologists are cautioned to expand their ideas of habitat improvement beyond the "in-stream" devices so popular in the 1930s when government agencies and enthusiastic sportsmen's groups, armed with unquestionable motives and little actual knowledge plunged into programs of habitat improvement—or should we say habitat alteration? Correct water flows and erosion and silt controls in the headwaters can be many times more valuable than a dam or deflector installed downstream. This is not to say that traditional "in-stream" devices are of no value, but only that their use should be kept in proper perspective.

Pollution

Fishery biologists need to be constantly concerned over changes in the chemical, physical, and biological properties of natural waters.

Pollutants may affect aquatic life in many ways, either directly or indirectly, by:

1. An increase in osmotic pressure.
2. An increase in acidity.
3. A decrease in oxygen content of the water.
4. Specific toxicity.
5. Destruction of food organisms.

6. Destruction of spawning grounds.
7. Mechanical injury to gills from silt or other suspended materials.
8. Blocking migration channels.
9. Rendering organisms unfit for food.
10. Accelerating aging of lakes and ponds.
11. Thermal alteration (either heating or cooling).

Major categories of the pollutants that are most hazardous to aquatic life are:

1. Chlorinated hydrocarbons, such as the insecticides Dieldren, Endrin, Aldrin, Toxaphene, Heptachlor, and DDT, and the polychlorinated biphenyls or "PCB's."
2. Nutrients (that is, organic or inorganic sources of phosphorus and nitrogen, particularly).
3. Detergents (these have both nutrient and toxicant properties).
4. Oil and related petrochemicals.
5. Silt.
6. Acid mine wastes.
7. Heavy metals.
8. Heat.
9. Radionuclides.

Available information on these topics is extensive, and excellent reviews are provided by authors such as Johnson (1968), McCauley (1966), McKee and Wolf (1963), Tarzwell (1968), and the Water Pollution Control Federation (1968–1972, inclusive), to name only a few.

Exact determination of the effect of pollution is difficult and complex. One method has been to determine the effect of various concentrations of a pollutant on fish in an acute static toxicity test in the laboratory even though this exposes the researcher to all the problems of field versus laboratory work. Flow-through tests rather than static tests are of more value. Further, there is considerable emphasis now on chronic or long-term testing using parameters such as growth, maturation of viable sex products, and production of normal offspring. Chronic testing from the egg stage to the production of offspring requires many months, even years, in the laboratory with all the complications of maintaining controlled conditions over long periods of time, but such testing is necessary if the problem of sublethal effects is ever to be answered.

Since, in a stream or other body of water, the effect of any particular effluent is often changed for better or worse by interaction with other effluents or with varying amounts of silt, any laboratory determinations of the effect of pollutants should, if at all possible, be checked against actual field observations. And because of the varying composition of industrial effluents from the same plants, it is usually insufficient to synthesize an effluent for use in laboratory tests. It is better, therefore, to collect effluents at various times and various stages, and particularly to collect the composite effluent that is discharged when all processes are in operation. Concentration of an effluent at any given distance from the point of discharge usually varies considerably from time to time. Thus, statements concerning average concentration of an effluent based on volume of discharge and volume of a stream over any period of time may be meaningless in terms of damage. A heavy discharge for a short time may cause the concentration to rise far above the average.

Because of these various complications, laboratory tests are usually not sufficient to give a realistic picture of the effect of a pollutant. Day-by-day, even hour-by-hour changes in the concentration and composition of pollutants can occur at any one point and make it impossible to collect and analyze sufficient water samples to determine the degree of pollution at a number of stations. Consequently, many investigators have made use of biological indicators to estimate the degree of pollution at a particular site. First, because plants and animals vary widely in their susceptibility to a given pollutant; and second, because wherever biological or physical conditions for survival are severe, there is always a tendency toward the survival of a smaller number of species, usually accompanied by an increased abundance of those that remain.

In past years biologists understandably searched for single-species indicators, but experience has taught the lesson that all habitats should be sampled qualitatively and the entire fauna considered. Determination of the abundance of various organisms is impractical, and best results can be obtained by sampling that will indicate the number of species present and their habitat requirements. In recent years diversity indices have received much attention as pollution indicators, but current findings have revealed that these too are undependable.

Fish Kills Caused by Pollution in 1971, the Twelfth Annual Report of the Environmental Protection Agency (formerly Federal Water Pollution Control Administration), lists 73.7 million fish killed by pollution in the 46 states reporting. This number of fish is greater by 81 percent than the number reported in any previous year on record beginning in 1960. Eight hundred and sixty reports of pollution-caused fish kills is also a record. For the first time since the annual report began in 1960 more fish were reported killed in estuarine waters than in fresh or salt water. The estuarine kills were the result of 31.4 million fish reported in two localized bays, one in Florida and one in Texas. These totals are only indications as some of these numbers may simply reflect better reporting by a concerned public. Many kills are probably not even seen, and some are not recorded; in other instances it is not certain what killed the fish, and occasionally reports on fish kills have been made in pounds or give no number estimates.

Because of their specialized training and greater understanding of total ecosystem concepts, biologists have a particularly strong public obligation to work for the stabilization and the reduction of pollution problems. Furthermore, pollution controls is, increasingly often, the most powerful tool available to the fishery manager for improving the fish habitat and for utilizing the aid of the community at large.

(For convenience, the pollution references have been separated from the general habitat improvement literature at the end of this chapter.)

Removal of Obstructions

Obstructions that retard or stop fish migrations should be removed to improve fish habitat. Most streams and rivers during the course of years accumulate obstructions of one sort or another—from logjams to hydroelectric dams. Fishways should be provided in obstructions which cannot be removed, providing the cost can be justified, and obstructions no longer serving any useful purpose should be removed.

The impact of obstructions on migratory fishes is serious when you consider:

1. Low discharge may discourage adult fish from leaving the estuary or moving upstream into fresh water.
2. Water quality is altered by temperature changes, reduced flows, and the addition of pollutants.
3. Diversion of normal flow from natural channels may lead migrating adults or young into blind alleys or, at least, delay migration.
4. Even though fish passage facilities are provided, problems can arise if they are inadequate and poorly designed. Young fish, moving downstream, must pass over or through the dam frequently suffering high mortalities. Fish may eventually pass the obstruction but any delays can be costly, and many times, as on the Columbia River, a series of dams obstruct migration.
5. Reservoirs may flood-out spawning and nursery areas, the lake-type

environment created can influence the migration of young fish, and undesirable fish may become established.

6. Predation becomes a problem in the tailraces where weakened migrants are exposed to large concentrations of predators.

Care should be taken in recommending the removal of natural barriers, as these may be maintaining a balance between populations above and below the obstruction. Natural barriers may also prevent the spread of an undesirable fish into headwater areas. Beaver dams are a problem where they obstruct fish movement. The dams may also change the general ecology of the flowage area by warming the water, silting the bottom, and changing the aquatic fauna of the stream bed. Ordinarily a beaver colony remains in an area about three to five years depending on the availability of food and water. After the colony migrates, the dam, unattended, may wash away; occasionally, however, dams remain as stream obstructions.

Many lakes depend on inlet and outlet streams to provide the necessary spawning and nursery areas to hatch and rear the young fish that constantly restock the lake. And, if populations of anadromous and adfluvial fishes are to be maintained, the fish must be assured of easy access to their spawning grounds. Even a log, formation of a sand or gravel bar, or exceptionally low water may prevent fish from entering spawning and nursery areas.

Deflectors

Although much more complicated devices are used in stream improvement than the simple deflector, more and more projects are discarding the complicated, hard-to-build-and-maintain devices in favor of deflectors that are primarily designed to speed up the current, thus washing out silt and providing graveled riffle areas. Increasing the speed of the current gives the water less chance for long exposure to the sun and helps to maintain the usually desired low temperature. Some deflectors in general use are described below.

The *single boulder deflector* is intended for speeding up the current in wide, shallow pools. The boulders may be placed in rows, their proximity depending on the depth of water that is to be piled up in the area behind. The boulders are arranged in lines, or they may be placed haphazardly in the stream bed. In general, the boulders speed up the current and also produce eddy currents that tend to scour the bottom and provide more movement in the pools. They are also advantageous in preventing silting.

A *boulder deflector,* when large boulders are handy in an area, if placed correctly, is one of the easiest of the improvement devices to construct and one of the most rewarding. A trench is dug in the stream bottom so that the foundation layer will be secure in the bottom. A large boulder is used to anchor the stream end of the boulder deflector.

A *log deflector,* consisting of a main log or logs, mud sill, and brace, may be constructed where sufficient boulders are not present. Trenches are dug to receive the mud sill and the main log. The face and bank and stream ends are protected with boulders.

Cement blocks can be used as deflectors. In one instance, two-foot-square *cement blocks* weighing approximately 520 kg were cast at the site of the project in reusable wooden forms. An iron ring was anchored in each block for ease in handling. Although the weight of the blocks is a distinct disadvantage, they are adaptable to most situations. If the water is deep, they can be piled on top of one another. If, after the device is finished, evidence suggests the advisability of alterations, the blocks can be shifted easily.

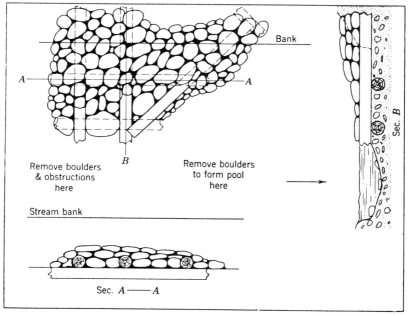

Figure 11-1. Log deflector. Courtesy of Lloyd L. Smith and John Moyle, 1944, Minnesota Dep. Conserv., Tech. Bull. 1:147–164.

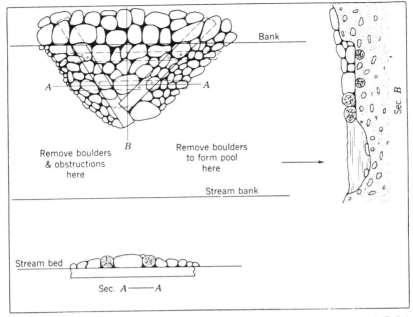

Figure 11-2. Triangular crib deflector. Courtesy of Lloyd L. Smith and John Moyle, Tech. Bull. 1, above.

The *triangular crib deflector* is a solidly built device consisting of a main log slanting at the desired angle downstream and anchored well back in the stream bank. A brace running from the back (downstream side) to the bank provides additional support and makes a crib which is filled with rocks or boulders. The front and back are also lined with rocks.

When sections of the stream are too wide to be affected by bank deflectors or in sections where silt deposits have built up, the *triangular rock-filled deflector* may be employed to speed up the current, wash out the silt, and in general narrow the width by dividing the current. In some cases several of these may be placed in a row in the center of the stream. Three logs are merely joined together in a triangle and securely fastened to the bottom. The entire structure is then filled with rocks.

The *underpass deflector* is intended to "blow out" the silt and the soft bottom in an effort to provide a pool. It is constructed with the main log set a few centimeters off the bottom so that the water will be forced to pass under the main log and in this manner dig the pool.

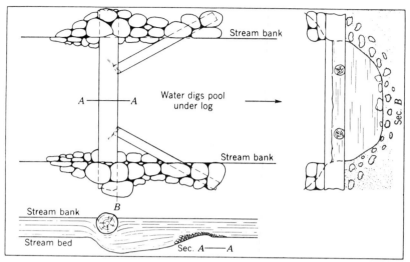

Figure 11-3. Underpass deflector. Courtesy of Lloyd L. Smith and John Moyle, Tech. Bull. 1, above.

The *double-winged deflector* has been developed in Michigan especially for use on larger streams with low gradient.

Dams

When the gradient of a stream is too steep, where ponds are desired, or where it is necessary to impound a large volume of water to insure a steady flow of water throughout periods of low rainfall, dams can be employed with success as improvement devices.

A *rock and boulder dam* takes advantage of the already existing supply of rocks and boulders. The fewer that are taken from the stream bed proper the better (except where a pool is indicated), since it would certainly be unwise to destroy already existing habitats in an attempt to create new ones. Frequently, advantage can be taken of an already existing large boulder which may be used as a keystone around which the dam can be built. Additional boulders are placed so that they interlock with the keystone and with each other. Care needs to be taken to protect the banks and insure against the water cutting around the ends of the dam.

The single-log dam is intended only for small streams not over 5 to 6 m in width; the construction makes it most adaptable for use on streams with soft bottoms. The backbone of this type of dam is the large log which extends across the stream and well into each bank, resting on log mud sills,

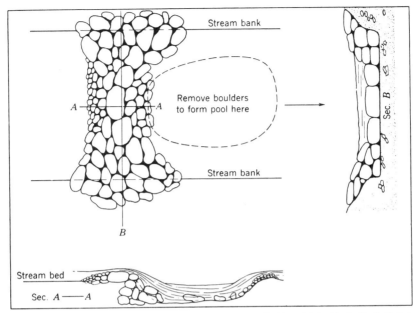

Figure 11–4. Rock and boulder dam. Courtesy of Lloyd L. Smith and John Moyle, Tech. Bull. 1, above.

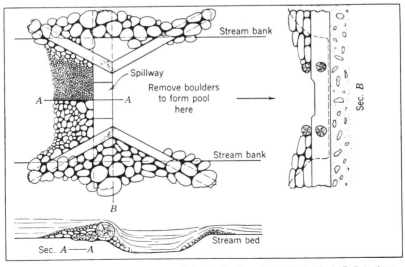

Figure 11–5. Single-log dam. Courtesy of Lloyd L. Smith and John Moyle, Tech. Bull. 1, above.

the number of mud sills depending on the width of the stream. A channel is dug across the stream for the main log and channels upstream for each of the mud sills. The mud sills are notched to receive the main log. After the main log has been fastened securely to the sills, heavy wire is stapled to the main log and sills, after which the dam is covered with a layer of heavy rocks and then topped off with earth, sand, gravel, or more boulders. The dam should have a spillway to take care of overflow of low water. Care needs to be taken, as with all installations, to make certain the banks are protected so that the water will not cut around the device.

K dams are designed for small streams with a hard or firm bottom. A trench is dug across the stream bed and extended well into each bank, 1½ to 2 m, on each side. The main log is fitted snugly into the trench. Braces are then attached to the downstream side and fastened to the log with drift bolts. The entire structure is covered carefully with rocks and boulders, including the spaces between the braces and banks. If the water is too deep for one log, then two logs are securely fastened together and used as the main log.

The pyramid dam derives its name from the three main logs extending across the stream. Two logs are placed in the carefully prepared trench and a third, forming the pyramid, is placed on top. All three are securely

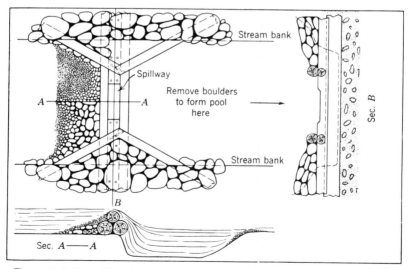

Figure 11–6. Pyramid-log dam. Courtesy of Lloyd L. Smith and John Moyle, Tech. Bull. 1, above.

fastened together. The construction, depending on whether the stream has a soft or hard bottom, follows the general plan of the two former types.

A *log and board dam* is intended for small streams and can be adapted to almost any stream except for those with a bedrock bottom. A trench is first dug lengthwise of the stream to accommodate about one-third of the mud sill. Then a trench is dug into both banks and crosswise across the stream deep enough for the main log to rest firmly on the mud sill. Braces are attached as indicated previously with earlier types of dams. The mud sill supports the main log several cm above the stream bottom thus providing shelter under the log and behind the boards. Five-centimeter lumber of any width available is then nailed directly to the main log extending upstream and with the upstream ends deep in a trench dug to receive them. As with the other dams, care should be taken that the water does not cut around the ends of the dam, and a spillway should be provided for periods of low water.

Although the *log brush dam* is intended for large streams, it is adaptable to almost any size. The usual trench is dug across the stream extending well into the banks and upstream a distance equal to the length of the brush which is used to fill the dam. The bottom of the trench is deeper on the upstream side. The main log is placed securely on the bottom in the crosswise trench. The first layer of brush is then placed in position with the butts nailed to the log and the tops extending upstream. This first layer of brush is well covered with rocks, earth, sand, or gravel. The next crosswise log is placed slightly behind the first. The boughs are attached as before, and covered with fill accordingly. The height of the dam is increased by continuing the same process. Care should be taken to protect the face of the dam and both banks with large boulders. The face of the dam should not be too steep; a pitch of 1:3 to 1:4 is recommended.

The *V dam* is intended for larger streams and is not recommended for those having a width less than 6 to 8 m. It is adaptable to streams with a hard bottom and consists of two logs joined together in the form of a *V* with the apex upstream. Depth is gained by using other logs on top. The usual trench is dug, and the logs are extended into each bank. Sloping the trench toward the center provides a spillway for low water; this allows for the principle that water, when passing over an obstruction, tends to take a course perpendicular to the obstruction and so is thrown to the center of the V dam. The upstream face of the dam is riprapped with small rocks and well shouldered at the banks. This type of dam may be modified for use in soft or silty bottoms by the provision of mud sills for the main logs to rest upon.

204 Principles of Fishery Science

The *truss dam* is a method of adapting some of the other styles to a wider stream by using two diagonal logs and one horizontal log. Angles of the cuts on the main logs are determined by the width of the stream. Woven wire is attached, and the dam is filled with rocks and completed with sand, gravel, and rubble. Care should be taken to protect the face of the dam and the sides so that there is no cutting around. A modification of the truss dam is the double truss which consists of a combination of two single dams for more strength and is for use on larger streams.

The *nonsilting dam* is designed to speed up the current and prevent silting. It is especially adaptable to an area of stream with a wide, shallow, slow flow. In general, the nonsilting dam narrows the channel and provides a board-planking overflow. The width of the board platform controls the volume of the flow. At normal flow a small pool should be backed up from the dam.

Gabions

Stream improvement devices are frequently constructed today using *gabions*. These are rectangular wire-mesh baskets filled with rocks, and wired shut. Gabion baskets are wired in series and used as building blocks. They are available in nine sizes ranging from 2 to 4 m long, and from ⅓ to 1 m high. Claims for the advantages of gabions are ease of installation, low cost (when rocks from the streambed are used, but this is a very poor idea), flexibility, durability, natural appearance, and permeability.

Cover, Shelter, and Attraction Devices

Cover, for protection and shelter, is sometimes provided by anchored trees, but they frequently become collectors of drift, form partial dams, and the increased pressure pulls them free. Carried downstream by the current, they are apt to lodge in some area where they are harmful or aesthetically displeasing.

Log and boulder devices consist of a log crib inclined upstream for a meter or so, leaving a large area for shelter underneath. The boulders in the crib keep the device from floating or moving downstream with the current. Log shelters may extend from the bank to provide cover and, if covered properly with boulders, are inconspicuous. As with all habitat-improvement devices, a major concern should be the aesthetics of the area.

The idea of placing shelters in lakes has received a great deal of attention. Brush shelters may be constructed by tying bushes together and

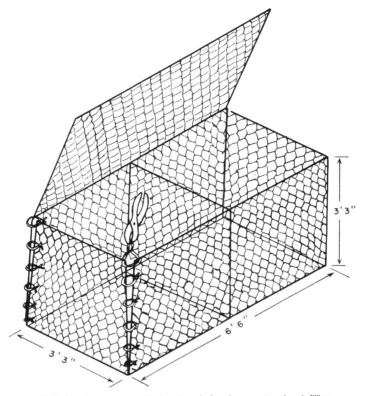

Figure 11-7. Gabion basket assembled and ready for placement and rock filling.

weighting them to sink. Some shelters have been more complicated, consisting of two frames with the inner frame less than 1 m square and the outer frame a little more than 1 m square. Bundles of brush are piled on the two frames, butt ends toward the center. Finished shelters are approximately 5½ m in diameter. Actually, what the shelter does in many cases is simply concentrate the fish for better availability by the angler.

The use of artificial reefs to improve sport fisheries in coastal marine waters has been developed, probably from the knowledge that fishing is good near sunken ships. The Sport Fishing Institute has provided excellent summaries on the building of reefs. The first recorded example of construction of a coastal fishing reef was in 1950 when broken-masonry building materials from New York City were deposited on the McAllister fishing grounds near Long Island, New York. Then again in 1953 a group of charter-boat captains dumped 14,000 concrete-weighted surplus wooden beer cases off Fire Island. No evaluation studies have been made of either of these reefs.

The first really substantial effort to construct artificial marine fishing reefs was in 1953 in the Gulf of Mexico by the Alabama Department of Conservation working with a charter boat club. Altogether, several thousand auto bodies were dumped in the Gulf floor off Alabama. Results were excellent. Snapper fishing became phenomenal, and there were snappers where there were none before. However, the auto bodies disintegrated in from three to five years. In Texas auto bodies were cabled together with concrete block weights in bundles of four and marked with lighted buoys, but these bodies disintegrated in the same time as for Alabama.

California set up evaluation studies to measure relative effectiveness, durability, and costs for a reef of auto bodies, one of streetcars, and one of artificial rocks. California reports that auto body reefs were highly attractive to fish—streetcars somewhat less—but both disintegrated into a pile of rubble in only three or four years, and fish attraction was greatly reduced. The reports indicate that concrete shelters were more successful than quarry rockshelters, although quarry rock is considerably less expensive. The concrete units were modified pontoons, originally designed to support boat slips and docks. Each shelter was 1½ × 2½ × 0.8 m with 8 to 10 38-cm-diameter holes. These 6-sided shelters cost $75 each. Including transportation, concrete shelters in units of 132, equal to 1,000 tons of quarry rock, cost between $8,000 and $8,700. Quarry rock in 2- to 3-ton chunks in 1,000-ton barge loads cost $4,800 dropped in three locations. An equivalent auto body reef would cost $3,000 but would have to be replaced in three to five years.

The Sandy Hook Laboratory of the National Marine Fisheries Service describes two tire units that can be carried offshore in any size boat. One unit is a single tire with a 7-kg concrete-ballast weight wedged between the sidewalls. The second is a 1.2-m-high stack of 7 to 8 tires held together with two 1.5 m lengths of 1 cm reinforcing rod projecting through aligned holes drilled in each tire. These are firmly anchored in concrete ballast that completely fills the space between the sidewalls of the base tire. Tires are readily available, and the laboratory reports that they do not decompose, corrode, rust, or give off toxic substances.

Any plans to use artificial reefs should include a check with the Coast Guard, to be sure that the location is safe and charted for ship traffic.

Some interesting work has been reported from Florida where the attraction of coastal pelagic fishes with artificial structures has been studied.

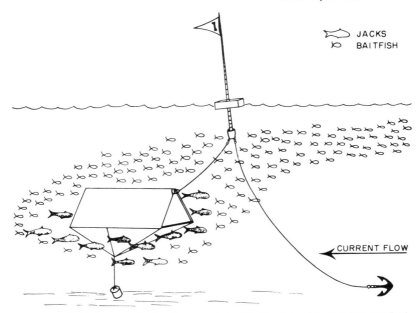

Figure 11-8. Underwater structure to concentrate fish. Courtesy of E. F. Klima and D. A. Wickham, 1971, Trans. Amer. Fish. Soc. 100(1):90.

Commercial and sport fishermen alike can confirm the concentrations of fish around floating debris. The simpler of the attracting devices (Figure 11-8) was constructed by using 2.5 by 5.0-cm wood. Right prisms, $1 \times 1 \times 1 \times 1\frac{1}{2}$ m, were covered on the upper two sides with vinyl cloth, creating a tentlike appearance. The structure was painted white. Commercial quantities of fish, estimated at up to 25 tons, were attracted to a structure on one occasion, and on six other occasions at least 5 tons of fish were attracted during the 20-day observation period. This experiment was largely a behavior study, and these preliminary results are encouraging and, it is hoped, will lead to more study.

Spawning Areas

Spawning channels are a very sophisticated method of creating a spawning area, either where one has been destroyed or where none previously existed. Survival from eggs to emergent fry can be increased by establishing stable and optimum physical conditions and by eliminating the major causes of heavy mortality: erosion, silting, and lack of adequate flow through the gravel. One of the first spawning channels to be constructed was the Robertson Creek channel, built by the Canadian Department of Fisheries. This channel is 1280 m long, although only 775 m is actually spawning channel, and the remainder is taken up by control structures, rearing ponds, experimental flume, bypass, and fish-transportation channels. Normal flow in the channel is 3 m^3/sec., maximum water depth is 0.8 m, average velocity and depth of water in the channel (at 3 m^3/sec.) is 0.5 m/sec. and 0.5 m deep. Total production of the channel in one year was 1,560,000 pink salmon. Total survival in the channel is reported as 91 percent, a figure which is more than favorable when compared with the 10 percent survival in a natural spawning area.

The Camanche spawning channel in California is constructed on the Mokelumme River just below Camanche Reservoir, where 20 km of good spawning area was flooded. The channel has a capacity for 15,000,000 salmon eggs. Segments of the channel are each 6 m wide at the bottom and 460 m long, making a total of 11,150 m^2 of nesting area. Since each female requires about 4 m^2, there is room for 3,000 adult females to dig their nests, fertilize, and cover their eggs. The bottom of Camanche Channel is constructed of a 1m layer of selected gravel from 2.5 cm to 10 cm in diameter. Seepage out of the channel is prevented by a layer of impervious clay. A flow of 1.7 m^3/sec. is required during the spawning period at which time the depth is about 46 cm and the velocity is 0.6 m/sec. After spawning is completed, the flow is reduced to 0.8 m^3/sec. which is enough for incubation, hatching, emergence, and transportation of fry to the river. Resting pools for adults are provided at 150-m intervals. The young fish remain in the gravel nests for a period of about 50 days.

Spawning channels are most valuable for those species of anadromous fishes that emerge from the gravel and travel almost immediately to the ocean. Long periods as fry and fingerlings in a stream nursery area before migration as a smolt would very much limit their production.

Spawning areas in lakes have been constructed by dumping quantities of large rocks at selected sites. The Maine Department of Inland Fisheries and Game placed 170 tons of rock in a lake trout lake to provide a spawning area. In this instance, the rocks were placed at 5-m depth, so that the 3-m normal lake fluctuation would not bother the spawning fish. Boxes of gravel, or the graveling of large areas, have been utilized in warmwater lakes as spawning aids for nest-building species such as smallmouth bass. Other spawning devices have been employed for minnows, such as cut brush along the shore, spawning slabs, and floating boards for those that lay their eggs underneath floating logs or other debris.

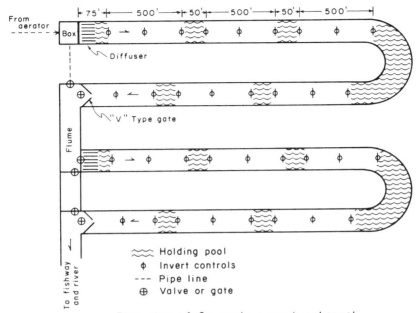

Plan view of Camanche spawning channel

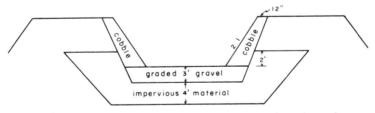

Typical cross section of Camanche spawning channel

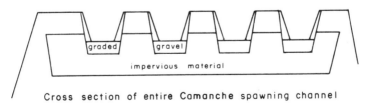

Cross section of entire Camanche spawning channel

Figure 11-9. Camanche spawning channel. Courtesy of California Dept. Fish and Game.

Artificial Destratification

Fish die-offs have been associated with problems relating to thermal gradients, stagnation, and ice-cover. Originally biologists thought of destratification as primarily a method to prevent winterkill, but more recently the effects on the total aquatic ecosystem are being considered. Eutrophication and new impoundments with associated water-quality problems emphasize the potential importance of artificial destratification. Fish distribution may be affected by severe thermal stratification; plankton blooms may cause recreational and commercial-use problems; drinking water quality deteriorates under anerobic conditions; and higher evaporation rates are associated with warm surface water.

Methods to recirculate and aerate water have ranged from the use of outboard motors to churn the water and arrangements to pump hypolimnion water to the surface to sophisticated systems involving compressors and plastic hoses or some type of aerohydraulic gun. Our experience has been with the Helixor, a one-piece polyethylene tube 2.7 m high and 46 cm in diameter (see Figure 11-10). A continuous spiral polyethylene coil inside

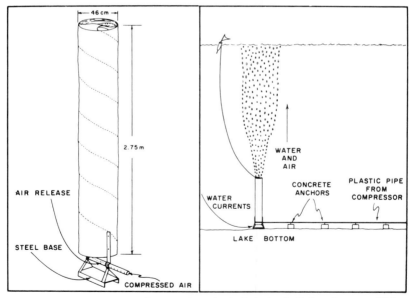

Figure 11-10. Artificial destratification and aeration facility using the Helixor. Courtesy of Robert T. Lackey.

the Helixor increases the surface interaction between bubbles and water about three times. The Helixor stands upright on the bottom of the lake. Compressed air, introduced at the bottom through a pipe with holes, travels up the Helixor to the surface of the lake. Two of these 3-m Helixors destratified a 25-ha lake with an average depth of 4.6 m. With treatment, the lake was 1 to 2.8 C colder during the winter and from 2.2 to 5 C warmer in the deep water in the summer. Surface-water temperatures stayed the same in summer for both treatment and control year. The dissolved oxygen loss in the bottom several feet, normally occurring in summer and winter, was eliminated.

Potentially useful applications of artificial destratification include plans affecting fish distribution, controlling the ecosystem so as to make it more desirable for valuable sport fish, altering outlet flows from large reservoirs, controlling algae blooms, improving recreation qualities, and improving drinking-water quality.

Fertilization

Work with small fish ponds and commercial fish growers has stimulated interest in the idea of fertilization of natural lakes and streams to increase fish production. Generally, fertilization has only been successful in very small water areas under intensive management.

Experimental work with the introduction of sucrose to a small Oregon stream resulted in a twofold increase in consumption of food and a more than sevenfold increase in trout production. These increases were largely the result of a greater abundance of aquatic food organisms, especially tendipedid larvae.

Productivity of waters is correlated directly with the productivity of the land they drain. A stream draining rich farmlands is very probably a productive stream, and a lake situated in a rich agricultural region is likely to produce more fish per acre than a high mountain lake or a lake in a drainage area of poor land. The rich fertilizers and top dressings now applied to farmlands find their ways into streams and lakes, enriching them in this manner. Nutrients are normally lost through drainage to the ocean, absorption into the bottom, removal of fish by anglers, or, as in the case of eels, by migration of the adults from the lake, when they take with them nutrients absorbed during their life cycle.

Perhaps before discussing fertilization further, some idea of its action in relation to the food chain should be mentioned. The addition of fertilizer to a body of water enriches it in certain inorganic nutrients of value in increasing the number of plankton organisms. Since most of the game fish in which we are interested are not plankton feeders, intermediate forms must transform the plankton into a form usable by the predator game fish. Intermediate forms may be insects on which trout will feed, or perhaps a small minnow which will feed on the plankton and in turn be fed upon by the larger and more desirable game fish. Most large fish feed on larger organisms than insects, so that the forage fish in the food chain is a vital necessity to the ultimate production of large fish.

The closer the feeding habits of the fish concerned are to the base of the food chain, the more the effects of fertilizer are reflected in the total weight of fish produced. The effect of fertilizer on the lower invertebrates is much more apparent than on the fish.

How can we be sure that the application of fertilizer will in fact improve the productivity? For example, where the restricted spawning area of a body of water is a limiting factor, then artificial fertilization will not change the numbers of fish produced. Unless there is some definite chemical deficiency in the water, the addition of nutrients may not increase the productivity. Particularly is this true if the slow growth is due to low temperature or some other variable. As with other improvement techniques, the first step before artificial fertilization is to collect adequate data showing whether or not fertilization will improve some basic deficiency of the habitat.

An interesting experiment on the fertilization of a trout lake was published by M. W. Smith. Much of the following has been taken from his paper. It was believed that the poor production in eight lakes of Charlotte County, New Brunswick, could be attributed primarily to the small amounts of nutrient material, such as phosphorus, in the waters. Commercial fertilizer was added to one of the lakes, Crecy, in a effort to improve trout production. One ton of ammonium phosphate (11–48–0) and 272 kg of potassium chloride were spread over the surface of the lake, concentrating the greater portion near the shores. It was calculated that this would give, providing it was adequately mixed, a phosphorus content of 0.39 ppm, which was 20 times the amount present before fertilization, and 0.21 ppm of nitrogen.

The highest total phosphorus content of the water (0.23 ppm) was determined a week after fertilization. There followed a rapid decrease in values, little above those found before fertilization, and by the next year the value had fallen to prefertilization levels. This drop occurred even though this particular lake had only a very slight flow of water from it. The phosphorus was evidently removed by other agencies, such as plant life or the bottom materials.

After two months a dense growth of microscopic plants (algae) produced a "bloom," although such a bloom had not been observed before in these waters. A year later no bloom occurred, but there were clumps of filamentous algae and rooted aquatic vegetation spread over an increased area of the lake bottom which could be taken as an indication that the bottom mud had been enriched. The zooplankters increased for only a short period within a month after fertilization. Actually, during the algal bloom there was a decrease in their number as a result of the changes created by the algae. The pH value during the bloom period increased from 6.6 to 8.2. In this particular case the bottom fauna, including immature insects, snails, and amphipods, made up the principal food of the trout. A year later a greater quantity of these foods in trout stomachs indicated a possible increase in the number of such fauna.

During the period from September 9, 1946, when planted, to May and June, 1947, fingerling trout increased in average length 14 cm. This increase was comparable to that made by the same stock in a hatchery rearing pond. Yearlings, when planted, showed an increase of 3.1 inches on an average. In this case, apparently, fertilization, coupled with stocking, was necessary to improve the fishing. The survival of 3.6 percent of the planted stock to the creel contributed 60 percent of the angling.

Even though Crecy was a natural lake, it should be realized that it was very small, 20 ha with a mean depth of 2.4 m. There is no permanent inflowing stream, and the outlet almost ceases to run in later summer and early fall. The shallow depth and ability to control the outlet are of prime importance in recommending a program of fertilization as practical.

One of the chief deterrents to fertilization in natural waters is the cost. Even a very small stream could require 1½ tons a year to produce any effect even for a short distance below the point of application. Costs have been estimated at from $15 to $20 per acre per year in shallow-water ponds with little outflow. What the cost would be for a square mile of lake of any depth and with a volume of water moving through it can hardly be considered as practical.

One danger of fertilization of large bodies of water is that more young fish will survive, and the trend may be toward stunted populations. Fertilization of lakes containing stunted populations will not always improve the fishing. Actually it may only provide additional food for more stunted fish.

Since the addition of fertilizer to a body of water increases the number of plants, the danger of winterkill in northern areas is also increased. As long as ample sunshine penetrates the water surface, and the surface of the lake is exposed to the air, plenty of oxygen will be present for the fish. In cold weather, however, when ice and then snow cover the lakes, all air and most of the light will be shut off from the water. Plants will continue to respire, using up the oxygen and producing carbon dioxide. With little or no sunlight, photosynthesis will be drastically reduced, and little or no oxygen will be contributed by the plants. Further, the increase of plant material from the fertilization and thus the increase of decaying organic material will also work to reduce the oxygen in the water. In addition, fertilization may sometimes cause an obnoxious growth of algae which is objectionable to fishermen, cottage owners, and swimmers. Such growths are unpredictable.

Miscellaneous Techniques for Habitat Improvement

Resource managers are continually searching for additional methods to improve water areas. The following examples may stimulate thinking about some better idea or a modification of present techniques.

The Forest Service, for example, has sponsored development of a machine for cleaning streambed gravels. The riffle sifter, as it is called, is a self-propelled amphibious machine using high-pressure underwater jets and a suction pump to remove sediment and make streambed spawning gravels clean and porous. The riffle sifter can clean about 4,000 square feet of streambed per hour. Initial tests in southeast Alaska reduced bottom-fauna population in each of the test streams, but within one year the populations apparently returned to pretreatment levels. A decrease of 30 percent in materials measuring less than 0.4 mm, and a decrease of 65 percent in materials measuring less than 0.4 mm were reported from two streams. The method is too new to assess the actual effects on the salmon populations.

Several agencies are investigating the practical effects of dredging lake bottoms to retard eutrophication and rejuvenate aging lakes. Disposal of the wastes, however, could create another problem here.

Weed control is a serious problem in the recreational use of lakes. Although some mechanical methods of weed control are available, aquatic herbicides usually produce the most effective control. But the introduction of chemicals to the ecosystem creates its own problems and should be approached with caution and after careful experimentation. All methods are temporary, particularly those that do not physically remove the weeds from the water area or reduce the sources of enrichment.

Chemical improvement, especially the application of lime, has been used in drastic situations to improve water quality and fish habitat. Applications of lime to some small Maine ponds by the Department of Inland Fisheries and Game resulted in doubling the oxygen to 8.0 ppm to depths of 3 m. At 4.6 m depth, oxygen was raised from 1.8 to 3.4 ppm.

U.S. Army Engineers are maintaining banks and bottoms of large rivers by using concrete mattresses, laid from barges, to prevent erosion and eating away of banks. Concrete blocks 7.6 cm thick, 36 cm wide, and 1.2 m long are fastened 2.5 cm apart on a screen of copper-coated steel fabric to form the mattress. The mattress may stretch as far as 152 m from shore and to a depth of 21 m. Width of a section is 43 m.

Fishing-pier improvements (habitat improvement for the fisherman) have been provided by the U.S. Army Engineers. Asphalt walkways capping rock jetties with 6-m-long sidewalks have made certain areas safer for fishermen. Some of these jetties support as much as 50,000 man-days of angling per year.

Plastic sheets have been used to convert soft, muddy bottoms for use by oyster growers. Thin sheets of polyethylene are laid down to support oyster shells or other set collectors, seed oysters, or spawners.

Since erosion is not as prevalent in areas where natural seaweed is present, the use of artificial seaweed made of tufts of polypropylene is being tested in Britain and the Netherlands in an attempt to stop coastal and seabed erosion. Bunches of polypropylene strands are anchored to the seabed a short distance offshore. Free ends of each tuft float upward like the fronds of natural seaweed, changing the wave pattern and trapping sand so there is a progressive build-up of material.

Wisconsin reports the prevention of winterkill by plowing the snow from ice-covered lakes. When oxygen dropped to 0.8 to 1.9 ppm, a snowplow worked two days plowing 18-m-wide strips on the main part of the lake. The area plowed on the 195-ha lake was equivalent to plowing 40 km of road. Two days later the oxygen read from 3.8 to 4.9 ppm under the areas plowed, while at the same time some snow-covered coves of the lake were still at 1.0 ppm of oxygen.

References (Pollution)

Anderson, R. B., and W. H. Everhart. 1966. Concentrations of DDT in landlocked salmon (*Salmo solar*) at Sebago Lake, Maine. Trans. Amer. Fish. Soc. 95(2):160–164.

Berkowitz, H. P., and J. C. Finnell. 1971. Power generation and environmental change. Symposium, Comm. Environ. Alteration, Amer. Assoc. Adv. Sci. M.I.T. Press, Cambridge, Mass. 440 pp.

Cooper, E. L., ed. 1967. A symposium on water quality criteria to protect aquatic life. Amer. Fish. Soc., Spec. Publ. 4. 37 pp.

Cordone, A. J., and D. W. Kelley. 1961. The influences of inorganic sediment on the aquatic life of streams. California Fish Game J. 47(2):189–228.

Davies, P. H., and W. H. Everhart. 1973. Lead toxicity to rainbow trout and testing application factor concept. Environ. Prot. Agency, Off. Res. Monit., Ecol. Res. Ser. EPA-R3-73-011c. 80 pp.

Dimond, J. B., A. S. Getchell, and J. A. Blease. 1971. Accumulation and persistence of DDT in a lotic ecosystem. J. Fish. Res. Bd. Can. 28(12):1877–1882.

Doudoroff, P., and M. Katz. 1950. Critical review of literature on the toxicity of industrial wastes and their components to fish. I. Alkalies, acids, and inorganic gases. Sewage Ind. Wastes 22(11):432–458.

———. 1951. Bio-assay methods for the evaluation of acute toxicity of industrial wastes for fish. Sewage Ind. Wastes 23(11):1380–1397.

———. 1953. Critical review of literature on the toxicity of industrial wastes and their components to fish. II. The metals, as salts. Sewage Ind. Wastes 25(7):802–839.

Environmental Protection Agency. 1971. Methods for chemical analysis of water and wastes, 1971. Natl. Environ. Res. Cent., Anal. Qual. Control Lab., Cincinnati, O.

Gaufin, A. R., and C. M. Tarzwell. 1952. Aquatic invertebrates as indicators of stream pollution. Public Health Rep., U.S. Dep. Health, Ed., and Welfare 67(1):57–64.

Hatfield, C. T., and J. M. Anderson. 1972. Effects of two insecticides on the vulnerability of Atlantic salmon (*Salmo salar*) to brook trout (*Salvelinus fontinalis*) predation. J. Fish. Res. Bd. Can. 29(1):27–29.

Henderson, C. 1957. Application factors to be applied to bioassays for the safe disposal of toxic wastes. *In* Tarzwell, C. M. Biological problems in water pollution. Trans. First Taft Sanit. Eng. Cent. Semin., Public Health Serv., U.S. Dep. Health, Ed., and Welfare, pp. 31–37.

Hynes, H. B. N. 1963. The biology of polluted waters. Liverpool University Press, Liverpool, England. 202 pp.

Johnson, D. W. 1968. Pesticides and fishes—a review of selected literature. Trans. Amer. Fish. Soc. 97(4):398–424.

Kittrell, F. W. 1969. A practical guide to water quality studies of streams. U.S. Dep. Inter., Water Pollut. Control Adm., Publ. CWR-5. 135 pp.

Kramer, R. H., and L. L. Smith, Jr. 1965. Effects of suspended wood fiber on brown and rainbow trout eggs and alewives. Trans. Amer. Fish. Soc. 94(3):252–258.

Keup, L. E., W. M. Ingram, and K. M. Mackenthun. 1967. Biology of water pollution. U.S. Dep. Inter., Water Pollut. Control Adm., Publ. CWA-3. 290 pp.

Mackenthun, K. M. 1973. Toward a cleaner aquatic environment. Environ. Prot. Agency. 273 pp.

McCauley, R. N. 1966. The biological effects of oil production in a river. Limnol. Oceanogr. 11(4):475–486.

McKee, J. E., and H. W. Wolf. 1971. Water quality criteria. 2nd ed. Resour. Agency, California Water Resour. Control Bd., Publ. 3-A. 548 pp.

Morton, J. M. 1977. Ecological effects of dredging and dredge spoil disposal: A literature review. U.S. Fish and Wildlife Service, Tech. Pap. 94. 33 pp.

Mount, D. I. 1968. Chronic toxicity of copper to fathead minnows (*Pimephales promelas,* Rafinesque). Water Res. 2(3):215–223.

Mount, D. I., and C. E. Stephan. 1967a. A method for establishing acceptable toxicant limits for fish—malathion and the butoxyethanol ester of 2,4-D. Trans. Amer. Fish. Soc. 96(2):185–193.

———. 1967b. A method for detecting cadmium poisoning in fish. J. Wildl. Manage. 3(1):168–172.

———. 1969. Chronic toxicity of copper to the fathead minnow (*Pimephales promelas*) in soft water. J. Fish. Res. Bd. Can. 26(9):2449–2457.

Office of Library Services. 1970. Mercury contamination in the natural environment. A cooperative bibliography. U.S. Dep. Inter., Washington, D.C. 32 pp.

Parsons, J. D. 1957. Literature pertaining to formations of acid-mine wastes and their effects on the chemistry and fauna of streams. Trans. Illinois Acad. Sci. 50:49–59.

Peters, J. C. 1967. Effects on a trout stream of sediment from agricultural practices. J. Wildl. Manage. 31(4):805–812.

Reish, D. J. 1956. An ecological study of lower San Gabriel River, California, with special reference to pollution. California Fish Game J. 42(1):51–61.

Rudd, R. L. 1964. Pesticides and the living landscape. University of Wisconsin Press, Madison, Wisc. 320 pp.

Ryder, R. A., and L. Johnson. 1972. The future of salmonid communities in North American oligotrophic lakes. J. Fish. Res. Bd. Can. 29(6):941–949.

Schouwenburg, W. J., and W. J. Jackson. 1966. A field assessment of the effects of spraying a small coastal coho salmon stream with phosphamidon. Can. Fish Cult. 37:35–43.

Sprague, J. B. 1969. Measurement of pollutant toxicity to fish. I. Bioassay methods for acute toxicity. Water Res. 3(11):793–821.

———. 1970. Measurement of pollutant toxicity to fish. II. Utilizing and applying bioassay results. Water Res. 4(1):3–32.

Subcommittee on Environmental Improvement, Committee on Chemistry and Public Affairs. 1969. Cleaning our environment. The chemical basis for action. Amer. Chem. Soc., Washington, D.C. 249 pp.

Sullivan, J. F., G. J. Atchison, D. J. Kolar, and A. W. McIntosh. 1978. Changes in predator-prey behavior of fathead minnows (*Pimephales promelas*) and largemouth bass (*Micropterus salmoides*) caused by cadmium. J. Fish. Res. Bd. Can. 35(4):446–451.

Tarzwell, C. M. 1960. Biological problems in water pollution. Trans. Second Taft Sanit. Eng. Cent. Semin., Public Health Serv., U.S. Dep. Health, Ed., and Welfare, Tech. Rep. W60-3. 285 pp.

———. 1965. Biological problems in water pollution. Trans. Third Taft Sanit. Eng. Cent. Semin., Public Health Serv., U.S. Dep. Health, Ed., and Welfare, Publ. 999-WP-25. 424 pp.

———. 1968. Section III. Fish, other aquatic life and wildlife. *In* National Technical

Advisory Committee. Water quality criteria. Water Pollut. Control Adm., U.S. Dep. Inter., pp. 27–110.

Tarzwell, C. M., and A. R. Gaufin. 1953. Some important biological effects of pollution often disregarded in stream surveys. *In* Proc. 8th Ind. Waste Conf., Purdue Univ., Eng. Ext. Bull. 83, pp. 295–316.

Vandermeulen, H. H. 1978. Recovery potential of oiled marine northern environments. Symposium. J. Fish. Res. Bd. Can. 35(5):499–795.

Warren, C. E. 1971. Biology and water pollution control. W. B. Saunders Co., Philadelphia, Pa. 434 pp.

Water Pollution Control Federation, Research Committee. 1968. A review of the literature of 1967 on wastewater and water pollution control. J. Water Pollut. Control Fed. 40(6):897–1219.

_____. 1969. A review of the literature of 1968 on wastewater and water pollution control. J. Water Pollut. Control Fed. 41(6):873–1251.

_____. 1970. A review of the 1969 literature on wastewater and water pollution control. J. Water Pollut. Control Fed. 42(6):861–1267.

_____. 1971. A review of the 1970 literature on wastewater and water pollution control. J. Water Pollut. Control Fed. 43(6):931–1417.

_____. 1972. 1971 water pollution control literature review. J. Water Pollut. Control Fed. 44(6):901–1294.

Wobeser, G., N. O. Nielsen, and R. H. Dunlop. 1970. Mercury concentrations in tissues of fish from the Saskatchewan River. J. Fish. Res. Bd. Can. 27(4):830–834.

References (Habitat Improvement)

Apmann, R. P., and M. B. Otis. 1965. Sedimentation and stream improvement. New York Fish Game J. 12(2):117–126.

Boussu, M. F. 1954. Relationship between trout populations and cover on a small stream. J. Wildl. Manage. 18(2):229–239.

Briggs, P. T. 1975. An evaluation of artificial reefs in New York's marine waters. N.Y. Fish Game J. 22(1):51–56.

Carbine, R. F., and O. M. Brynildson. 1977. Effects of hydraulic dredging on the ecology of native trout populations in Wisconsin spring ponds. Dep. Nat. Res. Madison, Wis. Tech. Bull. 98. 40 pp.

Carlisle, J. G., Jr. 1969. Results of a six-year trawl study in an area of heavy waste discharge: Santa Monica Bay, California. California Fish Game J. 55(1):26–46.

Clark, O. H. 1948. Stream improvements in Michigan. Trans. Amer. Fish. Soc. 75:270–280.

_____. 1953. The application of land use to fisheries management in Michigan. Prog. Fish-Cult. 15(2):64–71.

Davis, H. S., A. S. Hazzard, and C. MacIntyre. 1935. Methods for the improvement of streams. U.S. Bur. Fish. Memo. 1-133:1–27.

Ditton, R. B., et al. 1979. Access to and usage of offshore liberty ship reefs in Texas. Marine Fish. Rev., Nat. Marine Fish. Serv. 41(9):25–31.

Dunst, R. C., et al. 1974. Survey of lake rehabilitation techniques and experiences. Wisconsin Dep. Nat. Res. Tech. Bull. 75. 179 pp.

Ehlers, R. 1956. An evaluation of stream improvement devices constructed eighteen years ago. California Fish Game J. 42(3):203–217.

Fast, A. 1968. Artificial destratification of El Capitan Reservoir by aeration. California Dep. Fish Game. Fish Bull. 141, Part 1. 97 pp.

_____. 1971. Effects of artificial destratification on zooplankton depth distribution. Trans. Amer. Fish. Soc. 100(2):355–358.

Gard, R. 1961a. Effects of beaver on trout in Sagenhen Creek, California. J. Wildl. Manage. 25(3):221–242.

_____. 1961b. Creation of trout habitat by constructing small dams. J. Wildl. Manage. 25(4):384–390.

Gray, J. R. A., and J. M. Edington. 1969. Effect of woodland clearance on stream temperature. J. Fish. Res. Bd. Can. 26(2):399–403.

Hacker, V. A. 1957. Biology and management of lake trout in Green Lake, Wisconsin. Trans. Amer. Fish. Soc. 86:71–83.

Hasler, A. D., O. M. Brynildson, and W. T. Helm. 1951. Improving conditions for fish in brown-water bog lakes by alkalization. J. Wildl. Manage. 15:347–352.

Hazzard, A. S. 1940. Hunt Creek Fisheries Experiment Station. Michigan Conserv. 9(7):4–5.

_____. 1948. Stocking vs. environmental improvement. Michigan Conserv. 17(7):3, 14, 15.

Heman, M., R. S. Campbell, and I. G. Redmond. 1969. Manipulation of fish populations through reservoir drawdown. Trans. Amer. Fish. Soc. 98(2):293–304.

Herricks, E. E., and J. Cairns, Jr. 1974. Rehabilitation of streams receiving acid mine drainage. Virginia Water Resources Cent. Bull. 66. 284 pp.

Hooper, F. C., R. C. Ball, and H. A. Tanner. 1953. An experiment in the artificial circulation of a small Michigan lake. Trans. Amer. Fish. Soc. 82:222–241.

Hourston, W. F., and D. MacKinnon. 1957. Use of an artificial spawning channel by salmon. Trans. Amer. Fish. Soc. 86:220–230.

Hubbs, C. L., J. R. Greeley, and C. M. Tarzwell. 1932. Methods for the improvement of Michigan trout streams. Michigan Inst. Fish., Res. Bull. 1. 54 pp.

Hunt, R. L. 1971. Responses of a brook trout population to habitat development in Lawrence Creek. Wisconsin Dep. Nat. Resour., Tech. Bull. 48. 35 pp.

Hunter, G. W., III, L. M. Thorpe, and D. E. Grosvenor. 1940. An attempt to evaluate the effect of stream improvement in Connecticut. Trans. N. Amer. Wildl. Conf. 5:276–291.

Huntsman, A. G. 1948. Fertility and fertilization of streams. J. Fish. Res. Bd. Can. 7(5):248–253.

Klima, E. F., and D. A. Wickham. 1971. Attraction of coastal pelagic fishes with artificial structures. Trans. Amer. Fish. Soc. 100(1):86–99.

Lewis, S. L. 1969. Physical factors influencing fish populations in pools of a trout stream. Trans. Amer. Fish. Soc. 98(1):14–19.

Lister, D. B., and C. E. Walker. 1966. The effect of flow control on freshwater survival of chum, coho and chinook salmon in the Big Qualicum River. Can. Fish Cult. 37:3–22.

Lucas, K. C. 1960. The Robertson Creek spawning channel. Can. Fish Cult. 27:3–23.

Maccaferri Gabions of America, Inc. 1966. Maccaferri gabions—stream improvement handbook. 14 pp. (Several brochures are available from distributor at 55 West 42nd St., New York, N.Y. 10036.)

MacKinnon, D., L. Edgeworth, and R. E. McLaren. 1961. An assessment of Jones Creek spawning channel, 1954–1961. Can. Fish Cult. 30:3–14.

Madsen, M. J. 1938. A preliminary investigation into the results of stream improvement in the intermountain forest region. Trans. N. Amer. Wildl. Conf. 3:497–503.

Meehan, W. R. 1971. Effects of gravel cleaning on bottom organisms in three southeast Alaska streams. Prog. Fish-Cult. 33(2):107–111.

Parker, R. O., et al. 1979. Artificial reefs off Murrells Inlet, South Carolina. Marine Fish. Rev., Nat. Marine Fish. Serv. 41(9):12–24.

Parker, R. O., Jr., et al. 1974. How to build marine artificial reefs. Fishery Facts 10, Nat. Marine Fish. Serv. 47 pp.

Patriarche, M. H. 1961. Air-induced winter circulation of two shallow Michigan lakes. J. Wildl. Manage. 25(3):282–289.

Pollock, R. D. 1969. Tehama-colusa canal to serve as spawning channel. Prog. Fish-Cult. 31(3):123–130.

Prevost, G. 1957. Use of artificial and natural spawning beds by lake trout. Trans. Amer. Fish. Soc. 88:258–260.

Rasmussen, D. H. 1960. Preventing a winterkill by use of a compressed-air system. Prog. Fish-Cult. 22(4):185–187.

Richards, W. L. 1973. A bibliography of artificial reefs and other man-made fish attractants. Univ. North Carolina, Sea Grant Publ. UNC-SG-73-04. 21 pp.

Rodeheffer, I. A. 1939. Experiments in the use of brush shelters by fish in Michigan lakes. Pap., Michigan Acad. Sci. Arts Lett. 24:183–193.

———. 1940. The use of brush shelters by fish in Douglas Lake, Michigan. Pap., Michigan Acad. Sci. Arts Lett. 25:357–366.

———. 1945. Fish populations in and around brush shelters of different sizes placed at varying depths and distances apart in Douglas Lake, Michigan. Pap., Michigan Acad. Sci. Arts Lett. 30:321–345.

Scaratt, P. J. 1968. An artificial reef for lobsters (*Homarus americanus*). J. Fish. Res. Bd. Can. 25(12):2683–2690.

Shelton, J. M., and R. D. Pollock. 1966. Siltation and egg survival in incubation channels. Trans. Amer. Fish. Soc. 95(2):183–187.

Shetter, D. S., O. H. Clark, and A. S. Hazzard. 1949. The effects of deflectors in a section of a Michigan trout stream. Trans. Amer. Fish. Soc. 76:248–278.

Smith, L. L., Jr., and J. B. Moyle. 1944. Stream improvement on the north shore watershed. A biological survey and fishery management plan for the streams of the Lake Superior north shore watershed. Minnesota Dep. Conserv., Tech. Bull. 1:147–164.

Smith, M. W. 1948. Fertilization of a lake to improve trout angling. Fish. Res. Bd. Can., Note 105:3–6.

Steimle, F., and R. B. Stone. 1973. Bibliography on artificial reefs. Coastal Plains Cent., Mar. Devel. Serv., Wilmington, N.C., Publ. 73-2. 129 pp.

Stone, R. B., et al. 1979. A comparison of fish populations on an artificial and natural reef in the Florida Keys. Marine Fish. Rev., Nat. Marine Fish. Serv. 41(9):1–11.

Stroud, R. H. 1966. Artificial reefs as tools of sport fishery management in coastal marine waters. Sport Fish. Inst., Bull. 170. 8 pp.

Tarzwell, C. M. 1935. Progress in lake and stream improvement. Trans. Amer. Game Conf. 21:119–134.

———. 1937. Experimental evidence on the value of trout stream improvement in Michigan. Trans. Amer. Fish. Soc. 66:177–187.

———. 1938. An evaluation of the methods and results of stream improvement in the southwest. Trans. N. Amer. Wildl. Conf. 3:339–364.

Thomas, P. M., R. O. Legault, and G. F. Carpenter. 1968. Durability and efficiency of brush shelters installed in 1937 in Douglas Lake, Michigan. J. Wildl. Manage. 32(3):515–520.

Warner, K., and I. R. Porter. 1960. Experimental improvement of a bulldozed trout stream in northern Maine. Trans. Amer. Fish. Soc. 89(1):59–63.

Warren, C. E., J. H. Wales, G. E. Davis, and P. Doudoroff. 1964. Trout production in an experimental stream enriched with sucrose. J. Wildl. Manage. 28(4):617–660.

Waters, T. F. 1957. The effects of lime application to acid bog lakes in northern Michigan. Trans. Amer. Fish. Soc. 86:329–344.

White, R. J., and O. M. Brynildson. 1967. Guidelines for management of trout stream habitat in Wisconsin. Wisconsin Dep. Nat. Resour., Tech. Bull. 39. 65 pp.

12

Small Pond Management

The growing need for water-based recreation areas places increasing pressure on the utilization of *all* water areas, and one way of increasing the total amount of water available for fishing is to create impoundments. Small private ponds offer at least a partial answer to the problem.

A 1979 survey by the Soil Conservation Service state offices indicates that there are, in commercial food and baitfish production, 3,800 catfish ponds with a total area of about 22,500 hectares, 653 trout ponds with a total area of over 7,000 hectares, and 1,544 baitfish ponds with a total area of about 17,600 hectares. Fee fishing operations have 2,233 catfish ponds of about 16,000 hectares, 1,086 trout ponds with about 3500 hectares, and 5,555 "other" with about 27,200 hectares in total area.

History records the use of ponds in China in 500 B.C. for rearing carp; the Romans used farm ponds in the first century, and pond culture was common in Europe by the fourteenth century, chiefly for raising carp and pike. In the United States the use of mill ponds as fishing ponds began in colonial times with some attempts at pond fish stocking and management, but not much research was carried on in that period. Drs. H. S. Swingle and E. V. Smith, who conducted intensive studies in Alabama in the 1930s and 1940s, were the first in the United States to apply scientific methods to the development of pond management methods. As a result of their work, and by reason of plentiful water supplies, the highest concentration of successful fish ponds is in the Southeast.

Many state and federal government agencies are interested in small-pond construction and fish management, and information is available from the Soil Conservation Service (U.S. Department of Agriculture), state fish and game departments, state universities, U.S. Fish and Wildlife Service (U.S. Department of Interior), National Marine Fisheries Service (U.S. Department of Commerce), and private fishery consultants.

Many small ponds are not constructed primarily for recreational fishing. Other uses include commercial fish production, fee fishing, water storage for fire protection, livestock, irrigation, and erosion control, as well as swimming, skating, boating, waterfowl habitat, and the general aesthetic appeal of water and pond life near the home. This chapter is concerned with fish management for those ponds in which recreation is the primary objective. To simplify wording, the terms ''farm pond'' and ''small pond'' are used to denote any artificial pond (impoundment) between 0.1 and 2.0 ha in surface area.

Site Characteristics and Construction Features

An adequate water source is the most important requirement for small-pond management. Too much water can be as serious a problem as too little water. Once the pond is filled it is not ordinarily desirable to run a heavy flow through it, since this might lower the production by carrying away nutrients, cause losses of young fish, or produce erosion problems. Water shortages are most likely to occur in mid- to late summer. In some cases springs provide a dependable supply of cool water and hence are ideal for ponds to be managed for trout. The adequacy of a runoff water supply depends on the amount of precipitation and size of the watershed. For a 1-ha pond, a 5-ha watershed is required in New York State, while a 10-ha watershed is recommended in Alabama. A stream can be used to supply a pond, but to avoid flooding it is usually necessary to build the pond to one side of the main flow and control the water to be introduced into the pond with a head gate. A screening device may deter the introduction of undesired fish, but some of these will usually find their way into the pond eventually, if present in the stream.

Proper construction is essential to successful farm pond management. Local Soil Conservation Districts and County Agricultural Agents provide expert advice on farm pond construction, and there are excellent descriptions of these techniques in various bulletins. such as those of Dickson (undated) and Winkelblech (1955). Generally the best site for a farm pond is one not subject to flooding or silt runoff, on gently sloping ground (about 10:1), with relatively impervious soil (high clay content) at sufficient depths. The pond should have a maximum depth of about 2.5m, but in areas where high soil fertility, unusually severe winters, or other factors increase the danger of overfertilization leading to oxygen depletion and winterkills, a maximum depth of 3 to 4 m may be necessary. All sides of the pond should have 2:1 slopes out to a point where the water is always at least 1m deep. This minimizes the area of shallow water where weeds thrive, and lessens weed problems considerably. In trout ponds, shallow water is also undesirable because it warms up quickly during sunny weather, and warms the whole pond when mixed with the deeper water by wind action.

A vertical "trickle pipe" of 10- to 15-cm diameter provides an exit for excess water from the pond. This pipe is connected to a horizontal drain pipe in the bottom of the pond, as shown in Figure 12-1. An important asset in any fish pond, such an installation makes it possible to drain the pond to repair leaks or to make otherwise difficult adjustments to the fish population. To accommodate exceptional flows that the trickle pipe could not handle, a broad emergency spillway is made in undisturbed soil to one side of the earthen dike, at a level about 0.3m higher than the top of the trickle pipe. This prevents water from running over the top of the dike and cutting it. A thick grass cover should be quickly established and maintained on all sides of the earthen dike, the pond banks, and especially on the emergency spillway.

Fish Management

The selection of species for farm pond stocking is determined in part by owner preferences, maximum water temperatures expected, and pond size. Largemouth bass, bluegills and various other sunfish and their hybrids, golden shiners (as temporary forage for bass), channel catfish, rainbow and brook trout are the kinds of fish most frequently recommended for farm ponds. Because regional factors affect stocking recommendations, it is best to consult farm pond management bulletins for the area in question concerning the kinds and numbers to stock.

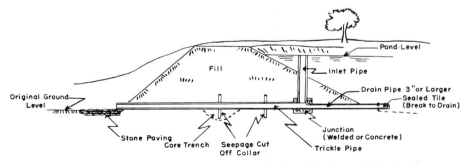

Figure 12-1. Cross section through farm pond dike showing trickle pipe and drain pipe. Courtesy of C. S. Winkelblech, 1955, New York State Coll. Agric. Life Sci., Cornell Ext. Bull. 949.

Survival of trout in farm ponds is influenced more by maximum summer water temperature than by any other factor. Although trout can withstand water temperatures as high as 27 C in farm ponds for periods of one or two days, prolonged periods of water temperature above 23 C will cause increasing mortality. Whether a farm pond will be suitable for trout depends chiefly on how long the coolest water, near the pond bottom, remains above 23 C for any one period. Satisfactory growth and reproduction of warm-water species such as largemouth bass and bluegills or golden shiners occur when the surface water of the pond becomes warm enough. For bass and shiners the temperature should be above 22 C and for bluegills 27 C for several weeks each summer.

In general, the larger the pond (within the size limits under consideration here), the easier it will be to manage. This is particularly true of the bass-bluegill combination, which is very likely to prove difficult to manage in ponds under ¼ ha in surface area, and even some under ½ ha.

Many state fish and game departments are unable to supply fish for private ponds. The U.S. Fish and Wildlife Service may supply fish for the initial stocking of ponds which meet certain criteria of size, construction features, use, and water quality. For all other circumstances, fish for stocking usually must be purchased from a commercial hatchery.

Trout-stocking recommendations derived from the Cornell University work in New York State provide some guidance on numbers and sizes of fish to stock. These figures apply only to brook trout and rainbow trout. Brown trout are not suitable for small ponds because they become increasingly difficult to harvest, and the old brown trout remaining in a pond prey heavily on any fingerlings introduced for restocking. Spring fingerlings, averaging 5 to 8 cm long after two or three months' growth, may be used for initial stocking at the rate of 4,900 per ha of pond surface. Fall fingerlings, usually 13 to 16 cm long after 7 or 8 months' growth, should be stocked at the rate of 1,500 per ha. Both reach catchable size the spring following stocking. Fall fingerlings give the most predictable results and the best return, especially for restocking. Mean natural mortality rates of farm pond trout in New York during various age intervals are given below.

Age interval	Natural mortality rate
0+ to 1+	0.60
I+ to II+	0.80
II+ to III+	0.80

Since few trout are left after two years, even without fishing, ponds should be restocked with fall fingerlings every other year. Trout usually do not reproduce successfully in farm ponds, at least not in New York. Standing crop will average about 120 kg per ha at the end of the first year after stocking with 1,500 fall fingerlings per ha, and 40 kg per ha at the end of the second year. These standing crop figures of course vary greatly between individual ponds.

The bass and golden shiner combination has proved satisfactory in warmwater ponds of the northeastern United States. An initial stocking of 250 bass fingerlings (3 to 5 cm long) and 1,000 golden shiner adults (usually 7 to 10 cm long) per ha is recommended. Although the shiners usually reproduce well, they tend to diminish in numbers as a result of bass predation until few if any are left four years after stocking (except in occasional ponds that are larger than 2 ha and have unusual amounts of protective cover for the shiners). Bass continue to survive and grow, however, although slowly. Bass should not be fished until they have reproduced successfully, which is usually two years after stocking in northern ponds, one year after stocking in southern ponds.

In general, it is recommended that the bass-bluegill combination be stocked at the rate of 250 bass fingerlings 3 to 5 cm long and 2,400 bluegill fingerlings 3 cm long, per ha. Both are usually stocked during midsummer. Because relative growth rates—and hence age at first reproduction—of both species vary greatly with climate, the techniques for, and the ease of, managing this species combination differ markedly with the geographical region under consideration, as well as with pond size and angler preferences. The basic idea of this species combination is that the young of the highly prolific bluegill provide forage for the larger bass. The bass and the larger bluegills provide more varied sport fishing (and eating) than could any single species. In many northern states bluegills reproduce the first year after stocking in ponds, but bass not until the second year. Thus the bluegills are off to a year's head start reproductively, and by the second summer after stocking their first year's hatch is likely to be too large to be eaten by the 25-cm bass then present. In Alabama, on the other hand, many of the stocked bass may reproduce in the first summer after stocking, and by the second summer may have reached a size large enough to consume the yearling bluegills. The preceding is somewhat oversimplified, but does illustrate why, in northern farm ponds, successful management of the bass-bluegill combination is more difficult to achieve and requires more intensive harvesting of bluegills. Many northern pond owners are not inclined to harvest bluegills intensively.

Probably the greatest lack in the general management of small ponds is firm research data, based on regional needs. Unfortunately, successful management practices in one locality, though they may suggest the general pattern for management in another, usually cannot be applied without further investigation and often some modification. Local conditions such as soil type, length of growing season, and endemic fish species are only a few of the obvious variables between regions. Factors such as these also cause large variations in the data from different ponds, as do variables such as calendar year, and the altitude, size, morphometry, water supply, and predators of the individual pond. The more variable the data, the more difficult and time-consuming it is to obtain valid management recommendations for even one region.

References

American Fisheries Society. 1969. Fish farming—a commercial reality. Proc., Tech. Sess. 4, 99th Annual Meet., Amer. Fish. Soc.; reproduced by Soil Cons. Serv., U.S. Dep. Agric., Portland, Ore.

Avault, J. W., Jr., R. O. Smitherman, and E. W. Shell. 1968. Evaluation of eight species of fish for aquatic weed control. Proc. World Symp. Warm-Water Pond Fish Cult. FAO Fish. Rep. 44(5):109–122. U.N.

Ball, R. C. 1949. Experimental use of fertilizer in the production of fish-food organisms and fish. Michigan State Coll., Agric. Exp. Stn., Tech. Bull. 210. 28 pp.

Ball, R. C., and H. D. Tait. 1952. Production of bass and bluegills in Michigan ponds. Michigan State Coll., Agric. Exp. Stn., Tech. Bull. 231. 24 pp.

Barrett, P. H. 1953. Relationships between alkalinity and adsorption and regeneration of added phosphorus in fertilized trout lakes. Trans. Amer. Fish. Soc. 82:78–90.

Beardon, C. M. 1967. Salt-water impoundments for game fish in South Carolina. Prog. Fish-Cult. 29(3):123–128.

Bennett, G. W. 1948. The bass-bluegill combination in a small artificial lake. Illinois Nat. Hist. Surv. Bull. 24(3):377–412.

_____. 1971. Management of lakes and ponds. 2nd ed. Van Nostrand Reinhold Co., New York. 375 pp.

Biological Sciences Communication Project, The George Washington University. 1971. Bibliography of aquaculture. Coastal Plains Cent., Mar. Devel. Serv., Wilmington, N.C., Publ. 71-4. 245 pp.

Blackburn, R. D. 1968. Weed control in fish ponds in the United States, Proc. World Symp. Warm-Water Pond Fish Cult. FAO Fish. Rep. 44(5):1–17. U.N.

Borell, A. E., and P. M. Scheffer. 1961. Trout in farm and ranch ponds. U.S. Dep. Agric., Farmers' Bull. 2154. 18 pp.

Boyd, C. E. 1979. Aluminum sulfate (alum) for precipitating clay turbidity from fish ponds. Trans. Amer. Fish. Soc. 108(3):307–313.

Bryan, R. D., and K. O. Allen. 1969. Pond culture of channel catfish fingerlings. Prog. Fish-Cult. 31(1):38–43.

Buck, D. H., and C. F. Thoits III. 1970. Dynamics of one-species populations of fishes in ponds subjected to cropping and additional stocking. Illinois Nat. Hist. Surv. Bull. 30(2). 165 pp.

Byford, J. L. 1970. Catfish production techniques manual. Univ. Georgia, Coop. Ext. Serv. 103 pp. Mimeogr.

Byrd, I. B., and D. D. Moss. 1957. The production and management of Alabama's state-owned public fishing lakes. Trans. Amer. Fish. Soc. 85:208–216.

Carlander, K. D. 1955. The standing crop of fish in lakes. J. Fish. Res. Bd. Can. 12(4):543–570.

_____. 1969. Handbook of freshwater fishery biology. Vol. I. The Iowa State Univ. Press, Ames, Ia. 752 pp.

Carlander, K. D., and R. B. Moorman. 1956. Standing crops of fish in Iowa ponds. Proc. Iowa Acad. Sci. 63:659–668.

Center for Continuing Education, University of Georgia. 1969. Proceedings of conference on commercial fish farming. Univ. Georgia, Coop. Ext. Serv., Inst. Community and Area Devel. 85 pp.

Cole, H., R. G. Wingard, R. L. Butler, and A. D. Bradford. (undated) Aquatic plants. Management and control in Pennsylvania. Pennsylvania State Univ. Coll. Agric., Nat. Resour. Ser., Spec. Circ. 79. 32 pp.

Crance, J. H., and L. G. McBay. 1966. Results of tests with channel catfish in Alabama ponds. Prog. Fish-Cult. 28(4):193–200.

Dickson, F. J. (undated) Georgia fish pond management. Georgia Game and Fish Comm. 89 pp.

Edminster, F. C. 1947. Fish ponds for the farm. Charles Scribner's Sons, New York. 114 pp.

Eicher, G. J., and G. A. Rounsefell. 1957. Effects of lake fertilization by volcanic activity on abundance of salmon. Limnol. Oceanogr. 2(2):70–76.

Eipper, A. W. 1964. Growth, mortality rates, and standing crops of trout in New York farm ponds. New York State Coll. Agric. Exp. Stn., Memoir 388. 68 pp.

Eipper, A. W., and H. A. Regier. 1972. Fish management in New York farm ponds. New York State Coll. Agric. Life Sci., Ext. Bull. 1089 (rev.). 40 pp.

Fessler, F. R. 1950. Fish populations in some Iowa farm ponds. Prog. Fish-Cult. 12(1):3–11.

Fielding, J. R. 1968. New systems and new fishes for culture in the United States. Proc. World Symp. Warm-Water Pond Fish Cult. FAO Fish. Rep. 44(5):143–161. U.N.

Flickinger, S. A. 1971. Pond culture of bait fishes. Colorado State Univ., Coop. Ext. Serv. Bull. 478A. 39 pp.

Forney, J. L. 1957a. Bait fish production in New York ponds. New York Fish Game J. 4(2):150–194.

_____. 1957b. Chemical characteristics of New York farm ponds. New York Fish Game J. 4(2):203–212.

_____. 1968. Raising bait fish and crayfish in New York ponds. New York State Coll. Agric. Life Sci. Ext. Bull. 986 (rev.). 32 pp.

Grizzell, R. A. 1967. Pond construction, water quality, and quantity for fish farming. Texas A. and M. Univ., Proc. Commer. Fish Farming Conf.:9–11.

Hansen, D. G., G. W. Bennett, R. J. Webb, and J. M. Lewis. 1960. Hook-and-line catch in fertilized and unfertilized ponds. Illinois Nat. Hist. Surv. Bull. 27(5):345–390.

Hasler, A. D. 1947. Eutrophication of lakes by domestic drainage. Ecology 28(4):383–395.

Hasler, A. D., and W. G. Einsele. 1948. Fertilization for increasing productivity of natural inland waters. Trans. N. Amer. Wildl. Conf. 13:527–552.

Henderson, C. 1949. Manganese for increased production of water-bloom algae in ponds. Prog. Fish-Cult. 11(3):157–159.

Holloway, A. D. 1951. An evaluation of fish pond stocking policy and success in the southeastern states. Prog. Fish-Cult. 13(4):171–180.

_____. 1952. Suggestions for the renovation of fish ponds. Prog. Fish-Cult. 14(1):27–29.

Howell, H. H. 1942. Bottom organisms in fertilized and unfertilized fish ponds in Alabama. Trans. Amer. Fish. Soc. 71:165–179.

King, W. 1960. A survey of fishing, in 1959, in 1,000 ponds stocked by the Bureau of Sport Fisheries and Wildlife. U.S. Fish Wildl. Serv., Fish Wildl. Circ. 86. 20 pp.

Knight, A., R. C. Ball, and F. F. Hooper. 1962. Some estimates of primary production rates in Michigan ponds. Pap. Michigan Acad. Sci. Arts Lett. 47:219–233.

Krumholz, L. A. 1949. Further observations on the use of hybrid sunfish in stocking small ponds. Trans. Amer. Fish. Soc. 79:112–124.

Lawrence, J. M. 1949. Construction of farm fish ponds. Alabama Polytech. Inst., Agric. Exp. Stn. Circ. 95. 55 pp.

_____. 1954. A new method of applying inorganic fertilizer to farm fish ponds. Prog. Fish-Cult. 16(4):176–178.

_____. 1962. Aquatic herbicide data. U.S. Dep. Agric., Agric. Handbook 231. 133 pp.

Maciolek, J. A. 1954. Artificial fertilization of lakes and ponds. U.S. Fish Wildl. Serv., Spec. Sci. Rep. Fish. 113. 41 pp.

McCarraher, D. B. 1959. The northern pike-bluegill combination in north-central Nebraska farm ponds. Prog. Fish-Cult. 21(4):188–189.

McConnell, W. J. 1966. Preliminary report on the Malacca *Tilapia* hybrid as a sport fish in Arizona. Prog. Fish-Cult. 28(1):40–46.

Meyer, F. P., K. E. Sneed, and P. T. Eschmeyer, eds. 1973. Second report to the fish farmers. U.S. Dep. Inter., Bur. Sport Fish. Wildl., Resour. Publ. 113. 123 pp.

Mortimer, C. H., and C. F. Hickling. 1954. Fertilizers in fish ponds. Colonial Off., London, Fish. Publ. 5. 155 pp.

Neely, W. W., V. E. Davison, and L. V. Compton. 1965. Warm-water ponds for fishing. U.S. Dep. Agric., Farmers' Bull. 2210. 16 pp.

Nelson, P. R., and W. T. Edmondson. 1955. Limnological effects of fertilizing Bare Lake, Alaska. U.S. Fish Wildl. Serv., Fish. Bull. 102:415–436.

Patriarche, M. H., and R. C. Ball. 1949. An analysis of the bottom fauna production in fertilized and unfertilized ponds and its utilization by young-of-the-year fish. Michigan State Coll., Agric. Exp. Stn., Tech. Bull. 207. 35 pp.

Pillay, T. V. R., ed. 1968. Proceedings of the world symposium on warm-water pond fish culture. FAO Fish. Rep. 44, Vol. 4. 492 pp. U.N.

Regier, H. A. 1962. On the evolution of bass-bluegill stocking policies and management recommendations. Prog. Fish-Cult. 24(3):99–111.

_____. 1963a. Ecology and management of largemouth bass and bluegills in farm ponds in New York. New York Fish Game J. 10(3):1–89.

_____. 1963b. Ecology and management of largemouth bass and golden shiners in farm ponds in New York. New York Fish Game J. 10(2):139–169.

_____. 1963c. Ecology and management of channel catfish in farm ponds in New York. New York Fish Game J. 10(2):170–185.

Schaeperclaus, W. 1933. Textbook of pond culture. (Transl. from German by Frederick Hund.) U.S. Fish Wildl. Serv., Fish. Leafl. 311. 261 pp.

Schultz, V. 1950. A selected bibliography on farm ponds and closely related subjects. Prog. Fish-Cult. 12(2):97–104.

Smith, M. W. 1953. Fertilization and predator control to improve trout production in Crecy Lake, New Brunswick. Can. Fish Cult. 13:33–39.

Smith, M. W., and J. W. Saunders. 1955. Fertilization and predator control to improve trout angling in natural lakes. J. Fish. Res. Bd. Can. 12(2):210–237.

Smith, R. H. 1955. Experimental control of water chestnut (*Trapa natans*) in New York State. New York Fish Game J. 2(2):173–193.

Surber, E. W. 1947. Variations in nitrogen content and fish production in smallmouth black bass ponds. Trans. Amer. Fish. Soc. 74:338–349.

_____. 1948. Increasing production of bluegill sunfish for farm pond stocking. Prog. Fish-Cult. 10(4):199–203.

_____. 1949. Control of aquatic plants in ponds and lakes. U.S. Fish Wildl. Serv., Fish. Leafl. 344. 20 pp.

_____. 1961. Improving sport fishing by control of aquatic weeds. U.S. Fish Wildl. Serv., Fish Wildl. Circ. 128. 49 pp.

Swanson, G. A., ed. 1952. Symposium on farm fish ponds and management. J. Wildl. Manage. 16(3):233–288.

Swingle, H. S. 1949. Experiments with combinations of largemouth black bass, bluegills, and minnows in ponds. Trans. Amer. Fish. Soc. 76:46–62.

_____. 1950. Relationships and dynamics of balanced and unbalanced fish populations. Alabama Polytech. Inst., Agric. Exp. Stn. Bull. 274. 74 pp.

_____. 1960. Comparative evaluation of two tilapias as pondfishes in Alabama. Trans. Amer. Fish. Soc. 89(2):142–148.

_____. 1970. History of warmwater pond culture in the United States. *In* N. G. Benson, ed. A century of fisheries in North America. Amer. Fish. Soc., Washington, D.C., Spec. Publ. 7, pp. 95–105.

Swingle, H. S., and E. V. Smith. 1947. Management of farm fish ponds. Alabama Polytech. Inst., Agric. Exp. Stn. Bull. 254. 30 pp.

Tanner, H. A. 1960. Some consequences of adding fertilizer to four Michigan trout lakes. Trans. Amer. Fish. Soc. 89(2):198–205.

Texas Agricultural Extension Service. 1969. Proceedings, the commercial bait fish conference. Texas A. and M. Univ., College Station, Texas. 52 pp.

Turner, W. R. 1960. Standing crop of fishes in Kentucky farm ponds. Trans. Amer. Fish. Soc. 89(4):333–337.

Weed Society of America. 1970. Herbicide handbook. Univ. of Illinois, Urbana, Dep. Agronomy, WSSA monogr. 3. 368 pp.

Winkelblech, C. S. 1955. Farm ponds in New York. New York State Coll. Agric. Life Sci., Ext. Bull. 949. 32 pp.

Woodford, E. K., and S. A. Evans, eds. 1965. Weed control handbook. Blackwell Scientific Publications, Oxford. 434 pp.

Zeller, H. D. 1953. Nitrogen and phosphorus concentrations in fertilized and unfertilized farm ponds in central Missouri. Trans. Amer. Fish. Soc. 82:281–288.

13

Fishways, Screens, and Guiding Devices

Fishways are water-filled locks, channels, a series of connected pools, or water-tight buckets by which fish may swim or be carried over and around an obstruction. Anadromous fishes must ascend streams to reach suitable and sufficiently extensive spawning grounds. Some fish use streams for spawning and nursery purposes, others use lake areas, and some use both. Failure to provide proper fishways over and around obstructions has destroyed natural populations and natural resources.

Basic requirements for any fishway are:

1. It should be suitable for and passible for all migratory species in the area.
2. It should operate at all water levels in the forebay and tailrace of a dam or above and below a natural obstruction.
3. It should operate at all volumes of stream flow.
4. Fish should ascend without injury or extreme exertion.
5. Fish must find the entrance, enter quickly, and pass through the fishway without delay.

General types of fishways are discussed below.

The *pool-type* fishway is a succession of pools connected by short rapids or low falls or by submerged orifices (Figure 13–1). The energy of the water is dissipated in each pool. A properly constructed pool-type fishway should not have any carry-over energy from one pool to the next. The pools should be deep enough to prevent excessive turbulence and long enough to prevent strong currents against the next weir.

The *Denil-type* fishway is a narrow chute carrying the water at a high velocity, especially at the surface. The force of this stream striking the closely spaced baffles aids in maintaining a low velocity near the bottom

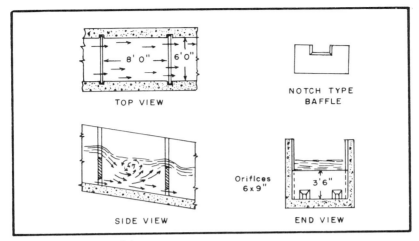

Figure 13-1. Pool-type fishway.

(Figure 13-2). The Denil type works on a different principle than the pool type. Whereas the pool type dissipates the energy of the falling water in each pool, the Denil type permits a very fast current. A portion of the energy generated by this fast current is dissipated against closely spaced marginal and bottom baffles and re-entrant angles. These angles cause

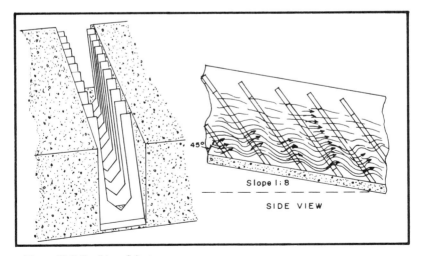

Figure 13-2. Denil-type fishway.

small backward jets that slow the current along the bottom. One can easily determine this for oneself by extending an arm or leg into one of these fishways and feeling the difference in velocity at the surface and at the bottom. Fish swim up the slower current nearer the bottom.

The Denil fishway requires less space and is cheaper to construct than the pool type; it is operable as long as there is water over the bottom of the baffles. It will operate, without adjustment, over a water fluctuation no greater than the depth of the baffles. The entrance to such a fishway must be submerged in the tailwater below the obstruction.

The *deep-baffled channel* or *vertical-slot* fishway is designed for a low gradient and is especially suitable to an extreme variation in flow and water level (Figure 13-3).

The *fish lock* depends on the rising water level or on some kind of moving screen within the lock to encourage the fish to rise in the lock and past the obstruction. Mechanical gates and either attendants or automatic devices are necessary to regulate water flows in and out of the lock (Figure 13-4).

The *fish-lift type* consists of some kind of water container into which the fish either swims or jumps and then is transported over the obstruction by some kind of lift or incline track (Figure 13-5).

Design and location of the fishway are basic to its success. Location is frequently complicated when a fishway must be added to an already constructed dam or industrial facility. Ideally, it is best to include the fishway as an integral part of the dam structure, built with the dam and even through the dam if this provides the most efficient facility. It should be kept

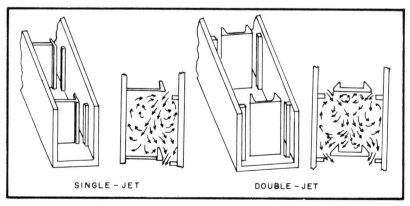

SINGLE – JET   DOUBLE – JET

Figure 13-3. Deep-baffle or vertical-slot fishway.

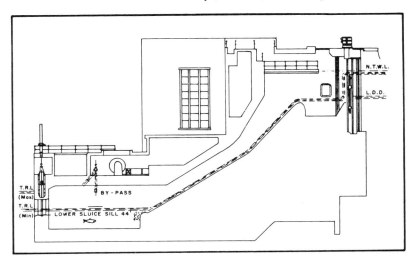

Figure 13-4. Borland fish lock.

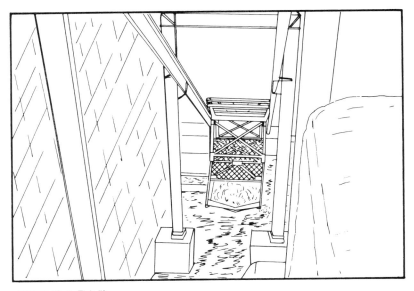

Figure 13-5. Fish lift.

in mind that one of the best ways to destroy a run of fishes is to settle for a "second best" fishway. If the dam owner has complied with the law, there is no legal recourse. At least as long as there is no fishway there is a chance to build a successful one.

Economics frequently preclude the construction of multiple fishways at obstructions, but for a salmon to find a single entrance to a fishway along a many-hundred-foot obstruction with spillways, turbine flows, and natural currents, the chances are slim. Thus, the most important feature of any fishway is the entrance. If the entrance is located too far downstream from the obstruction, it will be passed by the fish, which follow the main channel until stopped by a barrier. This is a more likely mistake in fishways constructed after the dam is built. Generally the entrance should be located close to the toe of the dam. If the dam happens to be built obliquely across a stream, it is usually best to locate the fishway at the upstream end. The primary requirement for attracting fish is that the entrance be located close enough to the main current from the spillway or tailrace so that the fish following the current will be led into the fishway entrance.

Fishway efficiency can be increased by providing multiple entrances with considerable flexibility of operation depending on spillway flows and turbine outfall. Thus, at Bonneville Dam there are 65 possible entrances over the powerhouse, with provisions for operating them in combinations up to 30. At the Dalles Dam, just above Bonneville on the Columbia, there are 138 possible entrances. Twenty-five are usually operated, although up to 30 can be used. Each entrance is 46 cm wide and 1.8 m deep. Using multiple entrances requires a transportation channel to carry fish from the entrance into the fishway.

Bonneville has as much as 68 m^3 of auxiliary flow, and Dalles has about 170 m^3. This auxiliary water increases the attraction by providing plenty of flow at the fishway entrances. One engineering problem is how to introduce this much auxiliary water without interfering with the fishes' movement up the main current in the fishway. Adding the auxiliary water at or near the entrance reduces the overall cost of the fishway which can then operate on much less flow.

Fishway exits should be provided with some type of trash rack and with some provision to accommodate for fluctuations in the forebay level. This may be accomplished with a single exit plus some means of manipulating several of the weirs in the top pools. Submerged-orifice fishways can be designed to split the difference in head. Six balancing pools can be designed so that at a forebay level of plus 1.8 m, each pool has a 0.3 m head difference. At a forebay level of plus 1 m, each pool has a 15-cm difference. At zero level there would be no head between any of the top six pools. Another way of providing for fluctuating forebay level is to have several exits in the fishway and simply open the one that provides for the best flow and the most efficient movement of the fish. The exit should be placed in water of sufficient depth but should not be located at the crest of the spillway or very close to any water diversion.

Few fish can swim faster than 11 to 13 km per hour and then only in short bursts of not very long duration. Fishways are not intended as obstacle courses, and velocities should be kept well below the maximum. As a general rule, velocities should be less than 1.5 m per second at any point in the fishway.

Pool-type fishways and deep-baffle fishways should be at least 3 m square. Water depths should be more than 1 m to insure energy is dissipated without excessive turbulence. Larger-size pools may be necessary where exceptionally large numbers of fish are expected.

The maximum drop in water surface between pools should be about 0.30 m. It is recognized that fish can jump higher than that, but the point is to provide for as rapid and easy migration as possible. Submerged orifices 15 cm wide and 23 cm high are usually sufficient.

In the straight-run fishway there is no advantage to staggering the position of the orifices. If the pools are the proper dimensions, there will be practically no carry-over of velocity from one pool to the next. If the orifices are in line, a fish will often swim through several in one fast rush. Crosscurrents created by staggering orifices make each pool less of a resting pool.

Dimensions of the Denil fishway are usually a maximum of 1.2 m in width with a gradient of 1:8. This means the fishway runs 2.4 m for every 0.3 m of height. Steeper gradients reduce the efficiency of the fishway because fish must climb entire runs either to resting pools or to the exit. Gradients exceeding 1:8 mean a longer fishway with increased costs and engineering problems.

No matter how well a fishway is engineered it is necessary to provide for proper operation and inspection of the facility.

Screens and Guiding Devices

Alterations in the aquatic environment result in the need for screens and guiding devices to help keep fish in favorable water areas. These facilities range from the simple bar screen to the artificial outlets on the salmon and steelhead streams in Oregon. Practical considerations with any screening or guiding facility are maintenance, problems with cleaning, whether or not water flow is restricted and impounded, and interference with boat passage. Screening and guiding are a fish-behavior problem, and species differences, age differences, size differences, and motivation for whatever movement is involved are confounded with the differences in water velocity, depth, turbidity, and color.

The need for screens and guiding devices is well illustrated by the entrapment and impingement of fish at power plant cooling water intakes. Losses of fish larvae and juveniles at the Connecticut Yankee Power Plant are estimated at 180 million and reports from other plants are in the one million range.

Fish guiding devices should be evaluated as to cost, efficiency, mechanical limitations, volume and flow, excessive head loss, survival of eggs, larvae, and fry, and maintenance and adjustment.

Simple bar screens are adaptable to conditions of low velocity where drift is not excessive. These screens are usually made of narrow round iron bars. Provision should be made for easy access for cleaning, particularly in high water when debris may be most serious and when clogging of the screen creates a damming effect and the threat of flooding. Many of these screens are made in removable panels for repair and for easy cleaning. Bar screens are used with some success for diverting and guiding large, adult fish but are almost useless in preventing movement of small fish.

Paddle-wheel screens of various designs have been used to guide fish. The screen pictured in Figure 13–6 is capable of passing about 1.1 m³/sec. of water. The paddle wheel powers a wiper blade which travels back and forth along the upstream side of the screen, taking advantage of the current to operate a cleaning device. Should the cleaning device fail and the screen become plugged, the float indicated on the diagram drops and releases the entire screen which falls flat on the stream bed. This screen is used in California for irrigation diversions.

Rotating vertical drum screens are used in irrigation projects to prevent fish from being pumped into the system and eventually ending up between rows of vegetables. One screen of this kind is a perforated drum 2.7 m in diameter and 1.2 m deep. It rotates from 3 to 9 m per minute and is powered by a small electric motor which in turn drives a hydraulic motor for considerable latitude in speeds and location. A stationary wiper-blade attachment keeps the screen clean.

Louver screens consist of a fence-like series of vertical steel slats with each slat at a 90-degree angle to the flow (Figure 13–7). Head loss may be less at lesser angles, but diversion is not as efficient. Fish approaching the louver screen "tail" along the front until they are picked up in the by-passes. The steel slats are about 6 cm in width with spacing between 2.5

Figure 13-6. Paddle-wheel screen.

cm and 5 cm. The louver screen at the Tracy pumping station in California has 2.5-cm spacing. Individual slat lengths can be varied to accommodate for irregularities of the bottom. A backup louver system may be necessary if too much water and too many fish pass through the primary screen.

Fishery biologists feel that the louver screen could be exceptionally efficient if there were some way to provide for movement. Moving the louver screen along in a matched velocity with the current would guide even more than the presently accepted figure of ''more than 80 percent.'' However, the cost and engineering problems of sealing the bottom of a moving screen and preventing damage to the fish are the subject of some controversy.

Artificial outlets are presently in operation at the power dams of the Portland and General Electric Company at North Fork on the Clackamas River, at the Pelton Dam on the Deschutes, and at the Round Butte Dam on the Deschutes River. The North Fork and Round Butte facilities are similar in design and considerably less expensive with fewer maintenance problems than the Pelton design.

The North Fork artificial outlet and fishway must operate in a forebay with a weekly fluctuation of 5.8 m. This is said to be the greatest range

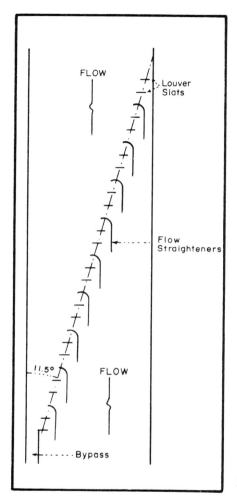

Figure 13-7. Louver screen. Courtesy of D. W. Bates and R. Vinsonhaler, 1957, Trans. Amer. Fish. Soc. 86:43.

resolved thus far in fishway exit design. Filure 13–8 demonstrates how the fishway and artificial outlet were carried through the spillway section of the dam at an elevation below minimum forebay level and continued through 22 4-m pools between high walls next to the shoreline of the spillway in a gently curving arc. The downstream migrant channel (artificial outlet) is 3 m wide and 9 m deep, parallel to the fishway.

Each of the 22 pools in this section of the fishway is equipped with a gate 0.6 meters deep and 1.2 meters long, opening into the adjacent migrant

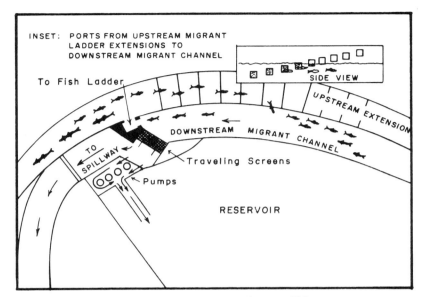

Figure 13–8. North Fork artificial outlet. Courtesy of George J. Eicher.

channel. Automatic controls open the gate of the pool, which has from 7.6 cm to 15.2 cm of head between it and the migrant channel, and 0.9 m³/sec. flows through. Fish ascend the ladder to the pool with the open gate, pass through it into the migrant channel and out the gate into the reservoir. The exit is far enough upstream so that fish are unlikely to be swept back down over the spillway, even in flood conditions.

Downstream migrant fish, both fingerling salmon and trout, and spent adult steelhead trout leave the reservoir through the artificial outlet. During normal river stages, four pumps near the downstream end of the migrant channel draw water at 5.7 m³/sec. through the outlet and return it through a 2.1-m pipe to a point deep in the reservoir. Two adjacent traveling water screens 3 m wide and 9 m deep, set at an angle to the channel, form a shunt to divert fish into a bypass where water moving at 0.3 m³/sec. overflows into a sump and passes through an orifice into the ladder.

Once the fish have been guided from the reservoir into the fish ladder, they must be separated out and guided into the bypass which will take them five miles downstream around two other hydroelectric facilities below. A fish separator (Figure 13-9) below the artificial outlet guides the downstream migrants into the bypass, while permitting the upstream migrants to pass on up the fishway. Downstream migrants are transported

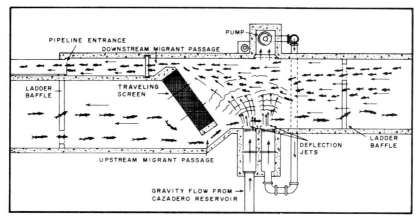

Figure 13-9. North Fork fish separator. Courtesy of George J. Eicher.

then for five miles in a smooth, low pressure, transite pipe 51 cm in diameter. The pipe level is graded so the head is largely balanced by friction loss, and negative pressures are avoided. Fish are moved along at from 1.2 to 1.8 m per second. They are introduced into the river below the lowest obstruction by means of a free fall from the end of the pipe 9 m above the water surface.

The Pelton-type artificial outlet (Figure 13-10) uses a movable perforated plate to strain out attraction water. The trash problem is greater at this installation, and clogging of the perforated plate alters hydraulics and reduces effectiveness of the facility. The North Fork design is less costly to build, easier to maintain, and cheaper to operate.

Chain screens composed of chain lengths dangling in the water have been proposed for guiding fish. The movement of the chains in the water is supposed to scare fish away and move them along the screen. Placing the entire chain screen on an eccentric permits the powershaking of the chains and results in some increase in effectiveness of guiding. A deflection of 79 percent of the sockeye yearlings was obtained when the screen was vibrated with a stroke of 10 cm and a rate of oscillation averaging 153 oscillations per minute. No success was obtained with coho yearlings under these conditions. Water velocity of even 0.03 m^3/sec. sharply reduced effectiveness, as did repeated experiences.

A *traveling cable deflector,* consisting of a curtain of 6 mm cables hanging vertically and traveling horizontally on an endless belt, has been studied in Canada. The upstream or advance row moves in the direction of the bypass. Since fish tend to hold position in a stream by taking a visual

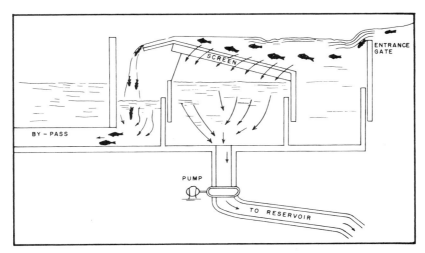

Figure 13-10. Pelton artificial outlet. Courtesy of George J. Eicher.

fix on near objects (rheotropism), those making visual reference with the front row of moving cables would be guided in the appropriate direction. The face of the cables was painted white to increase visual stimulus. Approach walls and doors of the entrance were covered with smooth plywood, painted dark brown, and lined partway along the side with mirrors. Water velocity was increased from 0.24 m/sec. at the entrance to 1.3 m/sec. at the narrowest point.

General conclusions as to the usefulness of this type of deflector were that the higher the rate of movement and the smaller the interval, the greater the deflection. An interval greater than 10 to 15 cm reduced efficiency. An average guidance of 89 percent of the sockeyes (range 86 to 94 percent) and 58 percent of the coho (range 36 to 81 percent) resulted.

Horizontal traveling screens have been developed essentially by substituting an endless screen belt in place of the cables in the traveling cable deflector. Reports on the effectiveness of the traveling screen indicate that matching the movement of the screen to the velocity of the water eliminates high mortalities by collecting the fish gradually and moving them directly into the safety of the bypass. One traveling screen installed in Stanfield Irrigation Canal near Echo, Oregon, utilized stretched nylon mesh with an effective open area of 72 percent with a small head loss and with a deflection percentage of 97 to 100 percent of the young steelhead and coho salmon.

Vertical traveling screens orient vertically to the water flow resulting in serious fish impingement problems. Research to improve survival of fish includes one design concept for a vertical traveling screen equipped with fish basket collectors. Fish are transported up and over the screen in baskets filled with water and then emptied automatically into the bypass. Fish impingement has also been reduced by implementation of high volume, low pressure spray jet wash systems.

Vertical traveling screens used to intercept migrants moving near the ceiling of a turbine intake are reported as successful in diverting 87 percent of the fish entering the intake into the gatewell slot. This installation takes advantage of the observed behavior of salmon and steelhead trout to rise in the water-filled gatewell slots.

Illumination has been studied as a method of guiding fish, but results have been conflicting. The most nearly universal response displayed by all sizes and species of salmon and trout in still water is to avoid the light. Thus, the area to be avoided should be brightly illuminated, and the area to be used should be darkened. Laboratory tests indicated that the smaller the angle of the barrier to the current, the smaller the number of fish that entered the lighted area. And fewer fish entered the lighted area as light intensity increased. Velocity of the water had the most effect, and depth difference between 7.6 and 61 cm did not have any significant effect. Field testing, using a 12-m-long light barrier, revealed the following significant results:

1. Definite guiding resulted with the use of the light barrier.
2. Efficiency of guidance was not increased either by shining lights upstream or downstream or by varying the levels of illumination.
3. Light barriers reduced the number trapped as compared with those trapped in equivalent dark periods.
4. Hatchery-reared and wild salmon exhibited no differences.
5. A 0.3-m-wide, 12-m-long lighted area was as effective as lighting the entire area behind or in front of barrier.

Repellents for fish guiding have attracted the attention of biologists and engineers for as long as we have known that washing your hands in a fishway will reduce or stop migration. Actually mammalian skin in a concentration of 1:80,000,000,000 will repel fish that may have just migrated through pulp-mill effluent of a million times greater concentration. Tests of a great many odors have shown that human skin, bear paw, deer hoof, dog paw, and sea-lion meat will produce a significant change in migration for periods as long as 15 minutes. One interesting field test of the bear paw is described from Canada where two bear paws on sticks were used to drive salmon into a net for collection.

Despite its effectiveness, the problems of controlling concentration and movement of repellent in an actual field condition have so far prevented use of this technique in guiding fish.

"Ski-jump" spillways have been utilized to concentrate flow over a spillway and deliver both water and fish so that the fish free-fall into a tailwater pool. This prevents injury to the fish from the surface of the dam and reduces the velocity at which they strike the river below.

Bubbles or curtains of bubbles have been used to guide fish in their migrations or to help concentrate them for fishing gear. General conclusions on these studies usually refer to some success, but velocity of water, turbidity, and different species behavior all seem to cloud the results.

Sound has been experimented with, but to date nothing has been developed that shows much promise. The problems of sound scattering, reflecting, and refracting have not been solved.

Electricity has been used to guide fish by suspending rows of positive and negative electrodes in the water, on the bottom, or by combining negative electrodes in the water with positive electrodes in the bank near the area where the fish are to be concentrated. Such electric guiding fences are effective. The most serious problem is one of maintenance to insure that electrodes do not become corroded or eroded, thus changing the current so as to be ineffective or to cause mortality in the fish population.

As our need for more and more fish for food and for sport fishing increases, the prospect of saving additional fish by protecting them from dangerous areas will continually pressure engineers and biologists into searching for better guiding methods.

References

Bates, D. W. 1970. Diversion and collection of juvenile fish with travelling screens. U.S. Fish Wildl. Serv., Fish. Leafl. 633. 6 pp.

Bates, D. W., and J. G. Vanderwalker. 1970. Traveling screens for collection of juvenile salmon Models I and II. U.S. Fish Wildl. Serv., Spec. Sci. Rep. Fish. 608. 15 pp.

Bates, D. W., and R. Vinsonhaler. 1957. Use of louvers for guiding fish. Trans. Amer. Fish. Soc. 86:38–57.

Bentley, W. W., and H. L. Raymond. 1968. Collection of juvenile salmonids from turbine intake gatewells of major dams in the Columbia River System. Trans. Amer. Fish. Soc. 97(2):124–126.

Brett, J. R., and D. F. Alderdice. 1954. Some aspects of olfactory perception in migrating adult coho and spring salmon. J. Fish. Res. Bd. Can. 11(3):310–318.

――――. 1956. Preliminary experiments using lights and bubbles to deflect migrating young spring salmon. J. Fish. Res. Bd. Can. 10(8):548–559.

――――. 1958. Research on guiding young salmon at two British Columbia field stations. Fish Res. Bd. Canada. Bull. 117. 75 pp.

Burner, C. J., and H. L. Moore. 1962. Attempts to guide small fish with underwater sound. U.S. Fish Wildl. Serv., Spec. Sci. Rep. Fish. 403:1–30.

Campbell, H. J. 1959. Field testing rotary fish screens in northwestern Oregon. Prog. Fish-Cult. 21(2):55–62.

Clay, C. H. 1955. A procedure for installation of fishways at natural obstructions. Can. Fish. Cult. 17:1–12.

_____. 1955. Downstream fish migration over dams. British Columbia. Prof. Eng., October. 4 pp.

_____. 1961. Design of fishways and other fish facilities. Dep. Fish., Ottawa, Can. 301 pp.

Cramer, F. R., and R. C. Oligher. 1964. Passing fish through hydraulic turbines. Trans. Amer. Fish. Soc. 93(3):243–259.

Deelder, C. L. 1958. Modern fish passes in the Netherlands. Prog. Fish-Cult. 20(4):151–154.

Ducharme, L. J. A. 1972. An application of louver deflectors for guiding Atlantic salmon (*Salmo salar*) smolts from power turbines. J. Fish. Res. Bd. Can. 29(10):1397–1404.

Eicher, G. J. 1958. Fish get novel treatment at PGE's North Fork project. Electr. Light Power, April 15:84–87.

_____. 1960. Fish bypass experience at PGE's new hydro project. Electr. Light Power, March 1. 4 pp.

_____. 1964. Round Butte Dam fish-handling costs 2.5% of total project outlay. Electr. World, February 10. 4 pp.

Fields, P. E. 1957. Guiding migrant salmon. Sci. Mon. 85(1):10–22.

Frischholz, D. 1924. Anlage und Betrib von Fischpassen (Construction and operation of fish passes). Handbuch der Binnenfischerei Mitteleuropas (Handbook of the Central European freshwater fisheries), Stuttgart 6(1). 137 pp.

Furuskog, V. 1945. En ny laxtrappa (A new fishladder). Svensk Fisk.-Tidskr. 11:236–239.

Hallock, R. J., and W. F. VanWoert. 1959. A survey of anadromous fish losses in irrigation diversions from Sacramento and San Joaquin Rivers. California Fish Game J. 45(4):227–296.

Hanson, C. H., J. W. White, and H. W. Li. 1977. Entrapment and impingement of fishes by power plant cooling-water intakes: An overview. Mar. Fish. Rev. 39(10):7–17.

Hourston, W. R., and J. W. Stokes. 1953. Moricetown Falls Biological Survey—1951. Can. Fish. Cult. 13:22–32.

John, K. R. 1954. A spaced-disc, self-cleaning fish screen. Prog. Fish-Cult. 16(2):70–74.

Kahn, R. A., and G. A. Rounsefell. 1947. Evaluation of fisheries in determining benefits and losses from engineering projects. U.S. Fish Wildl. Serv., Spec. Sci. Rep. Fish. 40. 10 pp.

Kupfer, G. A., and W. G. Gordon. 1966. An evaluation of the air bubble curtain as a barrier to alewives. Commer. Fish. Rev. 28(9):1–8.

Kupka, K. H. 1966. A downstream migrant diversion screen. Can. Fish. Cult. 37:27–34.

Lennan, B., and G. H. Bulik. 1966. Spill-pattern manipulation to guide migrant salmon upstream. Trans. Amer. Fish. Soc. 95(4):397–407.

McCabe, G. T., Jr., C. W. Long, and D. L. Park. 1979. Barge transportation of juvenile salmonids on the Columbia and Snake Rivers, 1977. Marine Fish Rev., Nat. Marine Fish. Serv. 41(7):28–34.

McLeod, A. M., and P. Nemenyi. 1941. An investigation of fishways. Univ. Iowa Stud., Eng. Bull. 24. 63 pp.

Menzies, W. J. M. 1934. Salmon passes—their design and construction. Fish. Scotland, Salmon Fish. 1934(1). 29 pp.

Monon, G. E., R. J. McConnell, J. R. Pugh, and J. R. Smith. 1969. Distribution of debris and downstream-migrating salmon in the Snake River above Brownlee Reservoir. Trans. Amer. Fish. Soc. 98(2):239–244.

Moore, H. L., and H. W. Newman, 1958. Effects of sound waves on young salmon. U.S. Fish Wildl. Serv., Spec. Sci. Rep. Fish. 172:1–19.

Moulton, J. M., and R. H. Backus. 1955. Annotated references concerning the effects of man-made sounds on the movements of fishes. Maine Dep. Sea and Shore Fish., Fish., Circ. 17.

Nemenyi, P. 1941. An annotated bibliography of fishways. Univ. Iowa Stud., Eng. Bull. 23. 64 pp.

Pryce-Tannatt, T. E. 1938. Fish passes. The Buckland Lectures for 1937. Edward Arnold and Co., London. 108 pp.

Raymond, H. L. 1956. Effect of pulse frequency and duration in guiding salmon fingerlings by electricity. U.S. Fish Wildl. Serv., Res. Rep. 43. 19 pp.

Rounsefell, G. A. 1944. Fishways for small streams. U.S. Fish Wildl. Serv., Fish. Leafl. 92. 55 pp.

Rowley, W. E., Jr. 1955. Hydrostatic pressure tests on rainbow trout. California Fish Game J. 41(3):243.

Schoeneman, D. E., R. T. Pressey, and C. O. Junge, Jr. 1961. Mortalities of downstream migrant salmon at McNary Dam. Trans. Amer. Fish. Soc. 90(1):58–72.

Slatik, E. 1970. Passage of adult salmon and trout through pipes. U.S. Fish Wildl. Serv., Spec. Sci. Rep. Fish. 592:1–18.

_____. 1975. Laboratory evaluation of a Denil-type steeppass fishway with various entrance and exit conditions for passage of adult salmonoids and American shad. Marine Fish. Rev., Nat. Marine Fish. Serv. 37(9):17–26.

Smith, J. R., and W. E. Farr. 1975. Bypass and collection system for protection of juvenile salmon and trout at Little Goose Dam. Marine Fish. Rev., Nat. Marine Fish. Serv. 37(2):31–35.

Warner, G. H. 1956. Report on the air-jet fish deflector tests. Prog. Fish-Cult. 18(1):39–41.

Webster, D. A. 1956. Modifications of the wolf-type fish trap. New York Fish Game J. 3(2):199–204.

Whalls, M. J., D. S. Shetter, and J. E. Vondett. 1957. A simplified rotary fish screen and an automatic water gate. Trans. Amer. Fish. Soc. 86:371–380.

Wolf, P. 1951. A trap for the capture of fish and other organisms moving downstream. Trans. Amer. Fish. Soc. 80:41–45.

14

Role of Hatchery-reared Fish

Early fish culturists were extremely optimistic in predicting that the world's streams, rivers, and lakes could be repopulated with hatchery-reared fish, and fishermen shared their enthusiasm. In the late 1930s, and particularly in the 1940s, a large number of studies demonstrated that the survival rate of trout planted as "fry" (3 cm or less in length), was usually too low to justify such stocking, and that plantings of "fingerling" trout (5 to 16 cm long) often provided a much higher proportionate yield to the angler. With increased knowledge also came the realization that stocking in general is no panacea and is seldom a satisfactory substitute for natural reproduction. Artificial propagation, like habitat improvement, is simply another tool of fishery management.

Policy makers and program planners for fishery agencies need to weigh the demands of fishermen, the enthusiasm of fish culturists, and the technical information from fishery biologists in assigning priorities to state and regional fishery programs. This is a complex and difficult, but important, task. As fishing waters become increasingly crowded, planners must accommodate not only greater fishing pressure, but also greater varieties of fishing demands and preferences.

Survival

The large body of data on the relative survival of wild versus hatchery-reared trout clearly indicates that the naturally produced fish have higher survival rates than do hatchery fish of the same species except in unusually favorable environments (including the hatchery). Webster (1963) summarized the reasons for this with the statement: "Review and evaluation of a now substantial amount of research on the performance of hatchery-reared fish under laboratory and natural conditions leads to an inevitable conclusion that the hatchery operation may significantly change the animal it produces: chemically, physiologically, behaviorally, and/or genetically." Most fish artificially propagated in North America have been salmonids, simply because most members of this group are more amenable to hatchery culture than are other groups of game fishes such as centrarchids, percids, and esocids.

Many of the characteristics of hatchery trout that decrease their potential for survival after stocking are genetically fixed, the result of deliberate or inadvertent selection of brood stock in the hatchery. This selection tends to produce progeny with characteristics that favor survival in a hatchery environment but mitigate against good survival in a natural environment. Such selected characteristics often include:

1. Adaptation to hatchery diet.
2. Adaptation to hatchery environment, both physically (lack of stamina, lack of adaptability to environmental stresses), and behaviorally (docility, reduced fright reactions, reduced territoriality).
3. Fast growth, early maturity (sometimes associated with shorter life span).

In addition, various conditions of hatchery existence such as diet, overfeeding, crowding, lack of exercise, protection, handling, and treatment with antibiotics tend to produce fish that are less well-adapted physically and/or chemically to withstand the stresses of a natural environment.

Two other causes of low survival in hatchery fish, not directly associated with the hatchery environment, can be:

1. Planting time—governed by the economics of the hatchery operation, rather than by the best time for fish survival.
2. Planting in unsuitable habitats to satisfy demands of fishermen.

Survival of trout fry to maturity or to the creel has been negligible in streams and seldom better in lakes. Fry can be utilized for introductions to barren or reclaimed lakes, or as an initial introduction in any suitable water body, but they will usually contribute little toward maintaining an established population. Trout fingerlings, larger at stocking, survive better than the fry, particularly in lakes with inadequate natural spawning areas or nursery areas, or in reservoirs where rapid and sustained water fluctuations are routine. The Colorado stocking program in lakes and reservoirs is a good example of the wise use of fingerling fish. Maine also makes good use of fingerling salmon and lake trout in its stocking programs. Survival of fingerlings does not approach that of fish stocked at a "catchable" size, but the fingerlings cost much less to produce than do the larger fish. Hence the total cost of stocking, per fish harvested, may sometimes be lower for fingerlings than for fish stocked at a "catchable" size. Another advantage in stocking fingerling trout is that they more nearly resemble naturally produced fish than do "catchables."

Uses

Hatchery fish are used to:

1. Supplement fish populations in streams or lakes incapable of producing natural populations large enough to satisfy a large angling demand, such as may exist near heavily populated areas. California's catchable trout program is a good example and will be discussed later.
2. Maintain fish populations in bodies of water that lack adequate spawning sites or nursery areas. Since the Columbia River is dammed by multiple hydroelectric facilities, thus reducing successful migration of natural populations, hatchery-reared fish are being given increasing credit for maintaining sport and commercial fishing in this once naturally produced fishery.
3. Add new species to a fish population complex. For example, rainbow trout have added another dimension to fishing in several of New York's Finger Lakes, apparently without having depressed the production of endemic species. Several agencies have reported success with "two-story fisheries," a program in which trout are added to a lake or reservoir that is thermally stratified during summer months, thus providing trout in the hypolimnion or thermocline, and endemic "warmwater" species in the epilimnion.
4. Restock waters after some natural or man-made catastrophe such as winterkill or pollution.
5. Restock lakes or streams after chemical reclamation programs (see Chapter 17).
6. Provide fish for laboratory research.
7. Provide fish for commercial purposes. Private hatcheries support a large bait-minnow industry, and others grow species such as rainbow trout and channel catfish for the restaurant trade and for fee-fishing ponds.

Stocking Catchable Trout

"Catchable-trout stocking" typically provides a short-term, artificial fishery by planting trout (18 to 20 cm long) in waters where they are expected to be caught within a short time. Survival for more than a year is not expected, and some of the waters stocked may not support trout for more than a few weeks in the spring. Stocking catchable trout has become "a dragon" to some state fishery agencies and maintaining the programs can require enormous effort by fish culturists, fishery biologists, and administrators, as well as being very costly. Unfortunately, the benefits from stocking catchable trout are not equitably distributed among customers because, in many sport fisheries, a small proportion of the anglers (usually 10 percent or less) take a large proportion (50 percent or more) of the total catch. There is something to be said too about the loss of aesthetics of a natural fishery, and the greediness and poor sportsmanship that frequently accompany the crowding of anglers.

On the other hand, some advantages of the catchable programs are that:

1. Fish are caught rapidly and thus may be planted in the spring in waters that will later become unsuitable habitat for trout.
2. Immediate fishing is provided.
3. Such stocking yields a high percentage return on investment, in that a large proportion of the fish stocked are caught.
4. Fishing can be provided near large population centers, where it can furnish enjoyable outdoor recreation for large numbers of people—including family groups—who would be unable to obtain it otherwise.
5. Fishing can be provided for large concentrations of tourists at certain parks, public campsites, and the like.

The California Fish Bulletin, "Catchable Trout Fisheries" (Butler and Borgeson, 1965) contains an excellent analysis of catchable trout programs, and the findings are supported by information from many other states. Much of the following is taken from that bulletin. Mean return of catchable rainbow trout from 7 California lakes was 83 percent, and mean return from 13 streams was 73 percent. The average time interval required to harvest 75 percent of the total catch was 11 days in lakes, 6½ days in streams. The principal source of mortality was obviously fishing, and the harvest was very rapid. About 50 percent more trout had to be stocked in streams than in lakes to provide the same number of angling days in both habitats. These data support a policy of giving higher priority to lakes than to streams in a catchable trout stocking program. Larger trout gave better returns. In one experiment 18- to 20-cm fish provided a mean return of 62 percent and fish 21 to 23 cm long provided a mean return of 72 percent.

Catchable trout are usually much more vulnerable to angling than are wild fish of comparable size. Also, fishing pressure is much higher in a stocked catchable fishery than in a wild trout fishery. Fishing pressure tends to be proportional to the number stocked, so that increasing the number of fish in a planting increases the number of anglers who harvest it, instead of increasing the catch per angler. A situation known as "truck following" sometimes develops where a heavy planting of catchable trout from the hatchery truck results in an immediate onslaught by a horde of eager fishermen and a rapid depletion of the stock, with little fishing available thereafter until the next visit by the hatchery truck, and so on. This syndrome is most acutely manifest in small streams. Managers of catchable-trout fisheries have found that this "pulsing" can be minimized by determining the frequency and size of plantings which will result in the least rapid drop in catch per hour during the intervals between plantings. Because angling effort is far more important in reducing the stocked trout populations than is time, all intervals between plantings should include, as nearly as possible, equal amounts of angling effort regardless of their length in days or weeks. The harvest of catchable trout may also be distributed more evenly by including more than one species (of different angling vulnerabilities) in each planting, and by distributing a given planting as evenly and over as wide a geographical area as possible. In addition, many states are now de-emphasizing the use of catchable-trout programs in the smaller streams.

It is often suggested that streams be closed after stocking to give catchable trout an opportunity to become acclimated and distributed. Experiments indicate, however, that much less movement than might be expected takes place in the short period after stocking. Further, in many states it would be impossible to keep up with the thousands of stockings by posting closures and then returning to take down the signs. Closure after stocking creates additional problems—which have been shared by many states. Anglers traveling many miles to fish a favorite stream are not pleased to find it closed for several days following a recent planting.

The problem of how to allocate costs of a catchable-trout stocking program needs solution. If the money is taken from the fishery agency's general budget, then the warmwater fishermen and the trout fishermen who seek out and fish only wild trout are paying part of the bill for something they do not use. To finance the stocking of catchable trout by means of a special fee in the form of a "trout stamp" is a practical and fair method in states such as New Jersey where there are few year-round trout waters and few naturally produced trout. But in many states, where the bulk of trout fishing still comes from wild populations, a system of "taxing" all trout fishermen to benefit a minority clearly is inequitable.

The practice of stocking catchable-size fish continues to increase despite the management problems involved, and soon may be extended to warmwater species. Figures from California, a leader in experience with the catchable-trout program, indicate that over 10 million catchable trout weighing over 2.5 million pounds were planted in the 1969–1970 fiscal year. Further, the cost per pound of catchable rainbow trout was $0.68 or just about $0.18 per fish.

In summary, "catchable" trout fisheries have their place. They can—and should—provide much recreational fishing of one particular kind. In waters that can provide wild fish populations by natural reproduction, or semiwild populations by stocking fingerlings, the fishery should be managed so as to satisfy the needs of anglers who find "catchable" fisheries unattractive.

Costs

The fishery manager is more often concerned with actual stocking costs to an agency, rather than with retail prices. Colorado has released the following data on that state's costs for hatchery trout. Unlike the cost information given by many states, these figures include costs of rearing, transportation, stocking, and amortization of capital investment:

Size in inches	Price per hundred fish		
	Trout (brook, brown, or rainbow)	Largemouth bass	Bluegills
1 to 2	$ 4.00	—	$25.00
3 to 4	16.50	$50.00	—
5 to 6	26.00	—	40.00
8 to 9	52.50	—	—

In 1970, private fish hatcheries in New York State were charging the following retail prices (per hundred fish) for stock at the hatchery (delivery charges extra):

Stock	*Cost*
Fry (1″), 2,500 per pound	$20.25 per pound = $ 0.81 per 100
Spring fingerlings (2″).	
300 fish per pound	$10.00 per pound = $ 3.33 per 100
Catchable size (8″-9″),	
4 fish per pound	$ .85 per pound = $21.25 per 100

To more fully evaluate a stocking program, the cost per fish caught (C) can be determined, as follows:

$$C = \frac{\text{Number planted} \times \text{cost per fish planted}}{\text{Total number caught}} = \frac{\text{Cost per fish planted}}{\text{Rate of exploitation}}$$

California biologists used this formula to assess the relative costs of planting fingerling-size versus catchable-size rainbow trout in streams, with the following results (Calhoun, 1966, p. 179):

	Stocked as fingerlings	*Stocked as catchables*
Cost per fish at stocking	$0.018	$0.165
Rate of exploitation	0.024	0.700
Cost per trout caught	0.75	0.24

Webster et al. (1959) found that over a period of 7 consecutive years in Cayuga Lake, New York, the survival of lake trout planted as yearlings averaged four times greater than the survival of fish planted as fingerlings. However, there was no economic advantage in planting yearlings, because they cost the state at least four times as much to produce as did fingerlings.

No appraisal of the benefits and costs of a stocking program can be considered complete if it does not include information on the degree to which the hatchery fish are increasing the quantity and quality of fishing that would be afforded in that water body if no stocking were done. There is often a strong possibility that adding some biomass of hatchery fish is subtracting some biomass from production within the native fish population.

Trends and Goals

The National Task Force for Public Fish Hatchery Policy Report submitted by Chairman Alex Calhoun to the U.S. Fish and Wildlife Service in

1974 contains 31 recommendations likely to affect federal and state fish production and stocking policies. Much of the report concentrates on the federal-state relationship with a general theme of greater control by the states over fish production and stocking of public waters. State fishery departments see the federal roles as limited to "support" production with a much increased federal research program in fish nutrition, diseases, and genetics.

Salmonid production, particularly trout, clearly dominates fish culture in the United States. Only 92 of the 505 state and federally operated hatcheries are producing warmwater fish. Major stocking programs for the Pacific salmon and steelhead on the West Coast and the salmon and lake trout in the Great Lakes illustrate the need and emphasis on the coldwater fishes. More than 26 million trout and salmon were planted in the Great Lakes in 1976 including about 6.5 million lake trout, the principal species. Most of the warmwater fish production is in the eastern two-thirds of the country, largely with bluegills, largemouth bass, and channel catfish, followed by walleye, northern pike, muskellunge, and striped bass.

Fish culturists are continuing work to improve the survival and other characteristics of hatchery-reared trout through better fish-cultural methods, improved diets, and selective breeding. Efforts are being made to develop genetic strains of trout particularly well suited to fulfilling specific management objectives. This approach has already achieved some noteworthy results and holds much promise for further development. The basic problem here is to determine more exactly the causes for lower survival of hatchery fish than wild fish of the same species in the same environment. Areas for investigation include physiological, nutritional, or behavioral differences, some of which might be genetically fixed. Another problem is to find measurable characteristics of the hatchery fish that are reliable predictors of their survival after planting. Much research interest also is being focused on the use of marine areas to rear fishes such as Atlantic salmon, pompano, striped bass, and Pacific salmon for commercial purposes.

References

Atkinson, N. J. 1932. A study of comparative results from stocking barren lakes with rainbow trout. Trans. Amer. Fish. Soc. 62:197–200.

Bams, R. A. 1970. Evaluation of a revised hatchery method tested on pink and chum salmon fry. J. Fish. Res. Bd. Can. 28(8):1429–1452.

Bregnlalle, F. 1963. Trout culture in Denmark. Prog. Fish-Cult. 25(3):115–120.

Briggs, J. C. 1953. Behavior and reproduction of salmonid fishes in a small coastal stream. California Dep. Fish Game, Fish Bull. 94. 62 pp.

Buss, K. 1959. Trout and trout hatcheries of the future. Trans. Amer. Fish. Soc. 88(2):75-80.

Buss, K., D. R. Graff, and E. R. Miller. 1970. Trout culture in vertical units. Prog. Fish-Cult. 32(4):187-191.

Buss, K., and J. E. Wright. 1958. Appearance and fertility of trout hybrids. Trans. Amer. Fish. Soc. 87:172-181.

Butler, R. L., and D. P. Borgeson. 1965. California "catchable" trout fisheries. California Dep. Fish Game, Fish Bull. 127. 47 pp.

Calhoun, A., ed. 1966. Inland fisheries management. California Dep. Fish Game, Sacramento. 546 pp.

_____. 1976. The 1974 report of a special task force on the public fish hatchery policy. Fisheries, Bull. Amer. Fish. Soc. 1(3):23-26.

Cobb, E. W. 1933. Results of trout tagging to determine migrations and results from plants made. Trans. Amer. Fish. Soc. 63:308-312.

Cooper, E. L. 1952. Rate of exploitation of wild eastern brook trout and brown trout populations in the Pigeon River, Otsego County, Michigan. Trans. Amer. Fish. Soc. 81:224-234.

_____. 1953a. Returns from plantings of legal-sized brook, brown, and rainbow trout in the Pigeon River, Otsego County, Michigan. Trans. Amer. Fish. Soc. 82:265-280.

_____. 1953b. Mortality rates of brook trout and brown trout in the Pigeon River, Otsego County, Michigan. Prog. Fish-Cult. 15(4):163-169.

Curtis, B. 1951. Yield of hatchery trout in California lakes. California Fish Game J. 37(2):197-215.

Flick, W. A., and D. A. Webster. 1964. Comparative first year survival and production in wild and domestic strains of brook trout, *Salvelinus fontinalis*. Trans. Amer. Fish. Soc. 93(1):58-69.

Foerster, R. E. 1938. An investigation of the relative efficiencies of natural and artificial propagation of sockeye salmon (*Oncorhynchus nerka*) at Cultus Lake, British Columbia. J. Fish. Res. Bd. Can. 4(3):151-161.

Green, D. M., Jr. 1964. A comparison of stamina of brook trout from wild and domestic parents. Trans. Amer. Fish. Soc. 93(1):96-100.

Greene, C. W. 1952. Results from stocking brook trout of wild and hatchery strains at Stillwater Pond. Trans. Amer. Fish. Soc. 81:43-52.

Hale, J. G., and L. L. Smith, Jr. 1955. Results of planting catchable-size brown trout, *Salmo trutta fario* L., in a stream with poor natural reproduction. Prog. Fish-Cult. 17(1):14-19.

Hansen, M. J., and T. M. Stauffer. 1971. Comparative recovery to the creel, movement and growth of rainbow trout stocked in the Great Lakes. Trans. Amer. Fish.Soc. 100(2):336-349.

Harkness, W. J. K. 1949. Catches of speckled trout from the plantings of hatchery raised fish in private waters of Ontario. Trans. Amer. Fish. Soc. 70:410-413.

Hatch, R. W. 1957. Success of natural spawning of rainbow trout in the Finger Lakes region of New York. New York Fish Game J. 4(1):69-87.

Hunt, R. L. 1969. Over winter survival of wild fingerling brook trout in Lawrence Creek, Wisconsin. J. Fish. Res. Bd. Can. 26(6):1473-1483.

King, W. 1942. Trout management studies at Great Smoky Mountains National Park. J. Wildl. Manage. 6(2):147–161.

King, W., and A. Holloway. 1952. Management of trout waters in the national forests of the southern Appalachians. Trans. Amer. Fish. Soc. 81:53–62.

Lachner, E. A., C. R. Robins, and W. R. Courtenay, Jr. 1970. Exotic fishes and other aquatic organisms introduced into North America. Smithsonian Contrib. Zool. 59, Washington, D.C. 29 pp.

Lane, J. P. 1957. Some angler preferences with respect to size and species of trout. California Dep. Fish Game, Inland Fish. Adm. Rep. 57-24. 19 pp. Mimeogr.

Lapworth, E. D. 1956. The effect of fry plantings on whitefish production in Easter Lake, Ontario. J. Fish. Res. Bd. Can. 13(4):547–558.

Leitritz, E. 1959. Trout and salmon culture. California Dep. Fish Game, Fish Bull. 107. 169 pp.

Loftus, K. H., convener. 1968. A symposium on introductions of exotic species. Ontario Dep. Lands For., Res. Rep. 82. 111 pp.

Lord, R. F. 1935. Trout harvest from Furnace Brook, Vermont's "test stream." Trans. Amer. Fish. Soc. 65:224–233.

———. 1946. The Vermont "test-water" study, 1935 to 1945 inclusive. Vermont Fish Game Serv., Montpelier, Fish. Res. Bull. 2. 110 pp.

McCrimmon, H. R. 1950. The reintroduction of Atlantic salmon into tributary streams of Lake Ontario. Trans. Amer. Fish. Soc. 78:128–132.

McFadden, J. T. 1961. A population study of the brook trout, *Salvelinus fontinalis.* Wildl. Monogr. 7. 73 pp.

Maciolek, J. A., and P. R. Needham. 1952. Ecological effects of winter conditions on trout and trout foods in Convict Creek, California, 1951. Trans. Amer. Fish. Soc. 81:202–217.

Miller, R. B. 1952. Survival of hatchery-reared cutthroat trout in an Alberta stream. Trans. Amer. Fish. Soc. 81:35–42.

———. 1959. Diet, glycogen reserves and resistance to fatigue in hatchery rainbow trout. J. Fish. Res. Bd. Can. 16(3):321–328.

Mottley, C. M. 1940. The production of rainbow trout at Paul Lake, British Columbia. Trans. Amer. Fish. Soc. 69:187–191.

Needham, P. R. 1939. Natural propagation versus artificial propagation in relation to angling. Trans. N. Amer. Wildl. Conf. 4:326–331.

———. 1949. Survival of trout in streams. Trans. Amer. Fish. Soc. 77:26–31.

Needham, P. R., and F. K. Kramer. 1943. Movement of trout in Convict Creek, California. J. Wildl. Manage. 7(2):142–148.

Needham, P. R., J. W. Moffett, and D. W. Slater. 1945. Fluctuations in wild brown trout populations in Convict Creek, California. J. Wildl. Manage. 9(1):9–25.

Needham, P. R., and D. W. Slater. 1944. Survival of hatchery-reared brown and rainbow trout as affected by wild trout populations. J. Wildl. Manage. 8(1):22–36.

———. 1945. Seasonal changes in growth, mortality, and conditions of rainbow trout following planting. Trans. Amer. Fish. Soc. 73:117–124.

Needham, P. R., and F. K. Sumner. 1942. Fish management problems of high western lakes with returns from marked trout planted in Upper Angora Lake, California. Trans. Amer. Fish. Soc. 71:249–269.

Nielson, R. S., N. Reimers, and H. D. Kennedy. 1957. A six-year study of the survival and vitality of hatchery-reared rainbow trout of catchable size in Convict Creek, California. California Fish Game J. 43(1):5–42.

Parsons, J. W. 1957. The trout fishery of the tailwater below Dale Hollow Reservoir. Trans. Amer. Fish. Soc. 85:75–92.

Pycha, R. L., and G. R. King. 1967. Returns of hatchery-reared lake trout in southern Lake Superior, 1955–1962. J. Fish. Res. Bd. Can. 24(2):281–298.

Ratledge, H. M., and J. H. Cornell. 1953. Migratory tendencies of the Manchester (Iowa) strain of rainbow trout. Prog. Fish-Cult. 15(2):57–63.

Rawson, D. S. 1941. The eastern brook trout in the Maligne River System, Jasper National Park. Trans. Amer. Fish. Soc. 70:221–235.

Schuck, H. A. 1945. Survival, population density, growth, and movement of the wild brown trout in Crystal Creek. Trans. Amer. Fish. Soc. 73:209–230.

Schuck, H. A., and O. R. Kingsbury. 1948. Survival and growth of fingerling brown trout (*Salmo fario*) reared under different hatchery conditions and planted in fast and slow water. Trans. Amer. Fish. Soc. 75:147–156.

Shetter, D. S. 1939. Success of plantings of fingerling trout in Michigan waters as demonstrated by marking experiments and creel censuses. Trans. N. Amer. Wildl. Conf. 4:318–325.

———. 1947. Further results from spring and fall plantings of legal-sized, hatchery-reared trout in streams and lakes of Michigan. Trans. Amer. Fish. Soc. 74:35–58.

———. 1950. Results from plantings of marked fingerling brook trout (*Salvelinus f. fontinalis* Mitchill) in Hunt Creek, Montmorency County, Michigan. Trans. Amer. Fish. Soc. 79:77–93.

Shetter, D. S., and A. S. Hazzard. 1941. Results from plantings of marked trout of legal size in streams and lakes of Michigan. Trans. Amer. Fish. Soc. 70:446–468.

Shetter, D. S., and J. W. Leonard. 1943. A population study of a limited area in a Michigan trout stream, September, 1940. Trans. Amer. Fish. Soc. 72:35–51.

Smith, L. L., and B. S. Smith. 1945. Survival of seven- to ten-inch planted trout in two Minnesota streams. Trans. Amer. Fish. Soc. 73:108–116.

Smith, O. R. 1947. Returns from natural spawning of cutthroat trout and eastern brook trout. Trans. Amer. Fish. Soc. 74:281–296.

Smith, S. B. 1957. Survival and growth of wild and hatchery rainbow trout (*Salmo gairdneri*) in Corbett Lake, British Columbia. Can. Fish Cult. 20:7–12.

Stockley, C. 1954. New method of artificially planting salmon eggs. Prog. Fish-Cult. 16(3):137–138.

Surber, E. W. 1937. Rainbow trout and bottom fauna production in one mile of stream. Trans. Amer. Fish. Soc. 66:193–202.

Thorpe, L. M., H. J. Rayner, and D. A. Webster. 1947. Population depletion in brook, brown, and rainbow trout stocked in the Blackledge River, Connecticut in 1942. Trans. Amer. Fish. Soc. 74:166–187.

U.S. Fish and Wildlife Service. 1968. National survey of needs for hatchery fish. Bur. Sport Fish. Wildl., Resour. Publ. 63. 71 pp.

Vincent, R. E. 1960. Some influences of domestication upon three stocks of brook trout (*Salvelinus fontinalis* Mitchill). Trans. Amer. Fish. Soc. 89(1):35–52.

Wagner, H. H. 1969. Effect of stocking location of juvenile steelhead trout, *Salmo gairdneri,* on adult catch. Trans. Amer. Fish. Soc. 98(1):27–34.

Wales, J. A. 1946. Castle Lake trout investigation. California Fish Game J. 32(3):109–143.

_____. 1947. Castle Lake trout investigations: 1946 catch and chemical removal of all fish. California Fish Game J. 33(4):267–286.

Webster, D. A. 1963. Improvements of survival in hatchery strains of trout. Cornell Univ., Dep. Nat. Resour. Mimeogr. 60-65. 5 pp.

Webster, D. A., W. G. Bentley, and J. P. Galligan. 1959. Management of the lake trout fishery of Cayuga Lake, New York, with special reference to the role of hatchery fish. Cornell Univ., New York State Coll. Agric. Exp. Stn., Memoir 357. 84 pp.

Wood, E. M. 1953. A century of American fish culture, 1853–1953. Prog. Fish-Cult. 15(4):147–162.

15

Stocking

Stocking policies must be based on extensive biological research and follow-up, and must be readily understandable by conservation agency personnel and by the public. Above all, the policies must remain flexible. A key objective of any stocking program should be to maximize fulfillment of the public's desire for recreational fishing. To formulate sound stocking recommendations, the fishery manager should have as much information as possible on the following:

1. Other management measures, aside from stocking, that may be undertaken to achieve the basic objectives of the fishery (other population manipulations, habitat improvements, regulatory measures).
2. Species for which the habitat is particularly well suited. Information is needed on physical, chemical, and biological characteristics of the habitat (see Chapter 1). Fish should never be stocked where populations of suitable fish are already utilizing the carrying capacity of the water body. Particular care must be taken to avoid introducing species that will prey on desired species already present, or compete with them for food, shelter, or spawning habitat. The introduction of the carp represents a classic violation of this principle, and almost every geographical area in the United States can provide an illustration of an unwise introduction.
3. Characteristics of fishing demands. These can vary widely with geographical region, population density, and various socioeconomic factors, such as the kind(s), relative abundance, and sizes of fish desired, the quality of fishing experience desired (for example, wilderness fishing vs. put-and-take), and the diversity of recreational fishing demands.

4. Costs of different stocking techniques.
5. Characteristics of the species and strain of fish being considered for stocking—their growth, longevity, reproductive capacity, vulnerability to angling—relative to characteristics of the habitat and the demand. To illustrate: for a put-and-take trout fishery subject to heavy exploitation in a low-quality stream, a trout strain with rapid hatchery growth and high vulnerability to angling is needed. Here longevity and reproductive capacity are of no importance, whereas they are likely to be paramount characteristics of the strain selected for stocking high-quality trout waters in remote areas.

Stocking Tables

Various systems have been and continue to be developed for estimating the correct number or weight of fish to stock in a given water body. Several of these are described below. In all of them the basic approach is to determine the number or weight of fish to stock that will provide the desired compromise between maximum population density and carrying capacity of the environment in question.

One of the first systems for formulating biologically sound stocking recommendations was that developed for trout streams by Dr. George C. Embody in 1927. This system combined information on the average stream width in feet, the relative size, depth, and frequency of pools, and the relative abundance of food, for use in a table to determine the numbers per mile of various size fingerling trout to stock yearly. An extension of Embody's stocking table is given in Table 15-1, taking into account the work of Leger (1910) which indicates that productivity per unit area decreases as stream width increases.

The number of 3-inch fingerlings to be planted yearly and per mile for streams of various widths and for different combinations of pool and food grade conditions can be read from this table. To calculate the number of 1-, 4-, 6-, or 10-inch trout to stock, multiply the tabular value by the appropriate "Factor" listed at the bottom of the table. This factor is based on the expected mortality at different sizes, listed below.

Size, in inches	1	2	3	4	6
Expected mortality, in percent	95	65	40	20	0

Pool conditions in the stream are classified A, B, or C, taking into account the following qualities:

Size

1. Pools having an average width more than double that of the stream.
2. Pools having a width about equal to that of stream.
3. Pools much narrower than stream.

Type

1. Deep (2 feet or more), exposed pools containing a great luxuriance of aquatic plants harboring a rich fauna, or deep pools with abundant shelter (overhanging banks, logs, roots, boulders), much drift or detritus, shaded by forest cover or shrubs.
2. Pools intermediate in depth, shelter, plant abundance, etc.
3. Shallow exposed pools without shelter and without plants; or scouring basins.

Frequency

1. More or less continuous pools.
2. Rather close succession of pools and rapids.
3. Pools infrequent with long stretches of swift water between them.

If we let S refer to size, T to type, and F to frequency, then a combination of $S1$-$T1$-$F1$ would receive the highest rating and $S3$-$T3$-$F3$ the lowest. Any combination in which the numerals add to 3 or 4 could be considered Grade A, any combination adding to 5, 6, or 7 could be considered Grade B, and any combination adding to 8 or 9 could be considered Grade C.

Food grades (columns 1, 2, and 3 in Table 15–1) are defined as follows:

Grade 1. Richest in food. Current moderate to sluggish; bottom of mud, silt, or detritus; extensive areas covered with watercress, moss, *Chara*, or other plants.

A rubble bottom in swift water with mayflies, stoneflies, and caddisflies occurring in great abundance.

A stream margined with willows, alders, button bush, and tall weeds such as goldenrod, with a large congregation of terrestrial insects.

Grade 2. Average quantity of food. Generally rapid current with bottom of coarse gravel and flat rocks, or sluggish with muck or silt bottom, with no submerged plants; exposed pools with few sheltering trees or shrubs.

Grade 3. Poor in food. Bottom of fine gravel, sand, hardpan, or clean bedrock; without vegetation; exposed pools; mayflies, stoneflies, caddisflies uncommon or rare.

Table 15-1. Extension of Embody's revised planting table for trout streams. (Embody, 1927)

Stream width, feet	Pool Grade A			Pool Grade B			Pool Grade C		
	1*	2	3	1	2	3	1	2	3
1	144	117	90	117	90	63	90	63	36
2	288	234	180	234	180	126	180	126	72
3	432	351	270	351	270	189	270	189	108
4	576	468	360	468	360	252	360	252	142
5	720	585	450	585	450	315	450	315	180
6	864	702	540	702	540	378	540	378	216
7	1008	819	630	819	630	441	630	441	252
8	1152	936	720	936	720	504	720	504	284
9	1296	1053	810	1053	810	567	810	567	324
10	1440	1170	900	1170	900	630	900	630	360
11	1584	1287	990	1287	990	693	990	693	396
12	1728	1404	1080	1404	1080	756	1080	756	432
13	1872	1521	1170	1521	1170	819	1170	819	468
14	2016	1638	1260	1638	1260	882	1260	882	504
15	2160	1755	1350	1755	1350	945	1350	945	540
16	2304	1872	1440	1872	1440	1008	1440	1008	576
17	2376	1930	1485	1930	1485	1039	1485	1039	594
18	2448	1989	1530	1989	1530	1071	1530	1071	612
19	2520	2047	1575	2047	1575	1102	1575	1102	630
20	2592	2106	1620	2106	1620	1134	1620	1134	648

For streams over 20 feet in width, use formula $\frac{1}{2}N_1 W + 8N_1 = X$, where N_1 = number fingerlings for stream 1 foot wide, W = average width, and X = number to be stocked per mile. The above table refers to 3-inch fingerlings only. To find the number of other sizes multiply the number of fish given for the stream width in question by the following factors (dependent on size):

Length, inches	1	3	4	6	10
	Fry	Fing.	Fing.	Legal	Adult
Factor	12	1	0.75	0.6	0.3

*The figures 1, 2, 3 at the heads of columns indicate the food grade.

The relative abundance of the various groups should be indicated on field banks by underscoring once if the animal is uncommon or rare, twice if common, and three times if abundant. At the same time a grade of 1, 2, or 3 should be recorded, indicating total richness of the whole stream in food.

Many variations of the basic stocking table have been introduced to reflect different environmental conditions, different fishing needs, and different management approaches. Examples of such modifications are described by Cooper (1948), Davis (1938), and Smith and Moyle (1944).

Other Stocking Systems

In West Virginia, each trout water is graded (essentially, ranked) according to its relative fishing pressure and its relative environmental suitability, and these grades are combined with acreage in a formula used to allocate the year's supply of catchable trout available for stocking among the state's trout waters. The stocking program is described in a small, easily read brochure available from the West Virginia Department of Natural Resources. West Virginia's stocking system does not apply to surplus stocking of fingerlings. These fish are stocked in accordance with recommendations made as a result of biological surveys to assess their best use. The stocking system for catchable-size trout (20 cm or longer) is based on three criteria:

1. Surface acreage factor—acres of trout water in each stream section or lake stocked.
2. Fishing pressure factor—graded into light (0.1), average (0.5), or heavy (1.0).
3. Water suitability factor—on a scale ranging from low (0.5) through high (1.0). A rating less than 0.5 is given to very low quality trout waters which contain catchable populations of warmwater fish. Most of these waters are located in the southern and northwestern parts of the state.

The three factors (acreage, pressure, and suitability) multipled together produce a rating factor. All rating factors, for each individual water, are added for a total figure which is divided into the total estimated production of catchable trout for the season, which in turn gives the planting factor for each water in the state. Multiplying the rating factor by the planting factor gives the total pounds of trout the water will receive during the stocking season. An example might be as shown below.

Estimated pounds of trout to be produced = 450,000
Total of all rating factors = 3,000
Planting factor = 450,000/3,000 = 150 pounds of trout
In a 10-acre stream-stocking section, the acreage factor = 10
Fishing pressure factor = 0.5 (average)
Water suitability factor = 1.0 (high)
$10 \times 0.5 \times 1.0 = 5$ rating factor
Thus, multiply $5 \times 150 = 750$ pounds stocking allotment for season

In West Virginia, stocking records are kept on keysort cards which can be easily punched for rapid sorting. These cards contain specific information as to the exact area to stock, the time it takes, species preference, the names of interested persons in the area to contact, the frequency of stocking, and the calculations used to arrive at allotment.

Colorado uses a lake-stocking system based on an estimate that combines data on carrying capacity and fishing pressure. Biologist W. C. Nelson of the Research Section of Colorado Division of Wildlife has de-

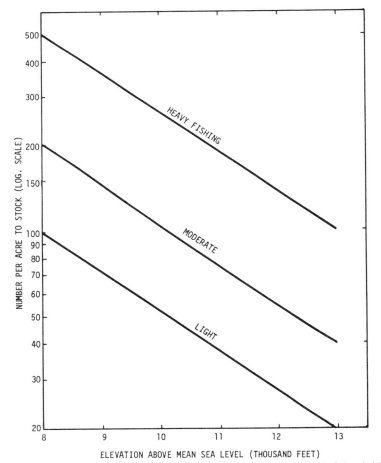

Figure 15-1. Proposed high-lake stocking schedule of fingerling (5 cm) trout in Colorado lakes for three levels of fishing pressure. Courtesy of W. C. Nelson, 1971, personal communication.

veloped and proposed a high-lake stocking schedule for fingerling (5 cm) trout. Based on research from several lakes, the schedule uses lake elevation as a convenient index of the environmental carrying capacity, on the assumption that the higher the lake the fewer fish it will support. The stocking rate proposed decreases by about 28 percent for every 1,000-foot increase in elevation. Stocking recommendations for conditions of light, moderate, and heavy fishing pressure are given in Figure 15–1 below. The figures given are intended to apply only to fingerling fish planted in July in areas where good survival has been demonstrated. The point is made and well taken that positive harm can result from overstocking of lightly fished lakes. This Colorado stocking schedule is an excellent example of research information being condensed into easily understood essentials.

Maine has developed a comprehensive system for stocking lakes that takes into account, besides surface acreage, the stocking purpose (introductory, maintenance or supplemental, experimental, or catchable), the size of fish planted, and the expected degree of competition and/or predation. To illustrate: landlocked salmon (*Salmo salar*) are stocked to supplement the salmon populations in certain lakes that are judged capable of supporting larger salmon populations than those maintained by natural reproduction. Salmon are stocked either as fall fingerlings, at rates up to 10 per acre (25 per hectare); as spring yearlings, up to 5 per acre (12 per hectare); or as fall yearlings, up to 2.5 per acre (6 per hectare). The more severe the expected competition or predation, the larger the number of salmon stocked.

Stocking to Provide a Mean Population Density or Mean Biomass

Often the biologist can estimate, on the basis of past information or experience, the approximate rates of total mortality and growth to expect in a fish population during a given period of time. In such cases one can estimate the number or weight of fish to plant that should provide the desired *mean* population density or biomass during that time period. A method is described below for making these determinations.

Basic assumptions in the procedure are, as described by Ricker (1958), that rates of growth, fishing mortality, and natural mortality are all exponential, and have parallel fluctuations with respect to time. Mean population density $(\bar{N})$ during the time (t) interval, $t = 0$ to $t = 1$, is a function of the initial population density (N_o), the total mortality rate $(1 - S = A)$, and the coefficient of total mortality (Z):

$$\bar{N} = N_o \cdot \frac{A}{Z}$$

The mean biomass $(\bar{P})$ is a function of the initial biomass (P_o), the coefficient of total mortality (Z), and the coefficient of growth (g):

$$\bar{P} = P_o \cdot \frac{e^{g-Z} - 1}{g - Z}$$

These formulae, plus the mortality relationship:

$$\frac{Z}{A} = \frac{F}{E} = \frac{M}{D}$$

as defined in Chapter 7, provide the basis for the following table and graphs used to estimate stocking and mortality rates. Table 15–2 gives the number of fish to stock (N_o) to obtain any desired mean population density at different levels of expected total mortality rate. Total mortality rate (or its complement, survival), may be estimated from catch-curve data, from recoveries of marked members in successive years, or by other methods described in Chapter 7.

Example of use. Green River has a rainbow trout fishery that is maintained entirely by stocking legal-size fish each spring. Total mortality rate is about 0.80 each year, and the problem is to determine how many fish should be planted so that their mean population density will be 250 per hectare during the year following stocking. Using the tabular data ($N = 250$, $A = 0.80$) we find that 503 per hectare should be stocked. (The "annual" mortality rate really denotes the mortality rate for the *period under consideration*—not necessarily a calendar year). So, as in the preceding example, if we expected the total mortality rate during the fishing season to be 0.65, and wished to provide a mean population density of 350 trout per hectare during the fishing season, we would stock 565 trout per hectare.

Figure 15–2 shows the weight of fish to stock to obtain a desired mean biomass during the time interval in question, when the expected coefficients of growth (g) and total mortality (Z) can be estimated.

Example of use: Most New York farm ponds can support brook trout populations of about 75 kilograms per hectare. Total mortality rate (A) averages 0.75 each year; therefore (from the Appendix Table to this chapter) $Z = 1.39$ per year. Trout stocked as 45-gram fall fingerlings average 225 grams one year later, so their growth coefficient (g) for the one-year period is:

$$\log_e \left[\frac{225}{45} \right] = 1.61$$

Table 15-2. Numbers of fish to stock (per unit area) to obtain a desired mean population density, at various levels of total mortality rate. (A. W. Eipper)

Mean population density N	Total mortality rate, A																		
	0.05	0.10	0.15	0.20	0.25	0.30	0.35	0.40	0.45	0.50	0.55	0.60	0.65	0.70	0.75	0.80	0.85	0.90	0.95
50	51	53	54	56	58	59	62	64	66	69	73	76	81	86	92	101	112	128	158
100	103	105	108	112	115	119	123	128	133	139	145	153	162	172	185	201	223	256	315
150	154	158	163	167	173	178	185	192	199	208	218	229	242	258	277	302	335	384	473
200	205	211	217	223	230	238	246	255	266	277	290	305	323	344	370	402	446	512	631
250	256	263	271	279	288	297	308	319	332	347	363	382	404	430	462	503	558	640	788
300	308	316	325	335	345	357	369	383	399	416	436	458	485	516	555	604	670	768	946
350	359	369	379	391	403	416	431	447	465	485	508	535	565	602	647	704	781	895	1104
400	410	421	433	446	460	476	492	511	531	555	581	611	646	688	739	805	893	1023	1261
450	462	474	488	502	518	535	554	575	598	624	653	687	727	774	832	905	1004	1151	1419
500	513	527	542	558	575	594	615	639	664	693	726	764	808	860	924	1006	1116	1279	1577
550	564	579	596	614	633	654	677	702	731	762	799	840	888	946	1017	1106	1228	1407	1734
600	616	632	650	669	690	713	738	766	797	832	871	916	969	1032	1109	1207	1339	1535	1892
650	667	685	704	725	748	773	800	830	864	901	944	993	1050	1118	1201	1308	1451	1663	2050
700	718	738	758	781	806	832	862	894	930	970	1016	1069	1131	1204	1294	1408	1562	1791	2207
750	769	790	813	837	863	892	923	958	996	1040	1089	1145	1211	1290	1386	1509	1674	1919	2365
800	821	843	867	893	921	951	985	1022	1063	1109	1161	1222	1292	1376	1479	1609	1786	2047	2523
850	872	896	921	948	978	1011	1046	1086	1129	1178	1234	1298	1373	1462	1571	1710	1897	2175	2680
900	923	948	975	1004	1036	1070	1108	1149	1196	1248	1307	1374	1454	1548	1664	1811	2009	2303	2838
950	975	1001	1029	1060	1093	1129	1169	1213	1262	1317	1379	1451	1534	1634	1756	1911	2120	2431	2996
1000	1026	1054	1083	1116	1151	1189	1231	1277	1329	1386	1452	1527	1615	1720	1848	2012	2232	2558	3153
1050	1077	1106	1138	1172	1208	1248	1292	1341	1395	1456	1524	1604	1696	1806	1941	2112	2344	2686	3311
1100	1128	1159	1192	1227	1266	1308	1354	1405	1461	1525	1597	1680	1777	1892	2033	2213	2455	2814	3469
1150	1180	1212	1246	1283	1323	1367	1415	1469	1528	1594	1670	1756	1857	1978	2126	2314	2567	2942	3626
1200	1231	1264	1300	1339	1381	1427	1477	1532	1594	1664	1742	1833	1938	2064	2218	2414	2678	3070	3784
1250	1282	1317	1354	1395	1438	1486	1539	1596	1661	1733	1815	1909	2019	2150	2310	2515	2790	3198	3942
1300	1334	1370	1408	1450	1496	1546	1600	1660	1727	1802	1887	1985	2100	2236	2403	2615	2901	3326	4099
1350	1385	1422	1463	1506	1553	1605	1662	1724	1794	1871	1960	2062	2180	2322	2495	2716	3013	3454	4257
1400	1436	1475	1517	1562	1611	1664	1723	1788	1860	1941	2033	2138	2261	2408	2588	2817	3125	3582	4415
1450	1488	1528	1571	1618	1669	1724	1785	1852	1926	2010	2105	2214	2342	2494	2680	2917	3236	3710	4572
1500	1539	1580	1625	1674	1726	1783	1846	1916	1993	2079	2178	2291	2423	2580	2773	3018	3348	3838	4730
1550	1590	1633	1679	1729	1784	1843	1908	1979	2059	2149	2250	2367	2503	2666	2865	3118	3459	3966	4888
1600	1641	1686	1734	1785	1841	1902	1969	2043	2126	2218	2323	2443	2584	2752	2957	3219	3571	4093	5045
1650	1693	1738	1788	1841	1899	1962	2031	2107	2192	2287	2396	2520	2665	2838	3050	3319	3683	4221	5203
1700	1744	1791	1842	1897	1956	2021	2092	2171	2258	2357	2468	2596	2746	2924	3142	3420	3794	4349	5361
1750	1795	1844	1896	1953	2014	2081	2154	2235	2325	2426	2541	2673	2826	3010	3235	3521	3906	4477	5518
1800	1847	1896	1950	2008	2071	2140	2215	2299	2392	2495	2613	2749	2907	3096	3327	3621	4017	4605	5676
1850	1898	1949	2004	2064	2129	2199	2277	2363	2458	2565	2686	2825	2988	3182	3420	3722	4129	4733	5834
1900	1949	2002	2059	2120	2186	2259	2339	2426	2524	2634	2758	2902	3069	3268	3512	3822	4241	4861	5991
1950	2000	2055	2113	2176	2244	2318	2400	2490	2591	2703	2831	2978	3149	3354	3604	3923	4352	4989	6149
2000	2052	2107	2167	2231	2301	2378	2462	2554	2657	2773	2904	3054	3230	3440	3697	4024	4464	5117	6307

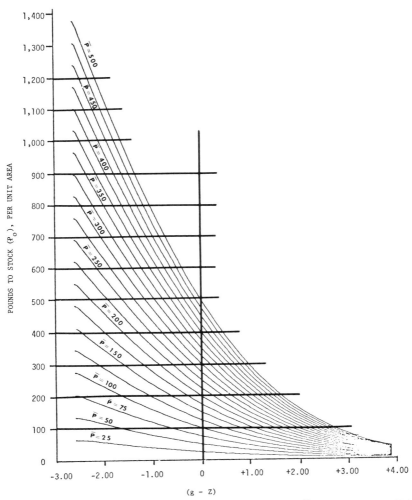

Figure 15-2. Graphic relationships between mean biomass desired ($\bar{P}$) and initial biomass (P_o) at various levels of difference between growth and mortality coefficient (g-Z). (A. W. Eipper)

Hence the value of $(g - Z) = 1.61 - 1.39 = + 0.22$. From the graph, we find that to provide a mean biomass of 75 kilograms per hectare during the year, we should stock 67 kilograms of the fall fingerlings. If the fish average 45 grams at stocking, this would mean planting 67,000/45 or 1,489 fingerlings per hectare.

In some fish populations the prevailing rate of natural mortality $(1 - e^{-M})$ may be approximately known (see estimation methods of Chapter 7), and the fishery manager may wish to determine what total mortality rates (A) will be produced by different rates of exploitation (E). In such cases A cannot be computed readily from the relationship $F/E = M/D = Z/A$, and the graph shown in Figure 15-3 can be useful. Points for these curves can be derived from the relationship:

$$1 - e^{-M} = 1 - S^{\frac{A - E}{A}}$$

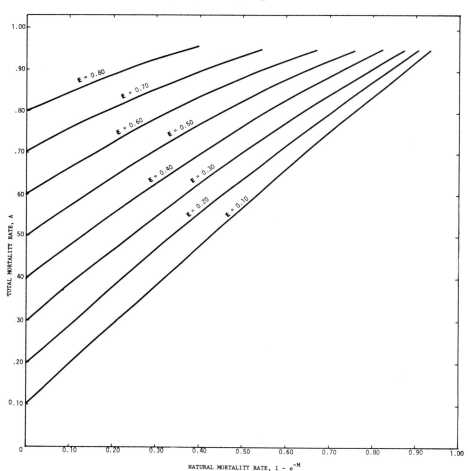

Figure 15-3. Relationships between total mortality rate, natural mortality rate, and exploitation rate (E). (A. W. Eipper)

Example of use: In Oxtooth Reservoir the natural mortality rate of wall-eyes (*Stizostedion vitreum*) is fairly constant at about 0.10 per year, and the annual rate of exploitation (*E*) averages 0.30. The State Fish and Game Department is contemplating regulatory changes that might double the catch and needs to know how this change would affect the total mortality rate. In Figure 15-3 we see that when the natural mortality rate is 0.10, changing *E* from 0.30 to 0.60 would increase total mortality from about 0.38 to 0.66. Such information, in turn, could be used with Table 15-2 or Figure 15-2 to determine the stocking rate needed if the catch were doubled. Alternatively, Figure 15-3 could be used to determine the maximum rate of exploitation to allow, given the maximum total mortality rate desired and an approximately known natural mortality rate.

The first step in formulating a stocking system is to provide a rational basis for determining the amount of fish to stock, by utilizing knowledge of the habitat, the fish population (or both), and the kinds and quantities of demand for fishing within all segments of the public—not just those who are already fishing. Survey techniques for making such unbiased estimates of angling demand are only now being developed. Stocking is an important fishery-management tool, but it is important to keep it in perspective. It is one member of that category of fish management measures called fish population manipulations. In terms of management priorities, habitat maintenance and improvement techniques take precedence over population manipulations. Habitat limits both the quality and the quantity of fishing that stocking can provide.

References

Bond, L. H. 1970. Guidelines to fish stocking in Maine. Maine Fish Game 12(3):22–25.

Cooper, G. P. 1948. Fish stocking policies in Michigan. Trans. N. Amer. Wildl. Conf. 13:187–198.

Davis, H. S. 1938. Instructions for conducting stream and lake surveys. U.S. Fish Wildl. Serv., Bur. Fish., Fish. Circ. 26. 55 pp.

Embody, G. C. 1927. An outline of stream study and the development of a stocking policy. Cornell Univ., Aquicultural Lab. 21 pp.

Fraser, J. M. 1972. Recovery of planted brook trout, splake, and rainbow trout from selected Ontario lakes. J. Fish. Res. Bd. Can. 29(2):129–142.

James, M. C., O. L. Meehan, and E. J. Douglass. 1944. Fish stocking as related to the management of inland waters. U.S. Fish Wildl. Serv., Cons. Bull. 35. 22 pp.

Kaesler, R. L., J. Cairns, Jr., and J. S. Crossman. 1974. Redundancy in data from stream surveys. Water Res. 8:637–642.

Larson, R., and J. M. Ward. 1955. Management of steelhead trout in the State of Washington. Trans. Amer. Fish. Soc. 84:261–274.

Leger, L. 1910. Principes de la methode rationnelle du peuplement des cours d'eau à salmonides. Univ. Grenoble, Trav. Lab. Piscic., fasc. 1:531–602.

Miller, R. B. 1954. Comparative survival of wild and hatchery-reared cutthroat trout in a stream. Trans. Amer. Fish. Soc. 83:120–130.

Mottley, C. M. 1942. Experimental designs for developing and testing a stocking policy. Trans. N. Amer. Wildl. Conf. 7:224–232.

Moyle, J. B. 1956. Relationships between the chemistry of Minnesota surface waters and wildlife management. J. Wildl. Manage. 20(3):303–320.

Newell, A. E. 1958. Trout stream management investigations of the Swift River watershed in Albany, New Hampshire. New Hampshire Fish Game Dep., Surv. Rep. 7. 40 pp.

Ratledge, H. M., and J. H. Cornell. 1952. The effect of trout stocking on the rate of catch. Prog. Fish-Cult. 14(3):117–121.

Ricker, W. E. 1958. Handbook of computations for biological statistics of fish populations. Fish. Res. Bd. Can., Bull. 119. 300 pp.

Smith, L. L., Jr., and J. B. Moyle. 1944. A biological survey and fishery management plan for the streams of the Lake Superior north shore watershed. Minnesota Dep. Conserv., Div. Game and Fish., Tech. Bull. 1. 228 pp.

Appendix Table. Corresponding values of survival (S), total mortality rate (A), and total mortality coefficient (Z)

S	A	Z	S	A	Z
0.01	0.99	4.605170186	0.5	0.5	0.6931471806
0.02	0.98	3.912023005	0.51	0.49	0.6733445533
0.03	0.97	3.506557897	0.52	0.48	0.6539264674
0.04	0.96	3.218875825	0.53	0.47	0.6348782724
0.05	0.95	2.995732274	0.54	0.46	0.6161861394
0.06	0.94	2.813410717	0.55	0.45	0.5978370008
0.07	0.93	2.659260037	0.56	0.44	0.5798184953
0.08	0.92	2.525728644	0.57	0.43	0.5621189182
0.09	0.91	2.407945609	0.58	0.42	0.5447271754
0.1	0.9	2.302585093	0.59	0.41	0.5276327421
0.11	0.89	2.207274913	0.6	0.4	0.5108256238
0.12	0.88	2.120263536	0.61	0.39	0.4942963218
0.13	0.87	2.040220829	0.62	0.38	0.4780358009
0.14	0.86	1.966112856	0.63	0.37	0.4620354596
0.15	0.85	1.897119985	0.64	0.36	0.4462871026
0.16	0.84	1.832581464	0.65	0.35	0.4307829161
0.17	0.83	1.771956842	0.66	0.34	0.415515444
0.18	0.82	1.714798428	0.67	0.33	0.4004775666
0.19	0.81	1.660731207	0.68	0.32	0.3856624808
0.2	0.8	1.609437912	0.69	0.31	0.3710636814
0.21	0.79	1.560647748	0.7	0.3	0.3566749439
0.22	0.78	1.514127733	0.71	0.29	0.3424903089
0.23	0.77	1.46967597	0.72	0.28	0.328504067
0.24	0.76	1.427116356	0.73	0.27	0.3147107448
0.25	0.75	1.386294361	0.74	0.26	0.3011050928
0.26	0.74	1.347073648	0.75	0.25	0.2876820725
0.27	0.73	1.30933332	0.76	0.24	0.2744368457
0.28	0.72	1.272965675	0.77	0.23	0.2613647641
0.29	0.71	1.237874356	0.78	0.22	0.2484613593
0.3	0.7	1.203972804	0.79	0.21	0.2357223335
0.31	0.69	1.171182982	0.8	0.2	0.2231435513
0.32	0.68	1.139434283	0.81	0.19	0.2107210313
0.33	0.67	1.108662625	0.82	0.18	0.1984509387
0.34	0.66	1.078809661	0.83	0.17	0.1863295782
0.35	0.65	1.049822124	0.84	0.16	0.1743533871
0.36	0.64	1.021651248	0.85	0.15	0.1625189295
0.37	0.63	0.9942522733	0.86	0.14	0.1508228897
0.38	0.62	0.9675840263	0.87	0.13	0.1392620673
0.39	0.61	0.9416085399	0.88	0.12	0.1278333715
0.4	0.6	0.9162907319	0.89	0.11	0.1165338163
0.41	0.59	0.8915981193	0.9	0.1	0.1053605157
0.42	0.58	0.8675005677	0.91	0.09	0.09431067947
0.43	0.57	0.8439700703	0.92	0.08	0.08338160894
0.44	0.56	0.8209805521	0.93	0.07	0.07257069283
0.45	0.55	0.7985076962	0.94	0.06	0.06187540372
0.46	0.54	0.7765287895	0.95	0.05	0.05129329439
0.47	0.53	0.7550225843	0.96	0.04	0.04082199452
0.48	0.52	0.7339691751	0.97	0.03	0.03045920748
0.49	0.51	0.7133498879	0.98	0.02	0.02020270732
			0.99	0.01	0.01005033585

16

Control of Undesirable Species

Sometimes, in accord with our preferences, populations of fish are described as undesirable. Certain species, the carp (*Cyprinus carpio*), for example, are usually considered undesirable by sport fishermen. The lamprey (*Petromyzon marinus*), as a parasite on valuable commercial and sport fishes in the Great Lakes, has been most undesirable. Populations of fish ordinarily considered popular become undesirable when they occur in such large numbers and such small size that they are not of commercial or sporting quality. Many people find the stunted populations of brook trout (*Salvelinus fontinalis*) in the cold streams of the Rockies undesirable and would prefer the native cutthroat. The abundance and predacious habits of the American eel (*Anguilla rostrata*) in the Atlantic salmon rivers of Canada brand this fish as undesirable. Heavy infections of parasites may ruin the aesthetic qualities of a fish population and make total control necessary. Fish present in hatchery water supplies are potential troublemakers and should be removed. Sometimes, as when Pacific salmon crowd their spawning grounds and the superimposition of redds results, it may be necessary to divert part of the same population to other areas. Thus, we find many reasons to consider a particular fish population undesirable and to recommend control.

Manufacturers and mail-order houses are marketing fish toxicants with such appealing slogans as ''Fishing Unlimited,'' ''Bigger Fish,'' ''How to obtain more fun, more utility from your body of water,'' and ''How to increase your income $250.00 per year.'' Fish toxicants are available to everyone, but only fishery biologists with considerable knowledge of the water areas involved, and with experience in the use of the chemicals, should ever use them.

Chemical reclamation has the most utility of the methods available for controlling fish populations. Ideally, a fish toxicant should:

1. Be nontoxic to any biological organism except the fish to be controlled.
2. Be easy and safe to apply.
3. Provide complete control.
4. Spontaneously decompose to a nontoxic condition.
5. Be unaffected by temperature.
6. Be unaffected by varying water chemistries and qualities.
7. Be low in cost.

Rotenone and antimycin are the two most-used fish toxicants at the present time. Squoxin, a new chemical, is selective for squawfish, and TFM(3-trifluormethyl-4-nitrophenol), a lampricide, is selective for lampreys. Many substances are toxic to fish, including copper sulfate, cresol, calcium, hypochlorite, and toxaphene.

Introducing toxicants into the ecosystem is a heavy responsibility, particularly for the resource manager who must at the same time take a strong stand against pollution. Biologists should be familiar with all the various products available, know their advantages and disadvantages, and study the proper concentrations.

Despite the effectiveness of the chemicals, complete kills and complete recovery of the dead fish are very difficult to achieve. Recovery of the dead fish is especially important because biologists frequently plan to use the recovered fish as the basis for conclusions about the dynamics of the fish population. Likely problems are that:

1. Chemicals may not reach all areas, and large springs may present enough flow to reduce the chemical to harmless concentrations in their immediate vicinity.
2. Fish may escape into the outlets or inlets of a lake and thus may not be treated completely.
3. Swampy areas or overhangs of aquatic vegetation may prevent the thorough distribution of the chemical.
4. Small fish are difficult to locate and are often too numerous for complete kills.
5. Smaller fish are sometimes eaten by larger fish in the first stages of chemical application.
6. Beds of emergent vegetation may hold fish which have become enmeshed during their death struggles.

7. Insects, snakes, turtles, birds, and mammals may, by eating the dead and dying fish, lower the number recovered.

Rotenone

Natives in widely separated tropical and subtropical countries have long used the crushed roots and extracts of plants containing rotenone as an aid to obtaining fish for food, Known also as tuba, cube, timbo, and derris, rotenone has been found in six genera of the family Leguminosae. In recommended doses, it is reasonably specific for fishes; it upsets the utilization of oxygen by the fish which suffocates even though there is plenty of oxygen in the water. Humans and animals with the notable exception of swine, can drink the treated water with no ill effects; fish-eating birds can consume the poisoned fish; and the Indians of South America can even use the poisoned fish for food. It is not harmful to plants.

Rotenone is manufactured under several trade names, including some products with a sulfoxide synergist. A general application rate might be about 2 ppm., but fishery managers are cautioned to read the labels and apply at the recommended concentrations. Products vary as to the actual amount of rotenone, carrying agents, and emulsifying agents that are included (powdered rotenone is very difficult to mix with water and is rarely used in dry form).

The most popular way of applying rotenone is by spraying the solution from pumps mounted in boats. Other methods are by pouring the toxicant into the wake of an outboard motor, towing bags of toxicant designed to dissolve slowly, spraying from planes and helicopters. Backpack pumps are useful when dispersing the chemical in vegetation or over swampy areas. Deep-water areas are treated by forcing the chemical down hoses suspended behind the boat. Constant-drip installations are used on streams to introduce the toxicant at a constant rate.

Fish in a treated pond will be affected in a matter of minutes, with the young fish usually distressed first. Fish may be dying for as long as two days; this is particularly true for species like the bullheads that may resist full exposure by being buried in the bottom during the day and not becoming active until evening.

In the interest of good public relations, it is usually better to treat ponds in September after vacationers and cottage users have left the area, and after fisherman use has slackened off. By the time the recreationist returns the next summer the pond has detoxified, catchable trout have been planted, and the water user has been least inconvenienced. This more than balances the fact that rotenone is more effective at higher summer water temperatures and detoxifies more rapidly.

Other general characteristics of rotenone are:

1. It repels fish.
2. It does not kill fish eggs.
3. It affects other gill-breathing organisms.
4. It may take weeks to detoxify.
5. Action may be reversible if fish are transferred to clean water soon enough.

If it should become necessary to detoxify very quickly, potassium permanganate may be used. New York State has suggested a 5.5 percent solution of permanganate at a concentration of 1 ppm. Chlorine will also detoxify the rotenone, and chlorinated lime has been used as an easy way of handling the chlorine. A general recommendation is to use either the chlorine or the potassium permanganate in concentrations equal to the rotenone. The addition of tannic acid to the solution will increase the detoxifying action.

Antimycin

Antimycin was discovered in 1945 and was so named (antimycin = antifungus) because it was a deadly fungicide. In 1963, the Wisconsin Alumni Research Foundation reported its toxicity to fish, and the Fish Control Laboratories at LaCrosse, Wisconsin, began intensive study. Antimycin is absorbed through the gills of fish and kills by interfering with the respiration of the body cells.

Ayerst Laboratories have been licensed to produce and market antimycin under the trade name Fintrol, as approved by the Pesticide Regulations Division of the U.S. Department of Agriculture for use in freshwater fishery management. Fintrol is formulated on sand grains with Carbowax (polyethelene = glycol-6000, Union Carbide Company). Each 100 g of Fintrol-5 contains 1 g of antimycin formulated to release the active chemical uniformly in the water by the time the Fintrol-5 has sunk to a depth of approximately 5 feet. Fintrol-15 and Fintrol-30 are under consideration for future use.

Fintrol can be applied with commercial grass seeders or by simply punching holes in the cans and letting the sand tagged with antimycin run into the water as the boat moves along. Recommended concentrations of Fintrol-5 range from 0.5 to 1 ppm. for the control of many freshwater fishes. For shortnose gar, bowfin, and goldfish, higher concentrations of 1.5 to 2.5 ppm. may be necessary. Less Fintrol is needed when the pH is low and water temperature is high.

Some characteristics of antimycin are:

1. Kills the fertilized fish eggs.
2. Does not repel fish.
3. Permits rapid degradation (1 to 14 days' range, usually between 4 to 7 days; higher pH's shorten degradation time).
4. Has little effect on fish food.
5. Does not kill plants, insects, tadpoles, frogs, salamanders, turtles, water snakes, mallards, ringneck pheasants, pigeons, chickens, quail, mice, rats, rabbits, guinea pigs, dogs, and lambs *at recommended dosages.*
6. May not be toxic to humans to eat fish killed by antimycin, but not accepted yet.
7. Imparts no color or odor to the water.

Potassium permanganate, at 1 ppm., will detoxify the antimycin, and tables are available with the commercial product to indicate the number of treatment days necessary for detoxification. Fingerling rainbow trout surviving for at least 48 hours in a livecar indicate detoxification has occurred.

Although many successful reclamations are reported from various areas and from the LaCrosse Laboratories, it should be noted that the reclamation of Turquoise Lake in Colorado did not remove the competing fish, particularly suckers. And while applications of antimycin at concentrations of 0.5 ppb. in a Maine lake eliminated yellow perch and five species of cyprinids, only partial kills of brook trout and redbelly dace resulted, and white suckers and burbot showed no apparent effects.

Investigations in Fish Control, published by the Bureau of Sport Fisheries and Wildlife provides the following information. Antimycin-A was subjected to field trials as a fish toxicant in 20 ponds and lakes and 5 streams in the East, Midwest, and West of the United States. The formulations of toxicant included three on sand grains which are designed to release antimycin uniformly within certain depths, and one formulation in a liquid. Ten parts per billion or less of the toxicant were effective against most of the 54 species of fish encountered, including carp, suckers, and green sunfish. Differences in sensitivity among fish suggest possible uses of antimycin as a selective toxicant. The efficiency of the toxicant is influenced by pH and water temperature, with slightly higher concentrations necessary at high pH or in cold water. Antimycin degrades rapidly, usually within a week. Fish-killing concentrations have little or no effect on other aquatic animals.

Antimycin effectively and economically controlled heavy infestations of green sunfish and golden shiners from selected channel catfish ponds at a Mississippi fish farm. An initial application of 5 ppb. of antimycin in two ponds and 7.5 ppb. in the third pond eliminated nearly 99 percent of the scaled fish. A followup treatment of 10 ppb., four days later, further reduced these populations with no apparent effect on yearling catfish. Comparison of the adjusted yields of catfish from treated and untreated ponds, ranging in size from 0.4 to 0.6 ha, indicates that treated ponds produced an additional 460 kg of fish worth $507.50, while antimycin cost only $145.79—a net return of $2.48 for each dollar invested in toxicant.

More than fifty applications of Antimycin-A, a fish toxicant, have been made to control fish. It has been used for partial reclamations, and as a general or selective toxicant. It appears to be effective against fish in fresh and marine waters, in acid and alkaline waters, in cold and warm waters, and in flowing and static waters. The toxic action, respiratory inhibition, is irreversible in most fishes.

Squoxin

Idaho researchers, during an intensive chemical screening program, developed a squawfish-killing chemical. The chemical (1,1'-methelenedi-2-naphthol) is applied to water areas at 0.1 ppm, although even smaller concentrations may prove effective at high temperatures. At this concentration Squoxin kills the squawfish by acting as a vasoconstrictor preventing efficient use of oxygen and the proper function of the blood vessels. The squawfish are not repelled by the chemical. Salmon and trout are unharmed, although the toxin is reported to kill a "few dace and shiners." No effect on aquatic insects, other fish foods, humans, or terrestrial animals is reported. Squoxin becomes ineffective within hours. The first dead squawfish are observed about 3 hours after application, and other deaths occur for 24 hours. Application over a long period is necessary because of the toxin's short life. Reports from one test estimated that about 200,000 squawfish were killed in a 12-km section of stream.

Elimination or control of squawfish should result in a much higher survival of young salmon and trout released from fish hatcheries, and for smolt disoriented in tailwater pools from passage over a dam or through turbines.

Lampricide

Control of the lamprey in the Great Lakes was not realized until a suitable chemical was discovered after screening and testing approximately 6,000 chemicals. The lamprey is available for control during the spawning run of adults upstream and during the larval period in the stream and estuaries. TFM has been judged about 98 percent effective. Larvae of all sizes are affected, and the metamorphosing larvae and young adults ready to migrate downstream are especially susceptible. Used in recommended dosages, TFM does not affect trout and other game species, although some species of minnows show distress.

TFM works best in streams (where proper concentrations are best determined for each stream separately) and has not been successful in deep ponds or in estuaries of streams as they run into the Great Lakes. After the control program began, researchers discovered that larval lampreys are also present in the lakes around the mouths of inlets. The Canadian Department of Fisheries is credited with the application of a chemical, Bayer 73, developed for snail control in the tropics. Sprayed on sand the chemical settles to the bottom where it forces the lamprey to the surface. Distressed lampreys can be netted or simply left for the chemical to kill. Reservations about this new chemical include its adverse effects on other fish, a problem which will restrict its use to selected areas.

Other Methods of Fish Control

Biological control is tempting and is more in keeping with most resource managers' philosophies. Introductions of northern pike (*Esox lucius*) and walleyes are examples of using predator fish for reducing populations. However, quantitative results on these introductions are not available, and this method should be approached with extreme caution. The introduction of coho salmon to the Great Lakes may give some definite information, as biologists study the interaction of the salmon and the alewives. Other possible biological controls are stocking hybrids with little or no potential for breeding; altering the sex by chemicals; and producing fish that will never develop sexually (energy ordinarily expended in maturing sex products and spawning is channeled into growth).

Netting to remove undesirable fish may generally be considered impractical and uneconomical. The reproductive potential of a fish in a suitable environment precludes much control, by netting, unless unusual circumstances concentrate and make available almost the entire population. Many times the population to be controlled simply compensates for the loss of fish with a faster growth rate and earlier maturing.

Weirs of all kinds have been used for fish control, but even fish on spawning migrations and reasonably available to the weir cannot always be controlled. High water may flood a weir, blow-outs underneath or around may allow fish passage, and with electric weirs there are problems about the proper voltage to allow fish to pass. Also fish weirs guide all species to collecting boxes where faulty handling may result in dead desirable fishes or release of some to be controlled. More sophisticated electric weirs with pulsated direct current are becoming more effective. A distinct advantage to the electric weir is that the floating or hanging electrodes do not catch debris in quantities large enough to retard flow and act as a dam.

Where water-level control is practical, various fishes such as sunfishes, minnows, suckers, perch, pike, pickerel, and carp which spawn in the shallow waters of lakes may be controlled. Water level may be dropped exposing desirable spawning areas. Or dams at the outlets of lakes, particularly oligotrophic lakes, may raise water levels, resulting in a more than proportionate increase in the warm, shallow littoral zone, but this increases the habitat favorable to undesirable fishes and to the stunting of existing populations of warmwater fishes.

A policy statement adopted by the Great Lakes Fishery Committee December 1, 1975, makes an important contribution in describing the role of dams in an integrated sea lamprey control program. The policy statement reads as follows:

"Barriers, natural or man made, play an extremely important role in limiting the number of streams used by spawning sea lampreys or in restricting the potential spawning area within a river system. Since the sea lamprey population in the Great Lakes is dependent upon reproduction which takes place in only about 400 of the 5,750 tributaries entering the Great Lakes, the Commission regards construction of barriers as a valuable and practical supplement to lampricides in development of an integrated sea lamprey control program.

Among the major advantages to be realized through the installation of properly designed barrier dams in selected sea lamprey producing streams are:

1. more efficient control on streams where physical characteristics make lampricide treatment difficult, expensive, or ineffective;
2. savings in time, manpower, and related costs through a reduction in stream miles requiring periodic lampricide treatment;
3. reduced dependency on chemicals;

4. reduced lampricide purchases in the face of rising costs and a potentially limited supply;
5. reduced quantity of lampricides added to the environment; and
6. restoration and/or survival of non-target species in some streams.

The benefits from dams designed specifically for sea lamprey control far outweigh the disadvantages. Proper design and knowledgeable selection of streams and sites minimize possible adverse effects such as significant changes in aquatic invertebrate communities, increased water temperatures, silting, or interference with upstream movement of anadromous fish.

References

Alton, F. M. 1959. The invasion of Manitoba and Saskatchewan by carp. Trans. Amer. Fish. Soc. 88(3):203–205.

Applegate, V. C., and E. L. King, Jr. 1962. Comparative toxicity of 3-Trifluormethyl-4-nitrophenol (TFM) to larval lampreys and eleven species of fishes. Trans. Amer. Fish. Soc. 91(4):342–345.

Ball, R. C. 1948. Recovery of marked fish following a second poisoning of the population of Ford Lake, Michigan. Trans. Amer. Fish. Soc. 75:36–42.

Balser, D. S. 1964. Management of predator populations with antifertility agents. J. Wildl. Manage. 28(2):352–358.

Beyerle, G. B., and J. E. Williams. 1967. Attempted control of bluegill reproduction in lakes by application of copper sulphate crystals to spawning nests. Prog. Fish-Cult. 29(3):150–155.

Bills, T. D., and L. L. Marking. 1976. Toxicity of 3-trifluoromethyl-4-nitrophenol (TFM), 2's-dichloro-4'-nitrosalicylanilide (Bayer 73), and 98:2 mixture to fingerlings, of seven fish species and to eggs and fry of coho salmon. U.S. Fish Wildl. Serv., Invest. Fish Control 69. 9 pp.

Bowers, C. C. 1955. Selective poisoning of gizzard shad with rotenone. Prog. Fish-Cult. 17(3):134–135.

Buck, D. H., M. A. Whitacre, and C. F. Thoits, III. 1960. Some experiments in the baiting of carp. J. Wildl. Manage. 24(4):357–364.

Burress, R. M., and C. W. Luhning. 1969. Field trials of antimycin as a selective toxicant in channel catfish ponds. U.S. Bur. Sport. Fish. Wildl., Invest. Fish Control 25. 12 pp.

Cahoon, W. G. 1953. Commercial carp removal at Lake Mattamuskeet, North Carolina. J. Wildl. Manage. 17(3):312–317.

Charles, J. R. 1957. Final report on population manipulation studies in three Kentucky streams. Kentucky Dep. Fish Wildl. Resour., Fish. Bull. 22. 45 pp. Mimeogr.

Clemens, H. P. 1952. An aid in the application of rotenone in pond reclamation. Prog. Fish-Cult. 14(1):31–32.

Clemens, H. P., and M. Martin. 1953. Effectiveness of rotenone in pond reclamation. Trans. Amer. Fish. Soc. 82:166–177.

Cushing, C. E., Jr., and J. R. Olive. 1957. Effects of toxaphene and rotenone upon the macroscopic bottom fauna of two northern Colorado Reservoirs. Trans. Amer. Fish. Soc. 86:294–301.

Finucane, J. H. 1969. Antimycin as a toxicant in a marine habitat. Trans. Amer. Fish. Soc. 98(2):288–292.

Foerster, R. E., and W. E. Ricker. 1941. The effect of reduction of predaceous fish on survival of young sockeye salmon at Cultus Lake. J. Fish. Res. Bd. Can. 5(4):315–336.

Foye, R. E. 1956. Reclamation of potential trout ponds in Maine. J. Wildl. Manage. 20(4):389–398.

_____. 1964. Chemical reclamation of forty-eight ponds in Maine. Prog. Fish-Cult. 26(4):181–185.

_____. 1968. The effects of a low-dosage application of Antimycin A on several species of fish in Crater Pond, Aroostook County, Maine. Prog. Fish-Cult. 30(4):216–219.

Fukano, K. G., and F. F. Hooper. 1958. Toxaphene (chlorinated camphene) as a selective poison. Prog. Fish-Cult. 20(4):189–190.

Gilderhus, P. A., B. L. Berger, and R. E. Lennon. 1969. Field Trials of Antimycin A as a fish toxicant. U.S. Bur. Sport Fish. Wildl., Invest. Fish Control 27. 21 pp.

Grice, F. 1958. Effect of removal of panfish and trashfish by fyke nets upon fish population of some Massachusetts ponds. Trans. Amer. Fish. Soc. 87:108–115.

Hayes, F. R., and D. A. Livingstone. 1955. The trout population of a Nova Scotia lake as affected by habitable water, poisoning of the shallows, and stocking. J. Fish. Res. Bd. Can. 12(4):618–635.

Hemphill, J. E. 1954. Toxaphene as a fish toxin. Prog. Fish-Cult. 16(1):41–42.

Hodges, J. W. 1972. Downstream migration of recently transformed sea lampreys before and after treatment of a Lake Michigan tributary with a lampricide. J. Fish. Res. Bd. Can. 29(8):1237–1240.

Hooper, F. F., and A. R. Grzenda. 1957. The use of toxaphene as a fish poison. Trans. Amer. Fish. Soc. 85:180–190.

Howland, R. M. 1969. Interaction of Antimycin A and rotenone in fish bioassays. Prog. Fish-Cult. 31(1):33–34.

Jackson, C. F. 1956. Control of the common sunfish or pumpkinseed, *Lepomis gibbosus,* in New Hampshire. New Hampshire Fish Game Dep., Tech. Circ. 12. 16 pp.

Jeppson, P. 1957. The control of squaw fish by use of dynamite, spot treatment, and reduction of lake levels. Prog. Fish-Cult. 19(4):168–171.

Johnson, W. C. 1966. Toxaphene treatment of Big Bear Lake, California. California Fish Game J. 52(3):173–179.

Krumholz, L. A. 1948. The use of rotenone in fisheries research. J. Wildl. Manage. 12(3):305–317.

Lawrence, J. M. 1956. Use of potassium permanganate to counteract the effects of rotenone on fish. Prog. Fish-Cult. 18(1):15–21.

Lee, T. H., P. H. Derse, and S. D. Morton. 1971. Effects of physical and chemical conditions on the detoxification of antimycin. Trans. Amer. Fish. Soc. 100(1):13–17.

Lennon, R. E., and B. L. Berger. 1970. A resumé of field applications of Antimycin A to control fish. U.S. Bur. Sport Fish. Wildl., Invest. Fish Control 40. 19 pp.

Lennon, R. E., J. B. Hunn, R. A. Schnick, and R. M. Burress. 1971. Reclamation of ponds, lakes, and streams with fish toxicant: a review. FAO, U.N., Fish. Tech. Pap. 100.

Leonard, J. W. 1939. Notes on the use of derris as a fish poison. Trans. Amer. Fish. Soc. 68:220-280.

Linhart, S. B. 1964. Acceptance by wild foxes of certain baits for administering antifertility agents. New York Fish Game J. 11(2):69-77.

Loeb, H. A. 1955. An electrical surface device for carp control and fish collection in lakes. New York Fish Game J. 2(2):220-231.

MacKay, H. H., and E. MacGillivray. 1949. Recent investigation on the sea lamprey, *Pretromyzon marinus*, in Ontario. Trans. Amer. Fish. Soc. 76:148-159.

MacPhee, G., and R. Ruelle. 1969. A chemical selectively lethal to squawfish (*Ptychocheilus oregonensis* and *P. umpquae*). Trans. Amer. Fish. Soc. 98(4):676-684.

Manion, P. J. 1969. Evaluation of lamprey larvicides in the Big Garlic River and Saux Head Lake. J. Fish. Res. Bd. Can. 26(11):3077-3082.

McLain, A. L. 1957. The control of the upstream movement of fish with pulsated direct current. Trans. Amer. Fish. Soc. 86:269-284.

Menzie, C. M., and J. B. Hunn. 1976. Chemical control of the sea lamprey: the addition of a chemical to the environment. *In* Environmental quality and safety: global aspects of chemistry, toxicology and technology as applied to the environment, Vol. 5. Academic Press, New York. 14 pp.

Miller, R. B. 1950a. The Square Lake experiment: an attempt to control *Triaenophorus crassus* by poisoning pike. Can. Fish Cult. 7:3-18.

_____. 1950b. A critique of the need and use of poisons in fisheries research and management. Can. Fish Cult. 8:30-33.

Miller, R. B., and R. C. Thomas. 1957. Albert's "pothole" trout fisheries. Trans. Amer. Fish. Soc. 86:261-268.

Patriarche, M. H. 1953. The fishery of Lake Wappapello, a flood-control reservoir on the St. Francis River, Missouri. Trans. Amer. Fish. Soc. 82:242-254.

Pintler, H. E., and W. C. Johnson. 1958. Chemical control of rough fish in the Russian River drainage, California. California Fish Game J. 44(2):91-124.

Post, G. 1955. A simple chemical test for rotenone in water. Prog. Fish-Cult. 17(4):190-191.

Powers, J. E., and A. L. Bowes. 1967. Elimination of fish in the giant grebe refuge, Lake Atitlan, Guatamala, using the fish toxicant, Antimycin. Trans. Amer. Fish. Soc. 96(2):210-213.

Prevost, G. 1960. Use of fish toxicants in the Province of Quebec. Can. Fish Cult. 28:13-35.

Priegel, G. R. 1971. Evaluation of intensive freshwater drum removal in Lake Winnebago, Wisconsin. Wisconsin Dep. Nat. Resour., Tech. Bull. 47. 29 pp.

Rawson, D. S., and C. A. Elsey. 1950. Reduction in the longnose sucker population of Pyramid Lake, Alberta, in an attempt to improve angling. Trans. Amer. Fish. Soc. 78:13-31.

Rose, E. T., and T. Moen. 1953. The increase in game-fish populations in East

Okoboji Lake, Iowa, following intensive removal of rough fish. Trans. Amer. Fish. Soc. 82:104–114.

Saila, S. B. 1954. Bioassay procedures for the evaluation of fish toxicants with particular reference to rotenone. Trans. Amer. Fish. Soc. 83:104–114.

Schnick, R. A. 1972. A review of literature on TFM (3-trifluormethyl-4 nitrophenol) as a lamprey larvicide. U.S. Bur. Sport Fish. Wildl., Invest. Fish Control 44. 31 pp.

Shields, J. T. 1958. Experimental control of carp reproduction through water drawdowns in Fort Randall Reservoir, South Dakota. Trans. Amer. Fish. Soc. 87:23–33.

Siegler, H. F., and H. Pillsbury. 1949. Progress in reclamation techniques. Prog. Fish-Cult. 11(2):125–129.

Smith, M. W. 1950. The use of poisons to control undesirable fish in Canadian fresh waters. Can. Fish Cult. 8:17–29.

_____. 1955. Control of eels in a lake by preventing the entrance of young. Can. Fish Cult. 17:13–17.

Stanley, J. G. 1976. Reproduction of the grass carp (*Ctenopharyngodon idella*) outside its native range. Fisheries, Bull. Amer. Fish. Soc. 1(3):7–10.

Stringer, G. E., and R. G. McMynn. 1958. Experiments with Toxaphene as fish poison. Can. Fish Cult. 23:39–47.

_____. 1960. Three years' use of Toxaphene as a fish toxicant in British Columbia. Can. Fish Cult. 28:37–44.

Stroud, R. H. 1951. Use of a wetting agent to facilitate pond reclamation. Prog. Fish-Cult. 13(3):143–145.

Swingle, H. S., E. E. Prather, and J. M. Lawrence. 1953. Partial poisoning of overcrowded fish populations. Alabama Agric. Exp. Stn., Circ. 113. 15 pp.

Tompkins, W. A., and J. W. Mullan. 1958. Selective poisoning as a management tool in stratified trout ponds in Massachusetts. Prog. Fish-Cult. 20(3):117–123.

U. S. Fish and Wildlife Service. 1966. Investigations in fish control. 3–8. U.S. Bur. Sport Fish. Wildl., Resour. Publ. 7–12, inclusive. (Papers describe use of toxaphene.)

Vanderhorst, R., and S. P. Lewis. 1969. Potential of sodium sulfite catalyzed with cobalt chloride in harvesting fish. Prog. Fish-Cult. 31(3):149–154.

Webster, D. A. 1954. A survival experiment and an example of selective sampling of brook trout (*Salvelinus fontinalis*) by angling and rotenone in an Adirondack Pond. New York Fish Game J. 1(2):214–219.

Weier, J. L., and D. F. Starr. 1950. The use of rotenone to remove rough fish for the purpose of improving migratory waterfowl refuge areas. J. Wildl. Manage. 14(2):203–205.

Zilliox, R. G., and M. Pfeiffer. 1956. Restoration and brook trout fishing in a chain of connected waters. New York Fish Game J. 3(2):167–190.

17

Regulations and Their Effects

Fisheries have been regulated on the basis of politics, social pressure, gear competition, prejudice, whim, and sometimes for biological reasons. Efforts have been made over a long period to manage fisheries by legislation. And a few laws have been helpful, but of many it can only be said that the fishery survived in spite of the regulation, not because of it. Emotions run high at hearings to discuss fishing regulations, and fishery biologists and administrators face difficult times in trying to keep discussions and decisions as biologically oriented as possible. It is particularly trying to stand by while wise resource management plays a subordinate role, for example, to international diplomacy, as fisheries become the pawn in a larger game, or, on a smaller scale, to local pressure from commercial fishing-camp owners to close lakes to ice fishing in the hope that more fish will then be available for the paying, nonresident summer visitor.

Fishing regulations need to be as simple and as few as possible. Every region has a standing joke which runs something like this, ''To go fishing requires a lawyer in attendance and if you intend to eat the fish you have to carry a cook to fry them as fast as they are caught.'' Those who advocate complex and numerous regulations would do well to browse through regulation books and note how many species fishermen must identify, how many size limits and bag limits they must know, and how many areas must be recognized. Fishing whether for fun or profit should not be unnecessarily complicated by indefensible regulation.

Fishing regulations are usually aimed at control of the fishermen with little concern given to the biological health of the fish. In general, fishing regulations simply confirm ownership of the fish by the state and assign responsibility for their care to the state fish and game agency. Regulations then require the fisherman to have a fishing license which legally permits him to acquire ownership of the fish, providing he fishes by legally described methods in the areas open to fishing at the legal times of year, month, and day. The license usually limits the fisherman to certain kinds, sizes, and amounts of fish.

Most regulations are based on the theory that the fewer fish caught now, the more will be available for future fishing. Such a "logical" theory has found wide public acceptance, even if not always true. It is generally believed that it is necessary to have a fairly large number of older fish for a spawning stock. This may be true for a few species, such as salmon, in which the large number of eggs is fairly well protected, and where a connection between spawners and progeny can be established. However, for many extremely prolific species, whose eggs are at the mercy of environmental conditions, it has not been possible to establish such a connection. For such species, maintenance of a large stock of older fish may represent a loss. This is not to discount the possibility of any connection, but rather to suggest that there may be an upper limit beyond which additional spawning stock is of no advantage. At lower levels the connection may be so obscured by the varying effects of environmental factors that it cannot be shown without a long series of data encompassing both spawning stock and ecological factors.

The fishing public has, for the most part, accepted the belief that all smaller fish should be fully protected—since they will grow into big fish—but this idea is also suspect, especially in sport fisheries. Many lakes and streams suffer from overpopulation by sunfish, bass, perch, and trout so stunted from competition for insufficient food they never reach a respectable size. Brook trout in the very cold ponds or cold streams of New England or the Rockies are good examples.

Another theory is that fish should be protected during the spawning season. Protecting a stock of mature fish may be justified when numbers are extremely low, or where the fish are particularly vulnerable to capture during the spawning period. But if fish are not, in fact, more vulnerable during the spawning period, little is gained by extra protection during that season that could not just as well be gained by reducing the catch during the remainder of the year, except that fish are likely to be in poorer condition during spawning season.

Man through his regulations can affect a species only by causing some changes in the predation he practices, unless one includes the indirect consequences of man preying also upon a competitor or predator of the species. Man may also affect a species with regulations regarding parasites and diseases. Exploitation of a species can be modified in two ways:

1. The total quantity can be reduced or limited.
2. The catch can be taken from selected portions of the population (or certain portions can be protected against capture).

A limitation or reduction in the total catch is achieved by:

1. A limitation of the efficiency of the individual fishing units.
2. A limitation on the number of fishing units permitted to operate.
3. A limitation (quota) on the total quantity of fish that can be captured.

Regulations designed to protect selected portions of the population may or may not result in overall reduction in catch. Such regulations include:

1. Restrictions or modifications on the gear used to lessen catches of sizes or groups it is desired to protect.
2. Closure of certain fishing areas.
3. Restriction of fishing to selected seasons.
4. Restrictions on the sizes, condition, or quantities that can be marketed.
5. Protection of individual fish based on sex or condition.

Restrictive regulations to the efficiency of the fishing unit are a commonly used method to reduce the size of the catch, and some typical examples are discussed below.

Restriction on the size and type of the fishing vessel is not as popular today as it was in the past when fishing boats were sometimes denied the use of power in certain areas or when actual size limits were imposed. Some sport fishing regulations prohibit the use of a motor or restrict the size of the motor. Most of these regulations draw the line at 10-horsepower outboards, with the result that manufacturers have developed motors with 9.5 or 9.8 horsepower. Restrictions on sport-fishing boats are oftentimes the result of social pressures or concern for aesthetics.

Regulating the size and type of commercial fishing vessels is highly artificial and, like most gear regulations, simply increases the costs of operation. This puts the commercial fisherman at a disadvantage when competing with fishery products from other regions or with other protein sources.

Restrictions on the type of gear used are a universally used form of control. The arguments favoring this kind of regulation are largely social, although a highly efficient gear in a localized area may take too large a proportion of the fish population and thus require some control. Restrictions on the type of gear permitted in commercial fisheries usually result in less efficiency in taking fish, thus increasing their cost.

Sport fishing has been the frequent target for regulations restricting the type of gear. Restrictions have included fly-fishing only, single-hook lures, size of hooks, and number of hooks on lures. Social pressures are common here, complicated by a desire to reduce the competition for fishing pools or to provide an environment free of other fishermen. Fish are just as dead when taken by a fly, a single-hook lure, or a worm. Arguments as to which is the better way to fish must be left to the fishermen themselves. Purist bait fishermen have as many rights as purist fly-fishermen.

Bag limits are a favorite regulation to restrict gear efficiency for sport fisheries and are considered a way of helping to share the catch among more fishermen. But bag limits are not always necessary or useful. In many of the warmwater fisheries and in some coldwater fisheries, stunted populations of brook trout might or might not show increased growth if bag limits were removed, but at least they would be harvested. Work on smallmouth bass populations in Maine did not indicate significant growth increases following a complete liberalization of the bag and size limits. If bag limits are to be successful they must be set low enough to reduce the catch to a specified level. Reducing the bag limit from 20 fish to 10 fish, if most people don't catch 10 fish, is obviously meaningless.

Regulations restricting the use of some types of gear in certain areas are usually based on either the high efficiency of the gear or on the actual or alleged destruction effected by the gear. Commercial fishermen are banned from fishing inshore to preserve those areas for local fishermen or sport fishermen. Herring seiners are prohibited from fishing in the estuaries of Atlantic salmon streams to prevent commercial exploitation of the populations. Lakes and streams may be closed to all kinds of fishing except fly-fishing. Many states are now closing sections of streams to certain types of fishing, thus creating fly-fishing sections, lures-only sections, and no-restriction areas. Naturally, the bait fishermen are now beginning to suggest bait-fishing-only sections.

Restrictions on the size of units of gear are also used to reduce efficiency. Lengths and depths and mesh sizes of commercial fishing gear have been regulated. Bait dealers are frequently limited to a small-size minnow seine. One result is that the dealer simply fishes the smaller net harder and with actually more damage possible to the habitat. What usually controls a bait dealer is the size and quality of his holding facility and the market. Hook size has been used in both commercial and sport fisheries to regulate the number and size of fish taken.

Regulations limiting the number of fishing units can limit the catch without necessarily restricting the efficiency of each unit of gear. Thus a commercial troller may be limited to four lines, or a sport fisherman to a single rod. The effect of such regulation may simply be that the fisherman will fish longer with the limited number of units and actually catch just as many fish. Weir sites are family property in the choice areas along the Atlantic coast of Canada, a situation which amounts to a limitation on fishing units. New England villages still restrict fishing for alewives to town-operated or town-leased fishing sites. (Bucksport, Maine, is known as the village where the alewives built the school.) The number of tip-ups permitted for sport ice-fishing or the number of lines permitted through the ice for commercial fishing are other examples of ways to limit the fishing units.

Another way to restrict the total catch without limiting either the efficiency of the individual unit of fishing gear or the number of fishermen is to impose a limit on the total quantity that can be taken or to set a quota. This method is suitable only for species about which enough biological knowledge has been accumulated to permit making a forecast of their expected abundance sufficiently in advance of the fishing season. Such a forecast has to be made within fairly narrow limits for the method to be practical. The reliability of the forecast depends on several factors, including adequate annual sampling of the population. Thus, if the adult population comprises fish of many age groups, the annual recruitment usually makes up a lesser proportion of the total population. For species in which the annual success of spawning exhibits wide variation, the forecasts will usually be less reliable. The quota system has been used almost entirely in commercial fisheries. Actually, the idea of a quota system for freshwater lakes to insure a continuing fishable population of acceptable-size fish is not unrealistic. Agencies setting quotas must be prepared to expend considerable effort obtaining information on stock size and production potential.

Certain portions of populations have been protected by gear restrictions. Public sympathy for the protection of small-sized fish, for example, makes such regulations easy to pass, even when unsupported by biological evidence. The advantage or disadvantage of protecting small fish must be weighed wholly by what happens to such fish and what losses may be incurred in extending such protection. If the group of young fish under consideration are growing with sufficient rapidity so their continuous loss in total weight from natural causes of mortality is much more than compensated for by gain in their total weight as a result of the increased size of the remaining individuals, protection assuredly pays, provided they are subject to recapture at some time in the future. The chief types of gear restrictions used to protect young fish are: minimum-sized meshes in the cod ends of otter trawls, minimum-sized meshes in trap and pound nets (this measure is also used in a mixed fishery to permit escape of smaller-sized species), minimum spacing between the bottom and the first lath in lobster pots, minimum-sized hooks in hook-and-line fisheries, and minimum-sized meshes in gill nets.

Protection of small fish should always be supported by evidence demonstrating its advantage and the sizes that need protection. In a fishery of low intensity the protection should be removed at a smaller size than in one of high intensity because it takes a certain period of time to harvest a group of fish. The lower the rate of capture the longer the period of time that will elapse between the time when protection is removed and the time when the small fish are recaptured. There should then be an adjustment between the maximum size protected and the period of time required to recapture them so they will be recaptured before they reach a size where their total gain in weight from growth is overbalanced by their total loss in weight from natural causes of mortality. If the fishing intensity is sufficiently low, no protection may be warranted. For example, the fishing pressure on sunfish, bluegills, perch, and other panfish in lakes is often too low to keep pace with their high reproductive capacity, and their abundance may be further augmented by the fishery removing large bass, pickerel, and other natural predators. As a result the numerous young fish outgrow their food supply. The solution in such cases is the removal of as many of the small fish of these species as can be captured by any means, until a balance is again achieved between the fish and the available food. In very cold ponds, and particularly in the small cold streams of the Northeast and the Rocky Mountain areas, populations of stunted brook trout dominate the fish populations. Protection of these small fish does nothing for the fishery.

Minimum-size limits are established to increase the sustained yield either by allowing more animals to mature and increase reproduction or by taking advantage of rapid growth periods offsetting losses through natural mortality. Size limits have been cited as the most effective means of increasing yield in lobster fisheries, but studies show that the catch of wild trout was "dramatically reduced" by a 23-cm limit. Colorado has a law in certain areas which requires all fish between 25 and 50 cm to be returned to the water immediately. The rationale behind this regulation is that the fisherman can keep the smaller, more numerous fish, catch but not keep the 25- to 50-cm fish, and, it is hoped, will catch an occasional trophy-sized fish.

Certain areas are often closed to fishing to achieve one or more of the following results: to limit the total catch, to protect fish on their spawning grounds, to protect fish while migrating through areas of restricted extent where they are very vulnerable to capture, to protect young fish on nursery grounds or areas, or to prevent fishing during periods when the smaller fish are more vulnerable to the gear. Fishing areas are also closed to prevent harvesting and sale of shellfish contaminated by sewage pollution, or to prevent poisoning from mussels or other mollusks at times when routine tests show them to be dangerously toxic through the ingestion of certain plankton organisms.

Protection based on sex or condition is sometimes used in the crab or lobster fisheries where regulations may prevent taking softshell crabs or "berried" lobsters (female carrying eggs). Regulations based on sex have been more popular in wildlife regulations where hen pheasants or does are protected and only cocks or bucks can be harvested. Wildlife biologists rarely favor this type of regulation.

Regulations against the sale of undersized fish provide additional motivation to use savings gear. Enforcement problems are also made easier when the sale of small fish is discouraged.

Over the years commercial fisheries have tended more and more to move away from a single-species fishery. As market demand has increased beyond production limits for one species, fishermen have been forced to turn to other fisheries. Technological advances in fish processing at sea have made it possible for commercial fishermen to take fish like the hake that deteriorate rapidly without refrigeration. The production of new fishery-consumer products such as fresh-frozen fillets and fish sticks have made it possible to market fish once displeasing to the consumer because of appearance. This continuing trend toward versatility in fishing vessels poses an increasing problem for the fishery administrator. When protection is provided for one species, more fishing effort is expended on others. Thus fishing regulations should consider total effects and should not be based solely on single-species management.

A similar situation exists in freshwater sport fisheries where highly restrictive regulations may actually increase fishing pressure on a few waters instead of spreading it out over many.

References

Burdick, M., and O. Brynildson. 1960. Fly fishing only. Wisconsin Conserv. Bull. 25 (6). 4 pp.

Churchill, W. 1957. Conclusions from a ten-year creel census on a lake with no angling restrictions. J. Wildl. Manage. 21(2):182–188.

Clark, C. F. 1965. Importance of angling regulations: ''Liberalized Angling.''
Proc. 32nd Annu. Meet., Midwest Fish Game Commissioners:174–180.

Hart, J. L. 1958. Some sociological effects of quota control of fisheries. Can. Fish
Cult. 22:17–19.

Hunt, R. L., O. M. Brynildson, and J. T. McFadden. 1962. Effects of angling
regulations on a wild brook trout fishery. Wisconsin Conserv. Dep., Tech. Bull.
26. 58 pp.

Kerswill, C. J. 1958. Regulation of the Atlantic salmon fisheries. Can. Fish Cult.
22:7–12.

MacKenzie, W. C. 1958. Some economic aspects of control by quota. Can. Fish
Cult. 22:21–24.

Oehmcke, A. A., and D. W. Waggoner. 1956. How liberal can you get? Wisconsin
Conserv. Bull. 21(5). 4 pp.

Ricker, W. E. 1958. Some principles involved in regulation of fisheries by quota.
Can. Fish Cult. 22:1–6.

Rupp, R. S. 1955. Studies of the eastern brook trout population and fishery in
Sunkhaze Stream, Maine. J. Wildl. Manage. 19(3):336–345.

Saila, S. B. 1958. Size limits in largemouth black bass management. Trans. Amer.
Fish. Soc. 87:229–239.

Shetter, D. S. 1969. The effects of certain angling regulations on stream trout
populations. Michigan Dep. Conserv., Res. Dev. Rep. 153:333–353.

Shetter, D. S., and G. Alexander. 1966. Angling and trout populations on the
North Branch of the AuSable River, Crawford and Otsego Counties, Michigan,
under special and normal regulations, 1958–1963. Trans. Amer. Fish. Soc.
95(1):85–91.

Wales, J. H., and J. P. Lane. 1958. Public opinion analysis in trout fishery
management. Trans. Amer. Fish. Soc. 87:163–171.

Wilder, D. G. 1958. Regulation of the lobster fishery. Can. Fish Cult. 22:13–16.

Appendix

Review of Mathematics and Statistics

A.1 Introduction

The study of animal populations in a quantitative manner requires some mathematical ability. The application of population dynamics theory to management of resources such as fish and wildlife requires estimation of parameters, determination of sampling programs, and identification of meaningful differences between parameter estimates or management strategies. Thus statistics is a necessary and useful tool. The purpose of this Appendix is to provide a review of the background required to understand population ecology. Mathematics and statistics are difficult for many students of biology and, accordingly, a more detailed review is given to areas that apply to dynamics of animal populations.

A.2 Terminology and Notation

The surest way for a biology teacher to reduce the number of students taking a course is to write a mathematical formula on the blackboard at the beginning of the first class. The symbolism invokes a very predictable behavior pattern. However, to successfully study animal populations, one must evolve some method of keeping track of characteristics being observed or measured.

These concepts will be illuminated in the following example. A pond with an area of one hectare is divided into one hundred plots or quadrants for the purpose of counting the number of submerged aquatic plants. The number of plants is not known but we can let this number be represented by Y. It seems clear that Y is not constant for all ponds of one hectare or even the same pond at different times. Y is called a variable as it may have

1	2	3	4	5	6	7	8	9	10
11									
21									
31									
41									
51									
61									
71									
81									
91									100

Figure A.2.1. Quadrants in a pond

different values depending upon the pond or time. In this case the variable Y is discrete because the number of plants is countable by units. If T were defined as the length of time necessary to count the plants in the pond, T would be called a continuous variable because passage of time is continuous.

Figure A.2.1 is a representation of the pond divided into units equal to 1/100 hectare. The number of plants, Y, is equal to the sum of the number of plants found in each individual plot. This might be represented symbolically in several different but equivalent ways. Suppose each cell is identified by a unique number, starting with 1 in the upper left corner and, numbering by rows, ending with 100 in the lower right corner. The total number of plants in the pond equals the number in plot 1 plus the number in plot 2 and so on until the 100 plots are added. A way of expressing this in mathematical shorthand would be as follows:

$$Y = \sum_{i=1}^{100} y_i = y_1 + y_2 + \cdots + y_{100}$$

The Greek letter sigma, Σ, stands for "the summation of," i is a subscript that takes on the values $1, 2, 3, \cdots, 100$, which in turn are the plot numbers.* The lower case y designates the value corresponding to the number of plants counted in the corresponding plot.

---

*The notation $\cdots$ is used to indicate the continuation of the law governing the operation. The continuation may be to an extent as indicated, for example, 100, or it may be open-ended. In the first case there are a finite number of terms; in the latter an infinite number of terms.

The same number Y may be representative in a slightly different fashion. Figure A.2.1 could be thought of as having ten rows and ten columns. Rather than numbering the plots consecutively from 1 to 100, a designation of the row number and column number would uniquely identify each quadrant. Quadrant number 36 would be found in the 3rd row and 6th column. The value for the number of plants in this quadrant could be designated as y_{36} or in row-column subscripts a $y_{3,6}$. The total number of plants in the pond could be represented in row-column notation as follows:

$$Y = \sum_{i=1}^{10} \sum_{j=1}^{10} y_{i,j}$$

The first subscript i refers to the row, while the second subscript j refers to the column. There are now two subscripts instead of one and the upper limit for these subscripts is 10.

In general, a capital letter will stand for a variable, whereas the same letter in lower case will stand for a specific observation of the variable. Many students tend to form associations between a particular letter and a specific interpretation as, for example, the letter N and population size. This sooner or later results in difficulty since there are many more population variables than there are letters. For this reason it is good practice to define each variable in the context being used in a specific study. In the preceding paragraphs Y was defined as the total number of plants in a pond of one hectare area; in following paragraphs Y may be defined as some other measurement.

The greek letter delta, Δ, generally is used to represent the change in the value of a variable. The water temperature of a stream may be 15°C at 0800 hours and 18°C at 1100 hours. The change, ΔC, would be the difference between the first observation and the second, or +3°C. The change in the time of day (T) for the same observations was 3 hours, which corresponds to ΔT. Rate of change, which is very important in population analysis, is often expressed as the ratio of two incremental changes, for example, $\Delta C/\Delta T$.

The Greek letter pi, Π, conventionally is used to represent the product of a series of variables and is analogous to the use of Σ to denote a sum. Serial multiplication would be given symbolically as:

$$\prod_{i=1}^{n} x_i = x_1 x_2 x_3 \cdots x_n$$

The variable x is indexed by i which in this case goes from 1 to n. In this case, n, the upper limit of the index is also a variable. Multiplication might be carried over multiple indices in an analogous fashion as was noted for the summation operation.

A particular product is formed by multiplying all integers up to and including some number n as defined by the notation ! called factorial. Thus $n!$, read as "n factorial," defines the product:

$$n! = n(n - 1)(n - 2)(n - 3) \cdots 2 \cdot 1$$

By definition $0! = 1$.

A.3 Functions and Graphs of Functions

A variable, y, is called a function of a second variable, x, if for known values of x, the value of y may be determined. The variable y is called the dependent variable and x is called the independent variable. This relationship is generally expressed as $y = f(x)$ and read "f of x." Other letters may be used to designate different functions such as, for example:

$$y = f(x) = 2x$$

$$y = g(x) = x^2$$

$$y = h(x) = x + 1$$

The f, g, and h functions are all different functions, namely $f(x) = 2x$, $g(x) = x^2$, and $h(x) = x + 1$. The value of y for a given value of x will depend on which function is used to determine y.

A variable may be the function of more than one variable. The notation is $y = f(x,z)$ where the independent variables are separated by commas. The area of a square is a function of length and width: Area $= f(l,w) = lw$.

A graph of a function (Figure A.3.1) is a picture of the values of the dependent variable for values of the independent variable. The vertical axis is called the ordinate and represents the dependent variable. The horizontal axis is the abscissa and represents the independent variable. The scale or distance corresponding to a unit of measurement may be different for the two axes. A graph is a visual presentation of the functional relationships between variables and, in order to incorporate the interesting features of a relationship, it may be necessary to change scale on an axis.

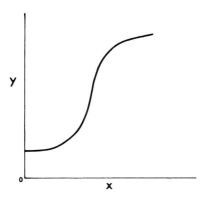

Figure A.3.1. Graph of a function $y = f(x)$

An asymptote (see Figure A.2.2) is defined as a fixed straight line that the tangent to a curve approaches as its limiting position.

If y is a function of x such that x can be expressed as some, generally different, function of y, the two functions are called inverse functions. An example would be $y = f(x) = x^2$ and $x = g(y) = \sqrt{y}$. The two functions $f(x)$ and $g(y)$ are inverse functions. Inverse functions of particular interest in population ecology are the logarithmic and exponential functions discussed in following sections.

A.4 Limit of a Function

A function $f(x)$ is said to approach the limit l if the difference between $f(a)$ and $f(x)$ becomes and remains arbitrarily small as x approaches a where a is some particular value on the abscissa. This is written symbolically

$$\lim_{x \to a} f(x) = l$$

The limit of the sum, product and quotient of two or more functions is respectively equal to the sum, product and quotient of their limits providing, in the case of the quotient, that the limit of the denominator is not zero.

A function $f(x)$ is said to be continuous at $x = a$ if 1) $f(a)$ has a defined value, and 2) the $\lim_{x \to a} f(x) = f(a)$. If either of these conditions is not satisfied, the function is said to be discontinuous at $x = a$. A function is continous in an interval $a \geq x \geq b$ if $f(x)$ is defined for every value of x within that interval.

The concept of limits of continuous functions is necessary for an under-

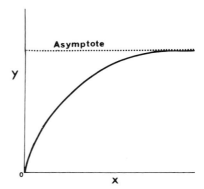

Figure A.3.2. Asymptote of a function $y = f(x)$

standing of the calculus. Limits are also important in certain sequences of numbers, as will be shown later.

A.5 Laws of Exponents

Let a be any number and n any positive integer. The product of n numbers, each equal to a, is equal to a^n, which is read the nth power of a. We call a the base and n the exponent. The following laws of exponents will be very important in understanding material presented later:

$$a^m \times a^n = a^{m+n}$$

$$(a^m)^n = a^{mn}$$

$$\frac{a^m}{a^n} = a^{m-n}$$

$$a^{-n} = \frac{1}{a^n}$$

$$(ab)^n = a^n b^n$$

$$\left(\frac{a}{b} \right)^n = \frac{a^n}{b^n}$$

$$a^o = 1$$

$$a^{\frac{m}{n}} = \sqrt[n]{a^m}$$

A.6 Logarithms

A logarithm is defined as the exponent to which a base would be raised to equal a given number. For three numbers a, x, and y such that $a^x = y$, the exponent x is called the logarithm, to the base a, of the number y. This is written as

$$x = \log_a y$$

For example, $2^3 = 2 \times 2 \times 2 = 8$ and the logarithm to the base 2 for the number 8 is 3 or equivalently $3 = \log_2 8$. If x is the logarithm to the base a, y is called the antilogarithm of x to the base a, for example, 8 is the antilogarithm of 3 to the base 2.

The following relationships for logarithms may be derived from the laws of exponents:

$$\log_a a = 1$$

$$\log_a 1 = 0$$

$$a^{\log_a x} = x$$

$$\log_a(mn) = \log_a m + \log_a n$$

$$\log_a \left(\frac{m}{n} \right) = \log_a m - \log_a n$$

$$\log_a (m^n) = n \log_a m$$

$$\log_a (\sqrt[n]{m}) = \frac{\log_a m}{n}$$

Logarithmic and exponential functions are inverse functions.

The two most common bases used with logarithms are the base 10 and the base e, generally referred to as common and natural logarithms respectively. The number e is an irrational number approximately equal to 2.71828.* Base 10 logarithms are more convenient when doing arithmetical operations, but the availability of calculators and computers has taken much of the drudgery from routine calculations. Operations involving the base e are fundamental to population dynamics. Logarithms to the base e

*The number e is defined as

$$\lim_{h \to o} [(1 + h)^{\frac{1}{h}}]$$

are frequently designated by *ln*. This notation will be followed in this book. Thus the logarithm to the base *e* of the number 5 would be *ln* 5.

Logarithms to the base *a* may be changed to logarithms to the base *b* by the following relationship:

$$\log_b n = \log_b a \, \log_a n$$

The number of $\log_b a$ is called the modulus of the base *b* system with respect to the base *a* system. The modulus of base *a* with respect to base *b* is the reciprocal of the modulus of base *b* with respect to base *a*, that is:

$$\log_b a = \frac{1}{\log_a b}$$

A.7 Series

A sequence of numbers is formed by some function $f(x)$ when x assumes only positive integer values. The sequence is called a finite sequence or finite series if $f(x)$ is defined for *n* terms; the sequence is called an infinite sequence or infinite series if $f(x)$ is defined for all integer values of *x*.

An arithmetic series (or progression) is a sequence of numbers formed by $f(x) = a + dx$, $x = 0, 1, 2, \cdots$ where *a* is the first term of the sequence and *d* is the common difference between any two consecutive numbers of the sequence. Defining *n* as the number of terms, *l* as the last term, and S_n as the sum of the first *n* numbers in a finite sequence, the following relationships exist:

$$l = a + (n - 1)d$$

$$S_n = \frac{n}{2}(a + l)$$

A geometric series is a sequence of numbers formed by $f(x) = ar^x$, $x = 0, 1, 2, \cdots$ where *r* is the (common) ratio between any two consecutive numbers in the sequence. The sum of the first *n* terms in a geometric series is:

$$S_n = a \left(\frac{r^n - 1}{r - 1} \right) = \frac{rl - a}{r - 1}$$

where a, l, and n are as defined above for the arithmetic progression. If r^2 < 1 the sum S_n approaches the limit S_∞ as n increases without limit. The limit, S_∞, is:

$$S_\infty = \frac{a}{1-r}$$

The sum of a geometric progression starting with some term, say k, other than the first term is given by:

$$S_n = \sum_{x=k}^{n} ar^x = \frac{ar^k - ar^{n+1}}{1-r}$$

for values of $o \leq r < 1$. Many of the procedures concerned with survival make use of geometric series.

A power series is a sequence of numbers formed by

$$f(x) = a_o + a_1x + a_2x^2 + \cdots = \sum_{i=o}^{\infty} a_ix^i \text{ for } x = 0, 1, 2, \cdots$$

Power series will be developed further as special cases. However, as an example of its use, e, the base of natural logarithms, may be expressed as

$$e = 1 + 1/1! + 1/2! + 1/3! + \cdots$$

Development of the special cases of power series requires the use of derivatives which will be reviewed next along with integration.

A.8 Differentiation

An understanding of the concepts of differential and integral calculus is necessary for effective study of animal population dynamics. Such basic processes as mortality, growth, and reproduction as well as interactions between different populations are normally expressed as rates of change. Only fundamental aspects of the calculus most related to population biology will be reviewed.

Differentiation is the operation of determining the derivative of a function. The derivative is the slope of the curve of the function and expresses the rate of change of the dependent variable with respect to the independent variable. The derivative of $y = f(x)$ with respect to x is defined as the limit of $\Delta y/\Delta x$ as Δx approaches 0. This is written as:

$$\lim_{\Delta x \to o} \frac{\Delta y}{\Delta x} = \frac{dy}{dx}$$

The notation of $f'(x)$ is also used equivalently with dy/dx. Derivatives may be derived by finding Δy dividing by Δx and taking the limit as Δx approaches zero. As an example assume that $y = f(x) = kx^n$ where k is a constant and n is a positive integer. dy/dx may be derived as follows:

$$y = kx^n$$

$$y + \Delta y = k(x + \Delta x)^n*$$

$$y + \Delta y = kx^n + knx^{n-1} \Delta x + \frac{kn(n-1)x^{n-2}}{2!} (\Delta x)^2 + \cdots + (\Delta x)^n$$

$$\Delta y = knx^{n-1}\Delta x + \frac{kn(n-1)x^{n-2}(\Delta x)^2}{2!} + \cdots + (\Delta x)^n$$

$$\frac{\Delta y}{\Delta x} = knx^{n-1} + \frac{kn(n-1)x^{n-2}}{2!} \Delta x + \cdots + (\Delta x)^{n-1}$$

$$\lim_{\Delta x \to o} \frac{\Delta y}{\Delta x} = \frac{dy}{dx} = knx^{n-1}$$

Because of the importance of functions involving logarithms, particularly natural logarithms, the derivation for the derivative of $y = f(x) = \log_a x$ is presented:

$$y = \log_a x$$

$$y + \Delta y = \log_a (x + \Delta x)$$

$$\Delta y = \log_a (x + \Delta x) - \log_a x$$

$$\Delta y = \log_a \left(\frac{x + \Delta x}{x} \right) = \log_a \left(1 + \frac{\Delta x}{x} \right)$$

*$(x + \Delta x)^n = (x + \Delta x)(x + \Delta x) \cdots (x + \Delta x)$ with the term $(x + \Delta x)$ occurring n times.

$$\frac{\Delta y}{\Delta x} = \frac{1}{\Delta x} \log_a \left(1 + \frac{\Delta x}{x}\right)$$

$$= \frac{1}{x}\frac{x}{\Delta x} \log_a \left(1 + \frac{\Delta x}{x}\right)$$

$$= \frac{1}{x} \log_a \left(1 + \frac{\Delta x}{x}\right)^{\frac{x}{\Delta x}}$$

$$\lim_{\Delta x \to o} \frac{\Delta y}{\Delta x} = \frac{1}{x} \lim_{\Delta x \to o} \left[\log_a \left(1 + \frac{\Delta x}{x}\right)^{\frac{x}{\Delta x}}\right]$$

$$\frac{dy}{dx} = \frac{1}{x} \log_a \left[\lim_{\Delta x \to o} \left(1 + \frac{\Delta x}{x}\right)^{\frac{x}{\Delta x}}\right]$$

let $z = \frac{x}{\Delta x}$, then if $\Delta x \to o$, $z \to \infty$

$$\therefore \frac{dy}{dx} = \frac{1}{x} \log_a \left[\lim_{z \to \infty} \left(1 + \frac{1}{z}\right)^{z}\right]$$

$$= \frac{1}{x} \log_a e$$

It follows that:

$$1) \quad \frac{d}{dx} \ln x = \frac{1}{x}$$

because $\ln e = 1$.

$$2) \quad \frac{d}{dx} \log_a u = \frac{1}{x}\frac{du}{dx} \log_a e$$

$$3) \quad \frac{d}{dx} \ln u = \frac{1}{x}\frac{du}{dx}$$

where $u = f(x)$.

An explicit function is a function in which $y = f(x)$, that is, y is defined in terms of x. If y is determined by x but is not defined as an explicit function of x, then y is said to be an implicit function of x. Implicit differentiation is the differentiation of an implicit function. An example will serve to demonstrate the technique. Let $y^2 = x^2 + 1$, an implicit function of x. The derivative of y with respect to x is found differentiating both sides of the equation with respect to x and then solving for dy/dx, for example

$$y^2 = x^2 + 1$$

$$2y\frac{dy}{dx} = 2x$$

$$\frac{dy}{dx} = \frac{x}{y}$$

The concept of implicit differentiation is necessary to obtain the derivative of exponential functions which will be considered next as the last example.

The exponential function is of the form $y = f(x) = a^x$. Differentiating y with respect to x by the rule for implicit functions gives the following:

$$y = a^x$$

$$\log_b y = x \log_b a$$

$$\frac{d \log_b y}{dx} = \frac{1}{y}\frac{dy}{dx} = \log_b a$$

$$\frac{dy}{dx} = y \log_b a = a^x \log_b a$$

If u is a function of x

$$\frac{d}{dx} a^u = a^u (\log_b a) \frac{du}{dx}$$

If a is equal to e, the base of natural logarithms, then

$$\frac{d}{dx} e^u = e^u \frac{du}{dx}$$

Higher-order derivatives are obtained by repeatedly taking derivatives of the function. If $y = f(x)$, then $dy/dx = f'(x)$, and $d^2y/dx^2 = f''(x) = \frac{d}{dx}$ $\frac{dy}{dx}$ which is the second derivative of y with respect to x; $f'''(x) = \frac{d}{dx}$

$\dfrac{d^2y}{dx^2} = \dfrac{d^3y}{dy^3}$ is the third derivative of y with respect to x. Higher-order derivatives may also be obtained for implicit functions.

Partial differentiation is the operation of differentiation when more than one independent variable is involved in the function. Let $z = f(x,y)$. The derivative of z with respect to x, with y treated as a constant, is called the partial derivative of z with respect to x. This partial derivative is written as $\partial z/\partial x$. Higher-order partial derivatives are obtained in a procedure analogous to higher-order total derivatives as discussed previously.

The differential of a function is equal to its derivative multiplied by the differential of the independent variable. This is written symbolically for $y = f(x)$ as:

$$dy = \frac{dy}{dx}\,dx = f'(x)\,\Delta x$$

In general, the formulas for derivatives become differential formulas by multiplying through by dx.

Equations involving derivatives and differentials are called ordinary differential equations; equations containing partial derivatives are called partial differential equations. It is frequently possible to express a problem as a differential equation and obtain an answer for the problem through solution of the differential equation. Solutions to differential equations are obtained by integration, the procedure reviewed next.

A.9 Integration

Integration is the inverse operation of differentiation, for example, given the derivative of a function, find the function. The symbol $\int$ is called the integral sign and indicates the operation of integration. If $F(x)$ is a function whose derivative is $f(x)$, then

$$F(x) = \int f(x)dx$$

The derivative of $F(x)$ must be equal to $f(x)$ and this may be used to check the´integration. However, because two functions that differ only by a constant have the same derivative, $\int f(x)dx$ is called the indefinite integral of $f(x)$.

The following fundamental integration formulas are of primary importance for material presented in this book:

$$\int u^n\, du = \frac{u^{n+1}}{n+1} + C \text{ providing } n \neq 1$$

$$\int \frac{du}{u} = \ln u + C$$

$$\int e^u\, du = e^u + C$$

$$\int a^u = \frac{a^u}{\ln a} + C$$

In many problems initial conditions will be known and then the constant of integration, C above, may be determined. The solution is then an exact solution. For example, let $y = \frac{1}{2}\, x + 2$, and therefore $dy/dx = \frac{1}{2}$. To find y, integrate

$$y = \int \tfrac{1}{2}\, dx = \tfrac{1}{2}\, x + C$$

If $x_1 = 0$, $y_1 = 2$ is given as an initial condition, C may be evaluated by substitution as follows:

$$y = \tfrac{1}{2}\, x + C$$

$$C = y - \tfrac{1}{2}\, x$$

in particular for the initial condition values of $x = 0$, $y = 2$

$$C = 2 - \tfrac{1}{2}\, (0) = 2$$

This gives the complete solution $y = \frac{1}{2}\, x + 2$.

The definite integral is shown by the symbol

$$\int_a^b f(x)\, dx$$

where b is called the upper limit of integration and a is called the lower limit of integration. The value of a definite integral is $F(b) - F(a)$, that is,

$$\int_a^b f(x)\, dx = [F(x) + C]_a^b = F(b) - F(a)$$

The constant of integration, C, is cancelled out by the subtraction.

Frequently, differential equations will be derived for which there is no apparent closed solution. In such cases it may be possible to obtain numerical solutions for the specific problem. The method of Euler consists of an approximation of dy/dx by the difference quotient $\Delta y/\Delta x$. Starting at some point y_o, the first Δy is by definition $y_1 - y_o$. The value for y_1 must be y_o plus $f(x_o)$ times Δx, that is, $y_1 = y_o + \Delta x f(x_o)$. This is a recursive relationship; $y_2 = y_1 + \Delta x f(x_1)$, $y_3 = y_2 + \Delta x f(x_2)$, $\cdots$. The process is carried forward to the desired end-point with results given in tabular or graphic form. This procedure gives an approximate solution which converges to the exact solution as $\Delta x \to 0$. However, as Δx gets smaller, the number of computations increases with an increasing round-off error. For most problems there is an acceptable Δx which will give a solution of sufficient accuracy. There are more accurate numerical techniques but they will not be reviewed here.

A.10 Taylor Series

If $f(x)$ and derivatives of $f(x)$ are continuous at the point $x = a$, within an interval including $x = a$, $f(x)$ may be expressed as a power series in $(x - a)$. Thus

$$f(x) = c_o + c_1 (x - a) + c_2 (x - a)^2 + c_3 (x - a)^3 + \cdots$$

Derivatives of the above expression are

$$f'(x) = c_1 + 2c_2 (x - a) + 3c_3 (x - a)^2 + \cdots$$

$$f''(x) = 2c_2 + 6c_3 (x - a) + \cdots$$

$$f'''(x) = 6c_3 + \cdots$$

$$.$$
$$.$$
$$.$$

Substituting $x = a$ in the above equations gives the following

$$f(a) = c_o$$

$$f'(a) = c_1$$

$$f''(a) = 2c_2$$

$$f'''(a) = 2 \cdot 3 c_3$$

.

.

.

Therefore the coefficients in the power series expansion of $f(x)$ are:

$$c_o = f(a)$$

$$c_1 = f'(a)$$

$$c_2 = f''(a)/2!$$

$$c_3 = f'''(a)/3!$$

.

.

.

$$c_n = f^n(a)/n!$$

The function $f(x)$ may then be expressed as

$$f(x) = f(a) + (x - a)f'(a) + \frac{(x - a)^2 f''(a)}{2!} + \cdots + \frac{(x - a)^n f^n(a)}{n!} + \cdots$$

or

$$f(x) = \sum_{i=0}^{\infty} \frac{(x - a)^i}{i!} f^i(a)$$

This series is called a Taylor series. The same concept may be applied to a function of more than one variable; consider a two-variable case $f(x,y)$ for which the function and its derivatives exist and are continuous at $x = a$, $y = b$. The function may be expressed as a Taylor series as

$$f(x,y) = f(a,b) + (x - a)\frac{\partial f(a,b)}{\partial x} + \frac{(x - a)^2}{2!} \frac{\partial^2 f(a,b)}{\partial x^2} + \cdots$$

$$+ (y - b)\frac{\partial f(a,b)}{\partial y} + \frac{(y - b)^2}{2!} \frac{\partial^2 f(a,b)}{\partial y^2} + \cdots$$

$$+ (x - a)(y - b) \frac{\partial^2 f(a,b)}{\partial x\, \partial y} + \cdots$$

A simple example is provided by $f(x,y) = xy$ to be determined at $x = 2.1$, $y = 3.1$. This function may be evaluated by Taylor series expansion around $x = 2$, $y = 3$. The derivatives are

$$\frac{\partial f(x,y)}{\partial x} = y \qquad\qquad \frac{\partial f(x,y)}{\partial y} = x$$

$$\frac{\partial^2 f(x,y)}{\partial x^2} = 0 \qquad\qquad \frac{\partial^2 f(x,y)}{\partial y^2} = 0$$

$$\frac{\partial^2 f(x,y)}{\partial x\, \partial y} = 1$$

Therefore $f(2.1,3.1)$ for $f(x,y) = xy$ expanded around $x = 2$, $y = 3$ is

$$f(2.1,3.1) = f(2,3) + (2.1{-}2) \times 3 + \frac{(2.1{-}2)^2}{2} \times 0 + \cdots$$

$$+ (3.1{-}3) \times 2 + \frac{(3.1{-}3)^2}{2} \times 0 + \cdots$$

$$+ (2.1{-}2)(3.1{-}3) \times 1 + \cdots$$

$$f(2.1,3.1) = 6.51$$

The result is exact in this case since all derivatives higher than the second one are equal to zero.

If $a = 0$ in a Taylor's series, then

$$f(x) = f(0) + xf'(0) + \frac{x^2}{2!} f''(0) + \frac{x^3}{3!} f'''(0) + \cdots$$

This series is called a Maclaurin's series. Consider, for example, finding $f(x) = e^x$:

$$e^x = 1 + x + \frac{x^2}{2!} + \frac{x^3}{3!} + \cdots \quad \text{(converges for all values of } x)$$

A.11 Matrix Algebra

An arrangement of elements in rows and columns is called a matrix. The elements may be numbers, variables, functions, or even other matrices. The dimensions of a matrix are given as the number of rows and the number of columns. A matrix of dimensions 3 by 4 is therefore composed of three rows and four columns, for example:

$$\mathbf{A} = \begin{bmatrix} 4 & 3 & 1 & 6 \\ 7 & 13 & 11 & 5 \\ 9 & 2 & 0 & 8 \end{bmatrix}$$

The size of a matrix is normally called its order; the order of matrix $\mathbf{A}$ is 3 by 4; usually written 3×4. A matrix is square if the number of rows equals the number of columns.

An array that has one row or one column is called a row vector or column vector respectively. A matrix or vector consisting of a single element is called a scalar. A scalar may be thought of as a 1×1 matrix for some purposes.

Subscript notation was reviewed in Section A.2. Matrix elements have two subscripts: the first referring to the row and the second to the column. Vector elements have a single subscript referring to the position of the element in the vector.

Arithmetic operations require conformability of the matrices; conformability differs depending upon the operation. Matrices of the same order are conformable to the operation of addition and subtraction. These operations are carried out on an element-by-corresponding-element basis. For example, if

$$\mathbf{A} = \begin{bmatrix} 1 & 3 \\ 7 & 9 \\ 13 & 17 \end{bmatrix} \text{ and } \mathbf{B} = \begin{bmatrix} 1 & 2 \\ 4 & 5 \\ 7 & 8 \end{bmatrix}$$

then the operation of addition is:

$$\mathbf{A} + \mathbf{B} = \begin{bmatrix} 1+1 & 3+2 \\ 7+4 & 9+5 \\ 13+7 & 17+8 \end{bmatrix} = \begin{bmatrix} 2 & 5 \\ 11 & 14 \\ 20 & 25 \end{bmatrix}$$

The operation of subtraction results in:

$$\mathbf{A} - \mathbf{B} = \begin{bmatrix} 1-1 & 3-2 \\ 7-4 & 9-5 \\ 13-7 & 17-8 \end{bmatrix} = \begin{bmatrix} 0 & 1 \\ 3 & 4 \\ 6 & 9 \end{bmatrix}$$

Matrix multiplication requires conformability in a different manner. Let **C** and **D** be the 2×2 matrices

$$\mathbf{C} = \begin{bmatrix} 1 & 3 \\ 5 & 7 \end{bmatrix}, \quad \mathbf{D} = \begin{bmatrix} 1 & 2 \\ 3 & 4 \end{bmatrix}$$

then multiplication is carried out as follows:

$$\mathbf{CD} = \begin{bmatrix} (1 \times 1)+(3 \times 3) & (1 \times 2)+(3 \times 4) \\ (5 \times 1)+(7 \times 3) & (5 \times 2)+(7 \times 4) \end{bmatrix} = \begin{bmatrix} 10 & 14 \\ 26 & 38 \end{bmatrix}$$

The element in the first row and first column of the matrix of the product is made up of the sum of the first row times the first column, on an element-by-corresponding-element basis. The analogous operation results in the other elements of the product matrix. The resulting matrix is the same order as **C** and **D**. These two matrices are also conformable to addition and subtraction.

If multiplication is attempted on the matrices **A** and **B**, the operation is not possible since there is not a one-to-one correspondence between the elements in the rows of **A** to the elements in the columns of **B**. However, if **A** is changed by making the first row the first column, the second row the second column, and the third row the third column, then the elements of first row of this new matrix do have a corresponding element in the first column of the matrix **B**. This new matrix formed by interchanging rows and columns of **A** is called the transpose of **A**, written as **A**′, for example:

$$\mathbf{A}' = \begin{bmatrix} 1 & 7 & 13 \\ 3 & 9 & 17 \end{bmatrix}$$

Multiplication of **A**′ times **B** is now possible since these two matrices are conformable for multiplications; the product is:

$$\mathbf{A'B} = \begin{bmatrix} 1 & 7 & 13 \\ 3 & 9 & 17 \end{bmatrix} \begin{bmatrix} 1 & 2 \\ 4 & 5 \\ 7 & 8 \end{bmatrix} = \begin{bmatrix} (1\times1)+(7\times4)+(13\times7) & (1\times2)+(7\times5)+(13\times8) \\ (3\times1)+(9\times4)+(17\times7) & (3\times2)+(9\times5)+(17\times8) \end{bmatrix}$$

$$\qquad\quad \text{2 by 3} \qquad \text{3 by 2} \qquad\qquad\qquad\qquad \text{2 by 2}$$

$$= \begin{bmatrix} 120 & 141 \\ 158 & 187 \end{bmatrix}$$

Notice that the order of the resulting product is different from either $\mathbf{A}'$ or $\mathbf{B}$. Notice also that $\mathbf{A}'$ and $\mathbf{B}$ are not conformable to the operation of addition and subtraction. It is seen that if the orders of the two matrices to be multiplied are placed in the order of the multiplication to be carried out, for example, 2 by 3 3 by 2, that conformability for multiplication requires the two inner numbers to be identical. The order of the resulting product is given by the two outside numbers, for example, 2×2. The product $\mathbf{BA}'$ would be set up as $3 \times 2\ 2 \times 3$, which is conformable but results in a matrix of order 3×3 and is therefore different than $\mathbf{A'B}$. The product of $\mathbf{AB}'$ is a matrix of order 3×3 but not the same as $\mathbf{A'B}$. However, $(\mathbf{A'B})'$ is the same as $\mathbf{B'A}$.

The multiplication of $\mathbf{AB}$ is called premultiplication of $\mathbf{B}$ by $\mathbf{A}$ or multiplication from the left by $\mathbf{A}$. Multiplication of $\mathbf{B}$ by $\mathbf{A}$ from the right is called postmultiplication of $\mathbf{B}$ by $\mathbf{A}$.

A scalar times a matrix is equal to the scalar times each element of the matrix. For example, the scalar 2 times matrix $\mathbf{C}$ is

$$2\mathbf{C} = 2 \begin{bmatrix} 1 & 3 \\ 5 & 7 \end{bmatrix} = \begin{bmatrix} 2 & 6 \\ 10 & 14 \end{bmatrix}$$

The same is true if the scalar is $1/2$ or $1/a$, for example:

$$\frac{1}{a}\mathbf{C} = \frac{1}{a} \begin{bmatrix} 1 & 3 \\ 5 & 7 \end{bmatrix} = \begin{bmatrix} \dfrac{1}{a} & \dfrac{3}{a} \\ \dfrac{5}{a} & \dfrac{7}{a} \end{bmatrix}$$

A square matrix is one in which the number of rows is the same as the number of columns. A symmetric matrix is a square matrix for which the transpose equals the original matrix, for example:

$$\mathbf{E} = \begin{bmatrix} 1 & 5 \\ 5 & 3 \end{bmatrix} = \mathbf{E}' = \begin{bmatrix} 1 & 5 \\ 5 & 3 \end{bmatrix}$$

A diagonal matrix is a square matrix in which all off-diagonal elements are zero. A special diagonal matrix consisting of 1 for each diagonal element is called the identity matrix. An identity matrix of order 3×3 is:

$$\mathbf{I} = \begin{bmatrix} 1 & 0 & 0 \\ 0 & 1 & 0 \\ 0 & 0 & 1 \end{bmatrix}$$

A matrix times an identity matrix is equal to the original matrix. For example:

$$\mathbf{EI} = \begin{bmatrix} 1 & 5 \\ 5 & 3 \end{bmatrix} \begin{bmatrix} 1 & 0 \\ 0 & 1 \end{bmatrix} = \begin{bmatrix} 1 & 5 \\ 5 & 3 \end{bmatrix}$$

The operation of division used to find the value of y in an algebraic expression such as $ay = c$, can be thought of as the process of multiplying each side of the equation by $1/a$ giving $\frac{1}{a} ay = \frac{1}{a} c$, which upon rearrangement becomes $1 \, y = a^{-1} c$. An analogous procedure is used in matrix algebra. An example of the procedure can be shown in the solution of simultaneous linear equation. The set of equations

$$2x + y + z = 1$$
$$x - 2y - 3z = 1$$
$$3x + 2y + 4z = 5$$

may be written in matrix form as:

$$\begin{bmatrix} 2 & 1 & 1 \\ 1 & -2 & -3 \\ 3 & 2 & 4 \end{bmatrix} \begin{bmatrix} x \\ y \\ z \end{bmatrix} = \begin{bmatrix} 1 \\ 1 \\ 5 \end{bmatrix}$$

The problem is to find the values of x, y, and z that satisfy this set of equations. Premultiplication on both sides of the matrix equation by some matrix that on the left-hand side would produce an identity matrix times the vector of variables would leave just the vector of variables as was shown above, for example, $\mathbf{IE} = \mathbf{E}$. Premultiplication on the right-hand side would provide the values for x, y, and z that satisfy this set of equations.

This is shown for the above example by premultiplying by the matrix

$$\frac{1}{-9} \begin{bmatrix} -2 & -2 & -1 \\ -13 & 5 & 7 \\ 8 & -1 & -5 \end{bmatrix}$$

since

$$\frac{1}{-9} \begin{bmatrix} -2 & -2 & -1 \\ -13 & 5 & 7 \\ 8 & -1 & -5 \end{bmatrix} \begin{bmatrix} 2 & 1 & 1 \\ 1 & -2 & -3 \\ 3 & 2 & 4 \end{bmatrix} \begin{bmatrix} x \\ y \\ z \end{bmatrix}$$

$$= \frac{1}{-9} \begin{bmatrix} -9 & 0 & 0 \\ 0 & -9 & 0 \\ 0 & 0 & -9 \end{bmatrix} \begin{bmatrix} x \\ y \\ z \end{bmatrix}$$

$$= \begin{bmatrix} 1 & 0 & 0 \\ 0 & 1 & 0 \\ 0 & 0 & 1 \end{bmatrix} \begin{bmatrix} x \\ y \\ z \end{bmatrix} = \begin{bmatrix} x \\ y \\ z \end{bmatrix}$$

for the left-hand side. Equality of the equation is maintained by premultiplying the right-hand side by the same component giving:

$$\frac{1}{-9} \begin{bmatrix} -2 & -2 & -1 \\ -13 & 5 & 7 \\ 8 & -1 & -5 \end{bmatrix} \begin{bmatrix} 1 \\ 1 \\ 5 \end{bmatrix} = \frac{1}{-9} \begin{bmatrix} -9 \\ 27 \\ -18 \end{bmatrix} = \begin{bmatrix} 1 \\ -3 \\ 2 \end{bmatrix}$$

which are the values for x, y, and z that satisfy the original set of equations. This may be checked by substituting these values in the original equations.

The matrix

$$\frac{1}{-9} \begin{bmatrix} -2 & -2 & -1 \\ -13 & 5 & 7 \\ 8 & -1 & -5 \end{bmatrix}$$

is called the inverse of matrix

$$\begin{bmatrix} 2 & 1 & 1 \\ 1 & -2 & -3 \\ 3 & 2 & 4 \end{bmatrix}$$

In a manner analogous to the scalar operation of $y = a^{-1}x$ as a solution to $ay = x$, the solution to the matrix expression $\mathbf{AY} = \mathbf{X}$ is $\mathbf{Y} = \mathbf{A}^{-1}\mathbf{X}$ where $\mathbf{A}^{-1}$ is the inverse of $\mathbf{A}$, providing $\mathbf{A}^{-1}$ exists. An inverse can exist only for a square matrix, but not all square matrices have inverses.

The process of finding the inverse of a matrix necessitates the use of determinants. A determinant is a scalar value of a matrix. A 1×1 matrix has as its determinant the value of that single element. The determinant of a second order (2×2) matrix is shown by example. Let

$$\mathbf{Z} = \begin{bmatrix} 4 & 3 \\ 1 & 2 \end{bmatrix}$$

then the determinant of $\mathbf{Z}$, written as $|\mathbf{Z}|$, is equal to:

$$|\mathbf{Z}| = (4 \times 2) - (1 \times 3) = 5$$

This could be viewed as the product of each element of a single row (or column) times a term called a cofactor. The cofactor is defined as $(-1)^{i+j}\,|\mathbf{M}_{ij}|$, where $|\mathbf{M}_{ij}|$ is the determinant of the matrix formed by crossing out the row and column of the chosen element. $|\mathbf{M}_{ij}|$ is called a minor of $|\mathbf{Z}|$. For example, in the matrix $\mathbf{Z}$ above, the element 4 has as its cofactor the matrix left after crossing out the row and column containing the element 4, that is:

$$\begin{bmatrix} \cancel{4} & \cancel{3} \\ 1 & 2 \end{bmatrix}$$

Applying this procedure to each element in the row (or column) and adding the results provides the value of the determinant, as shown in:

$$|\mathbf{A}| = \begin{vmatrix} 4 & 3 \\ 1 & 3 \end{vmatrix} = [4 \times (-1)^{1+1}\,|2|] + [3 \times (-1)^{1+2}\,|1|]$$

$$= (4 \times 2) + (-3) = 5$$

The same result is obtained if we use any row or column, for example, using the second column results in:

$$|\mathbf{A}| = \begin{vmatrix} 4 & 3 \\ 1 & 2 \end{vmatrix} = [3 \times (-1)^{1+2} \times |1|] + [2 \times (-1)^{2+2} \times |4|] = 5$$

The same principle may be applied to a higher-order square matrix. In the example set of equations, the matrix

$$\begin{bmatrix} 2 & 1 & 1 \\ 1 & -2 & -3 \\ 3 & 2 & 4 \end{bmatrix}$$

has the determinant, based on the first row, as given below:

$$2 \times (-1)^{1+1} \begin{vmatrix} -2 & -3 \\ 2 & 4 \end{vmatrix} + 1 \times (-1)^{1+2} \begin{vmatrix} 1 & -3 \\ 3 & 4 \end{vmatrix}$$

$$+ 1 \times (-1)^{1+3} \begin{vmatrix} 1 & -2 \\ 3 & 2 \end{vmatrix} = \text{determinant}$$

$$2 \times (1) \times [(-2) \times 4 - (-3) \times 2] + 1 \times (-1) \times [1 \times 4 - 3 \times (-3)]$$

$$+ \quad 1 \times (1) \times [1 \times 2 - 3 \times (-2)] \quad = \text{determinant}$$

$$[2 \times -2] \quad + \quad [1 \times -13] \quad + \quad [1 \times 8] \quad = -9$$

The inverse of a matrix is found by 1) obtaining the determinant, 2) calculating the values of the cofactors for all elements either by row or column, 3) replacing the row or column by its cofactors to form a matrix of cofactors, and 4) multiplying the reciprocal of the determinant by the transpose of the cofactor matrix. The solution for the example set of equations is shown below, with the determinant found using rows. The first-row cofactors were obtained above in finding the determinant. Cofactors of the second row are:

$$(-1)^{2+1} \begin{vmatrix} 1 & 1 \\ 2 & 4 \end{vmatrix} \qquad (-1)^{2+2} \begin{vmatrix} 2 & 1 \\ 3 & 4 \end{vmatrix} \qquad (-1)^{2+3} \begin{vmatrix} 2 & 1 \\ 3 & 2 \end{vmatrix}$$

$$-2 \qquad\qquad\qquad 5 \qquad\qquad\qquad -1$$

and of the third row:

$$(-1)^{3+1} \begin{vmatrix} 1 & 1 \\ -2 & -3 \end{vmatrix} \qquad (-1)^{3+2} \begin{vmatrix} 2 & 1 \\ 1 & -3 \end{vmatrix} \qquad (-1)^{3+3} \begin{vmatrix} 2 & 1 \\ 1 & -2 \end{vmatrix}$$

$$-1 \qquad\qquad\qquad 7 \qquad\qquad\qquad -5$$

The matrix of cofactors becomes:

$$\begin{bmatrix} -2 & -13 & 8 \\ -2 & 5 & -1 \\ -1 & 7 & -5 \end{bmatrix}$$

Multiplying the transpose of this matrix by the reciprocal of the determinant provides the inverse that was used in finding the solution to the example equation set.

It is not practical to work out by hand the inverse of a matrix of much higher order. Nearly all computing facilities have programs that can determine the inverse.

A.12 Probability

The previous material reviewed has been concerned with determining relationships where, when x was substituted in the function $f(x)$, values for y would be completely specified. The study of animal populations, on the other hand, involves observation or measurement of variables that, on an individual basis, do not have a known outcome. The number of plants for a given plot in a pond, the sex or the length of the next fish caught provide little information when considered singly. The subject of statistics was developed in order to describe or make inference about a population by using information obtained by taking a sample from the population being studied. A basic tool of statistics is probability. The review presented here will only present the concept of probability and will not review the subject for students not familiar with the terminology.

Classical or a priori probability is defined as the fraction resulting from the ratio of the number of ways an event having a given attribute can occur to the total number of equally likely and mutually exclusive ways the event can occur. This is best illustrated by example: a coin toss may result in a head or a tail, a cast die may result in a 1, 2, 3, 4, 5, or 6, and a card drawn from an ordinary playing card deck may be black or red. An a priori probability may be assigned to each of these situations: ½ for the coin toss, $1/6$ for the cast die, ½ for the card color.

The probability of an event occurring may also be observed empirically as the frequency of occurrence for a particular event in a large number of trials. This is a posteriori or frequency probability. An example would be the age frequency distribution in a sample of fish.

The probability of event A is designated as $P(A)$. Probabilities by definition will be numbers in the 0 to 1 interval. The probability of either event A or event B occurring, written $P(A \cup B)$, is the probability of A plus the probability of B minus the probability of both A and B occurring, that is, $P(A \cup B) = P(A) + P(B) - P(A \cap B)$. The meaning of the symbols $\cup$ and $\cap$ may be conveyed by referring to Figure A.12.1. If the area of the square is taken as 1, then the area of circle A represents the probability of event A, and the area of circle B represents the probability of event B. The cross-hatched area is a part of both A and B. This cross-hatched area is the probability of both A and B occurring and is represented by the symbol $\cap$, read the intersection of A and B. The area of circle A and circle B not included in the cross-hatched section is the probability of either event A or event B, or both. The unhatched area within circles A and B is represented by the symbol $\cup$, read the union of A and B.

The probability of both A and B occurring, $P(A,B)$, is equal to the conditional probability of B given that A has occurred, $P(B|A)$, times the probability of A, that is $P(A,B) = P(A) \cdot P(B|A)$. The concept of conditional probability is best shown by example. Suppose a box has 10 red balls numbered 1 to 10 and 10 black balls numbered 1 to 10 (see Table A.12.1). The probability of choosing a particular ball is $1/20$. The probability of a chosen ball being both red and 2, $P(R,2)$, is equal to probability of being red, $10/20$, times the probability of being a 2 given the ball is red, $P(2|R)$, that is, $P(R,2) = 10/20 \times 1/10 = 1/20$. As another example, consider the probability of choosing a black ball with a number greater than 5:

$$P(B,>5) = P(B)\ P(>5|B)$$

$$= \frac{10}{20} \times \frac{5}{10} = \frac{1}{4}$$

Notice that $P(B,>5)$ is also equal to $P(>5)\ P(B|>5)$.

$$P(B,>5) = P(>5)\ P(B|>5)$$

$$= \frac{10}{20} \times \frac{1}{2} = \frac{1}{4}$$

These ideas may be extended to situations involving more than two criteria.

Independence in a probability sense may be defined in terms of the conditional probability in that if $P(A|B)$ does not depend on B, the events are independent. Two events will be independent if any of the following conditions are met: $P(A|B) = P(A)$ or $P(B|A) = P(B)$ or $P(A,B) = P(A)\ P(B)$.

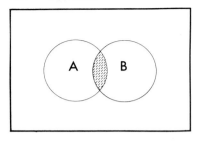

Figure A.12.1. Schematic representation of probability

A probability mass function, say $f(x)$, for the random variable x, is a function assigning a probability value for a specified value of x in the discrete case. A probability mass function satisfies the following:

$$f(x_i) \geq 0 \quad i = 1, 2, \cdots$$

and

$$\sum_i f(x_i) = 1$$

Any function that satisfies these conditions may be a probability mass function. Examples of discrete distributions are Poisson, binomial, multinomial, hypergeometric, and negative binomial, among others.

A probability density function assigns a probability value for a specified interval for x in the continous case and must satisfy the following:

$$f(x) \geq 0 \quad -\infty < x < \infty$$

and

$$\int_{-\infty}^{\infty} f(x)dx = 1$$

Table A.12.1. Frequency of numbered red and black balls in a box

| | Number | | | | | | | | | | |
	1	2	3	4	5	6	7	8	9	10	*Total*
Red	1	1	1	1	1	1	1	1	1	1	10
Black	1	1	1	1	1	1	1	1	1	1	10
Total	2	2	2	2	2	2	2	2	2	2	20

For all a and b $(a<b)$,

$$P\ (a<x<b) = \int_a^b f(x)dx$$

Examples of continuous distributions are the normal, uniform, beta, gamma, and Cauchy.

A.13 Expected Values

A sample space is that set consisting of all possible outcomes for which a probability density function is defined. If x is a real-valued function defined on that sample space, x is called a random variable. The expected value of a random variable x with probability density function $f(x)$ is defined as

$$E(x) = \sum_x xf(x)$$

for discrete variables, and as

$$E(x) = \int_{-\infty}^{\infty} xf(x)dx$$

for continuous variables. The expected value is synonymous with the mean.

Moments of a distribution are the expected values of powers of the random variable. The rth moment is generally written as:

$$E(x^r) = \sum_n x^r\ f(x) = u_r'$$

for discrete variables and as:

$$E(x^r) = \int_{-\infty}^{\infty} x^r\ f(x)dx = u_r'$$

for continuous variables. The first moment is the mean.

Moments about the mean are designated as u_r and are defined as:

$$E(x - u_1')^n = \sum_x (x - u_1')^r f(x) = u_r$$

or

$$E(x - u_1')^r = \int_{-\infty}^{\infty} (x - u_1')^r f(x)dx = u_r$$

The second moment about the mean is called the variance of x, an important parameter when comparing populations. The expected values of the moments of a random sample from a population are the population moments.

A.14 Statistics

Experimentation is the foundation of science but seldom is an experiment conducted such that the entire population is observed. More generally a fraction of the population, sample, is observed and then inferences concerning the population are based upon the sample observation. There is always doubt associated with an inference or an inductive process. If the experimentation is done in a prescribed manner, a level of probability associated with that doubt may be stated. The subject of statistics provides the framework for quantifying such probability levels.

A statistic is a function, not containing unknown parameters, of an observable random variable. The observed value of a statistic will influence or cause a certain decision to be made. If the statistic is a poor one, there may be a loss associated with the decision action. Statistics that have a small loss associated with them would intuitively seem to be worth pursuing, and many procedures for deriving estimators are based on minimization of a loss function. Adequate representation of a loss function is difficult in many cases, but an obviously appealing loss function would relate an observation and prediction of a random variable by some squared error function of the form Σ (Observation $-$ Prediction)$^2 \cdot g$ (Parameters).

A.15 The Normal Distribution

The normal distribution is of particular importance since many variables tend to be normally distributed and the distribution of sample means from other distributions tend to the normal distribution. This is stated in the central-limit theorem. If $f(x)$ is a density with mean u and variance σ^2 and $\bar{x}_n$ is a sample mean from n observations, then the quantity $\frac{\bar{x}-u}{\sigma}\sqrt{n}$ approaches a normal distribution with mean o and variance 1 as $n\to\infty$. The density function for the normal distribution is:

$$f(x) = \frac{1}{\sigma\sqrt{2\pi}}\, e^{-\frac{1}{2}\left(\frac{x-u}{\sigma}\right)^2}$$

The two parameters u and σ are unknown and in order to specify the distribution, as for example to define some characteristic or evaluate an experiment, estimates for these parameters are necessary. There are many different procedures for producing estimators such as method of least squares, method of Bayes, method of moments, and method of maximum likelihood. The method of maximum likelihood will be used to derive estimators for the parameters of the normal distribution and will serve as an illustration of the method.

A random sample of n observations of a variable x is taken from a population assumed to have a normal distribution with parameters u and σ. Then the distribution of x_1 is assumed to be $f(x_1;u,\sigma)$, x_2 is distributed as $f(x_2;u,\sigma)$, and for each $i \leq n$, x_i is distributed as $f(x_i;u,\sigma)$. The joint density, also called the likelihood function, of these observations is then:

$$L(u,\sigma) = f(x_i;u,\sigma)\, f(x_2;u,\sigma) \cdots f(x_n;u,\sigma)$$

Assuming a normal distribution for each x_i, this likelihood function becomes

$$L(u,\sigma) = \left(\frac{1}{2\pi\sigma^2}\right)^{n/2} e^{-\frac{1}{2\sigma^2}\sum_{i=1}^{n}(x_i-u)^2}$$

As the name of the method suggests, the procedure is now to choose estimators for u and σ such that the likelihood function has a maximum value. The maximum value for a function may be found by taking its derivative, setting the derivative equal to zero and solving for the parameters. Since the logarithm of a function will have its maximum at the same values of the parameters as the function itself, and because it is easier to take derivatives in some cases, the natural logarithm of the likelihood function will be used. This becomes:

$$ln\ L\ (u,\sigma) = L^*(u,\sigma) = -\frac{n}{2}\ ln\ 2\pi - \frac{n}{2}\ ln\ \sigma^2 - \frac{1}{2\sigma^2} \sum_{i=1}^{n} (x_i - u)^2$$

Taking derivatives of $L^*(u,\sigma)$ with respect to the parameters results in:

$$\frac{\partial L^*(u,\sigma)}{\partial u} = \frac{1}{\sigma^2} \sum_{i=1}^{n} (x_i - u)$$

$$\frac{\partial L^*(u,\sigma)}{\partial \sigma^2} = -\frac{n}{2\sigma^2} + \frac{1}{2\sigma^4} \sum_{i=1}^{n} (x_i - u)^2$$

Equating these to o and solving for parameter u and σ gives:

$$\hat{u} = \frac{1}{n} \sum_{i=1}^{n} x_i = \bar{x}$$

and

$$\hat{\sigma}^2 = \frac{1}{n} \sum_{i=1}^{n} (x_i - \bar{x})^2$$

The expected value of $\bar{x}$ is u, that is, $E(\bar{x}) = u$, and $\bar{x}$ is therefore an unbiased estimator of u. On the other hand σ^2 is a biased estimator of σ^2 since $E(\hat{\sigma}^2) \neq \sigma^2$, but $\frac{n}{n-1}\ \sigma^2 = s^2$ is unbiased. Many estimators discussed in the text will be maximum-likelihood estimators. The principle of their derivation is the same as discussed above.

A.16 Confidence Intervals

Estimators such as those derived above provide point estimates for parameters but provide no basis for the degree of uncertainty discussed at the beginning of the previous section. For example, $\bar{x} = 10$ may be an estimate of the population mean but, if a second sample from the population were drawn, would 20 be a reasonable value to expect for $\bar{x}$?

The central-limit theorem stated that for any density the variable $y = \frac{\bar{x}-u}{\sigma}\sqrt{n}$ approaches the normal distribution with $u = 0$ and $\sigma^2 = 1$. The density of y is

$$f(y) = \frac{1}{\sqrt{2\pi}} \, e^{-\frac{1}{2} y^2}$$

and therefore the probability that y is between two values may be computed. For example:

$$P(-1.96 > y > 1.96) = \int_{-1.96}^{1.96} f(y) dy = 0.95$$

Substituting for y in this expression gives:

$$P(-1.96 < \frac{\bar{x} - u}{\sigma/\sqrt{n}} < 1.96) = 0.95$$

which is equivalent to:

$$P\left(-1.96 \, \frac{\sigma}{\sqrt{n}} < \bar{x} - u < 1.96 \, \frac{\sigma}{\sqrt{n}}\right) = 0.95$$

Rearranging terms provides the following statement which is called a confidence interval:

$$P\left(\bar{x} - 1.96 \, \frac{\sigma}{\sqrt{n}} < u < \bar{x} + 1.96 \, \frac{\sigma}{\sqrt{n}}\right) = 0.95$$

This statement is interpreted as follows: If samples of size n were repeated exhaustively, the random interval of $\bar{x} - 1.96 \, \sigma/\sqrt{n}$ to $\bar{x} + 1.96 \, \sigma/\sqrt{n}$ would contain u 95 percent of the time. Any value of probability $(.90, .95, .99)$ may be used at the discretion of the experimenter. These values are found in tables for the cumulative normal distribution with zero mean and unit variance.

A.17 Binomial Distribution

If an event or outcome of an experiment can be classified into two categories, say occurrence or nonoccurrence, with some probability p of occurrence, then the resulting density is the point-binomial distribution

$$f(x) = \begin{pmatrix} 1 \\ x \end{pmatrix} p^x \, (1 - p)^{1-x} \qquad x = 0, 1 \qquad 0 \leq p \leq 1$$

A number, n, of independently repeated trials of such an experiment results in the binomial distribution with density

$$f(x) = \binom{n}{x} p^x (1 - p)^{n-x} \qquad x = 0, 1, 2, \cdots, n \qquad 0 \leq p \leq 1$$

The symbolism $\binom{n}{x}$ is read "n chose x" and is the number of ways n objects may be chosen x at a time. The expected value or mean of x is np. The variance of x is npq where $q = 1 - p$. The variance of $\hat{p}$ is pq/n. An obvious example of the applicability of the binomial distribution to population study is in the consideration of survival.

A.18 Poisson Distribution

As n approaches infinity and p approaches zero such that np approaches a constant, λ, the binomial distribution approaches the Poisson distribution with density

$$f(x) = \frac{e^{-\lambda}\lambda^x}{x!} \qquad x = 0, 1, 2, \cdots \qquad \lambda > 0$$

The mean and variance of the Poisson distribution are both λ. The distribution of infrequent or rare events, as, for example, the incidence of naturally missing fins on lake trout, are frequently distributed as a Poisson random variable.

A.19 Multinomial Distribution

If experimental results or events of a discrete nature can be categorized in more than two ways, the resulting distribution is a multinomial with density

$$f(x_1, x_2, \cdots, x_\kappa) = (x_1, {}^n x_2, \cdots, x_\kappa) \, p_1^{x_1} \, p_2^{x_2} \cdots p_\kappa^{x_\kappa}$$

where x_κ is specified as

$$n - \sum_{i=1}^{\kappa-1} x_i$$

and p_κ is specified as

$$1 - \sum_{i=1}^{\kappa-1} p_i.$$

With only two possible results, the multinomial reduces to the binomial. The p_i's are estimated by n_i/n with variance $p_i q_i/n$. Expected values for x_i would be np_i, analogous to the binomial case. Recaptures of numbered tags placed on fish provides an example of a multinomial distribution.

A.20 Hypergeometric Distribution

Another discrete distribution of importance in the study of animal populations is the hypergeometric. If the outcome of an experiment or event has two possible results, as with the binomial case, but a sample of size n is taken without replacement, then the number of occurrences of x is distributed as a hypergeometric random variable with density

$$f(x) = \frac{\binom{k}{x}\binom{m-k}{n-x}}{\binom{m}{n}} \qquad x = 0, 1, 2, \cdots, n$$

This distribution has mean kn/m and variance $nk(m-k)(m-n)/m^2(m-1)$. Certain types of mark-recapture methods of estimating population size are based on this distribution.

A.21 Negative Binomial

The negative binomial distribution is a discrete distribution that has been shown to provide a reasonable model for many biological situations. Distributions of counts that are like a Poisson but with a changing parameter from event to event give rise to a negative binomial. Counts of parasite per host, number of fish in a gill net, and certain methods of mark-recapture population estimation are examples of the negative-binomial distribution. One form of the density of this distribution is

$$f(x) = \binom{x+k-1}{x} q^x p^k \qquad x = 0, 1, 2, \cdots, k \qquad p + q = 1$$

This form is called the Pascal form of the negative binomial and represents the waiting time until the kth success or occurrence, that is, the variable x is the number of failures or nonoccurrences before the kth success. The mean and variance of this distribution are kq/p and kq/p^2 respectively.

An alternate form of the negative binomial is given by the density

$$f(x) = \binom{x + k - 1}{x} p^x q^{-x-k} \qquad q = 1 + p \qquad k > 0$$

This form is derived from a situation that would normally be Poisson but with a changing rather than fixed parameter λ. In this case k need not be an integer. Estimates of k may be obtained by method of moments, that is:

$$\hat{k} = \frac{\bar{x}^2}{s^2 - \bar{x}}$$

where $\bar{x}$ is the arithmetic mean and s^2 is $\dfrac{1}{n-1} \displaystyle\sum_{i=1}^{n} (x_i - \bar{x})^2$. A more accurate estimate of k may be obtained from the maximum-likelihood estimator

$$n \log_e \left(1 + \frac{\bar{x}}{k} \right) = \sum_{x=0} \frac{A_x}{\hat{k} + \bar{x}}$$

where A_x is the total count exceeding x and the summation limit is the value of x corresponding to the last observed frequency. At this point A_x becomes zero. The solution for $\hat{k}$ is obtained by iteration.

It is worthwhile to note the relationship of the mean and variance for the binomial, Poisson, and negative-binomial distributions. In the binomial $u > \sigma^2$, in the Poisson $u = \sigma^2$, and in the negative-binomial $u < \sigma^2$.

A.22 Linear Regression

One of the simplest functional relationships between two variables is that of a straight line. The linear relation is expressed by $y = a + bx$ where a and b are constants, y is the dependent variable, and x is the independent variable. In the context of linear regression, y is a random variable as opposed to being completely specified by $y = a + bx$. It is assumed that associated with each x there is a normally distributed y that has a mean or expected value of $a + bx$ and a variance σ^2.

The independent variable x is assumed to be fixed or controlled and the response or dependent variable y is observed or measured. As an example consider the weight of a hypothetical fish of known age as shown by the following data:

Age (yrs)	1	2	3
Weight (g)	41,44,44,47	54,56,57,59	62,63,64,66

Considering age as the independent variable, weight is a function of age. A plot of these data is shown in Figure A.22.1. The figure suggests that as age increases, weight increases, at least for the younger ages. It might be of interest to predict the weight of fish at age 1½ or 4½, for example. The method of linear regression provides the techniques for making such predictions. In order to make a prediction from linear regression, the coefficients, a and b, must be estimated from the observed data.

One method of estimation is to choose the coefficients a and b so that the squared distance from the observations to the line $y = a + bx$ is as small as possible. This is the method of least squares. If S represents the sum of squares of deviations from the observation to the line, then this may be expressed as:

$$S = \sum_{i=1}^{n} (Y_i - a - bX_i)^2 = \sum_{i=1}^{n} \epsilon_i^2$$

Treating this as a minimization problem of calculus results in the following equations:

$$\frac{\partial S}{\partial a} = -2 \, \Sigma (Y_i - a - bX_i)$$

$$\frac{\partial S}{\partial b} = -2 \, \Sigma \, X_i \, (Y_i - a - bX_i)$$

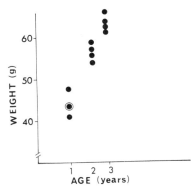

Figure A.22.1. A plot of fish weight against age

Setting these equations equal to zero provides the following normal equations:

$$\sum_{i=1}^{n} Y_i - na - b \sum_{i=1}^{n} X_i = 0$$

$$\sum_{i=1}^{n} X_i Y_i - a \sum_{i=1}^{n} X_i - b \sum_{i=1}^{n} X_i^2 = 0$$

The solution for b from the above equations is taken as the estimate for that coefficient

$$\hat{b} = \frac{\Sigma X_i Y_i - (\Sigma X \Sigma Y)/n}{\Sigma X_i^2 - (\Sigma X_i)^2/n}$$

The estimate of a is provided by:

$$\hat{a} = \frac{\Sigma Y_i}{n} - \hat{b} \frac{\Sigma X_i}{n}$$

The summation in the above expressions is from 1 to n, the number of observations in the sample.

The observations y_i from a sample of size n may be written:

$$
\begin{aligned}
y_1 &= a + bx_i + \epsilon_1 \\
y_2 &= a + bx_2 + \epsilon_2 \\
y_3 &= a + bx_3 + \epsilon_3 \\
&\vdots \\
y_n &= a + bx_n + \epsilon_n
\end{aligned}
$$

Expressed in matrix terms this becomes:

$$\mathbf{Y} = \mathbf{X}\beta + \boldsymbol{\epsilon}$$

where $\mathbf{Y}$ is the column vector of y_i observations

$$\mathbf{Y} = \begin{bmatrix} y_1 \\ y_2 \\ \cdot \\ \cdot \\ \cdot \\ y_n \end{bmatrix}$$

X is the matrix

$$\mathbf{X} = \begin{bmatrix} 1 & x_1 \\ 1 & x_2 \\ 1 & x_3 \\ \cdot & \\ \cdot & \\ \cdot & \\ 1 & x_n \end{bmatrix}$$

$\boldsymbol{\beta}$ is the column vector of coefficients

$$\boldsymbol{\beta} = \begin{bmatrix} a \\ b \end{bmatrix}$$

and ϵ is the column vector of normal, random error, ϵ_i

$$\boldsymbol{\epsilon} = \begin{bmatrix} \epsilon_1 \\ \epsilon_2 \\ \epsilon_3 \\ \cdot \\ \cdot \\ \cdot \\ \epsilon_n \end{bmatrix}$$

The normal equations in matrix terms are expressed as:

$$\mathbf{X'X}\,\boldsymbol{\beta} = \mathbf{X'Y}$$

Estimates of $\boldsymbol{\beta}$ are provided by:

$$\hat{\boldsymbol{\beta}} = (\mathbf{X'X})^{-1}\,\mathbf{X'Y}$$

The variance-covariance matrix of $\hat{\beta}$ may be calculated by:

$$\mathbf{V}(\hat{\beta}) = (\mathbf{X'X})^{-1} (\mathbf{Y'Y} - \hat{\beta}'\mathbf{X'Y})/n-2$$

where $(\mathbf{Y'Y} - \hat{\beta}'\mathbf{X'Y})/n-2$ is an estimate of σ^2. An estimate of the variation accounted for by regression is given by:

$$\frac{\hat{\beta}'\mathbf{X'Y} - n\bar{Y}^2}{\mathbf{Y'Y} - n\bar{Y}^2}$$

which will have a value between 0 and 1, with 1 meaning that all variation is accounted for by regression. A predicted, $\hat{Y}_k$, at $\mathbf{X}_k$ is given by $\hat{Y}_k = \mathbf{X'}_k \hat{\beta}$ with variance

$$\mathbf{V}(\hat{Y}_k) = \mathbf{X'}_k (\mathbf{X'X})^{-1} \mathbf{X}_k \sigma^2$$

A linear relationship that passes through the origin does not have an intercept value; that is, the relationship is $y = bx$ rather than $y = a + b_x$. The normal equations may be derived in the same manner as above but with the model $y = bx$. The matrix $\mathbf{X}$ will not have the lead column of 1's and the expression for variation accounted for by regression reduces to:

$$\frac{\hat{\beta} \mathbf{X'Y}}{\mathbf{Y'Y}}$$

The appropriate model should be used, however, regardless of a greater decrease in variability when an inappropriate model is used. If a relationship is known to pass through the origin, that is, $y = 0$ when $x = 0$, that model should be used. If it is not known to pass through the origin, the more general model may be used followed by testing for the contribution of a general mean. This may be done by an analysis of variance:

Source	Sum of squares	Degrees of freedom	Mean square	F
Mean (a)	$(\Sigma Y)^2/n$	1		
Regression (b|a)	$\hat{\beta} \mathbf{X'Y} - n \bar{Y}^2$	1		
Residual	$\mathbf{Y'Y} - \hat{\beta} \mathbf{X'Y}$	n-2	estimate of σ^2	
Total	$\mathbf{Y'Y}$	n		

The mean square values are determined by dividing the component sum of squares by their respective degrees of freedom. The F-value is used as the test statistic and is calculated as the ratio of the component mean square to the residual mean square. The resulting F-value is then compared to values from the F distribution with respective degrees of freedom and desired level of error. These values have been tabulated and are found in most books dealing with statistics. Thus testing for an intercept value is done by dividing the mean mean square by the residual mean square and comparing to the tabular value of F with 1 and $n - 2$ degrees of freedom.

The extension to more than one independent variable is straightforward and is easily handled in matrix notation; the matrix $\mathbf{X}$ may represent any number of variables. The estimates of $\boldsymbol{\beta}$ will be provided by the same solution,

$$\hat{\boldsymbol{\beta}} = (\mathbf{X}'\mathbf{X})^{-1}\,\mathbf{X}'\mathbf{Y}$$

provided $(\mathbf{X}'\mathbf{X})^{-1}$ exists. If there are p unknown parameters in the equations and the p normal equations are independent, then $(\mathbf{X}'\mathbf{X})^{-1}$ exists and a solution may be obtained.

A number of models may be transformed to the linear model and solutions derived from the linear model. A few examples of model and transformations are shown below:

Model	*Transformation*
$y = ax^b\epsilon$	$ln\ y = ln\ a + b\ ln\ x + ln\ \epsilon$
$y = ax_1^b\,x_2^c\,x_3^d\,\epsilon$	$ln\ y = ln\ a + b\ ln\ x_1 + c\ ln\ x_2 + d\ ln\ x_3 + ln\ \epsilon$
$y = e^{a+bx_1+cx_2+\epsilon}$	$ln\ y = a + bx_1 + cx_2 + \epsilon$
$y = \dfrac{1}{1 + e^{a+bx_1+cx_2+\epsilon}}$	$ln\left(\dfrac{1-y}{y}\right) = a + bx_1 + cx_2 + \epsilon$

Not all nonlinear relationships can be transformed to the linear form and other methods must be used to estimate parameter values. One such method is discussed in the next section.

A.23 Fitting Nonlinear Regression by Least Squares

A function of the form $y = ae^{bx} + \epsilon$ is nonlinear and cannot be put in linear form by transformation as was $y = ae^{bx}\epsilon$ shown in the preceding section. The concept of least squares can be applied to the problem using a Taylor series approximation of the sum of squares.

Using the model $y = ae^{bx} + \epsilon$ as an example, let $s(a,b)$ be the sum of squares of the differences between the observed random variable and its prediction from the proposed model based on parameters a and b. For some choice of values for a and b, the sum of squares should have a minimum value; the problem is to find those values. The sum of squares, $s(a,b)$, is given by:

$$s(a,b) = \sum_{i=1}^{n} (y - y^*)^2$$

where y^* is the predicted value given by the model, for example, $y^* = ae^{bx}$. Approximating $y^* = f(x;a,b)$ by Taylor series gives:

$$y^* = f(x;a_o,b_o) + (a - a_o)\frac{\partial f(x;a_o,b_o)}{\partial a} + (b - b_o)\frac{\partial f(x;a_o,b_o)}{\partial b}$$

For our specific example,

$$y^* = a_o e^{b_o x} + (a - a_o)e^{b_o x} + (b - b_o)\,axe^{b_o x}$$

Letting $\mathbf{C} = [(a - a_o)(b - b_o)]'$ and

$$\mathbf{D} = \begin{bmatrix} \dfrac{\partial f(x_1;a_o,b_o)}{\partial a} & \dfrac{\partial f(x_1;a_o,b_o)}{\partial b} \cdot \\ \cdot & \cdot \\ \cdot & \cdot \\ \cdot & \cdot \\ \dfrac{\partial f(x_n;a_o,b_o)}{\partial a} & \dfrac{\partial f(x_n;a_o,b_o)}{\partial b} \end{bmatrix}$$

The original model may then be rewritten as:

$$y - y^* = \mathbf{D}\,\mathbf{C} + \epsilon \text{ equivalent to the linear model } \mathbf{Y} = \mathbf{X}\boldsymbol{\beta} + \boldsymbol{\epsilon}.$$

Therefore the vector $\mathbf{C}_o = (\mathbf{D}'\mathbf{D})^{-1}\mathbf{D}'(y - y^*)$ will minimize the sum of squares,

$$\sum_{i=1}^{n}\left[y_i - y^* - \mathbf{C}\left[\frac{\partial f(x_i;a_o,b_o)}{\partial a} \quad \frac{\partial f(x_i;a_o,b_o)}{\partial b} \right] \right]^2$$

The elements of the vector $\mathbf{C}_o$ are in effect correction factors which may be added to the original value assigned the parameters a_o and b_o. These new values, say a_1 and b_1, may then be used in repeating the process. This procedure is carried forward until the values of a_i and b_i differ from a_{i+1} and b_{i+1} by some preselected small amount.

Variances for parameter estimates may be estimated by determining the variance, σ^2, which is the sum of squares of the differences between observations and prediction based on fitted model, and a matrix of sums of partial derivatives of the model with respect to each parameter. The variance σ^2 is given by:

$$\sigma^2 = \sum_{i=1}^{n} (\text{Observation-Prediction})^2/n - p$$

where p is the degrees of freedom; one degree of freedom is used for each parameter estimated. If M represents the model, the matrix of sums of squares of partial derivatives based on parameter a, b, and c would be, say

$$\mathbf{D'D} = \begin{bmatrix} \Sigma\left(\dfrac{\partial M}{\partial a}\right)^2 & \Sigma\left(\dfrac{\partial M}{\partial a}\right)\left(\dfrac{\partial M}{\partial b}\right) & \Sigma\left(\dfrac{\partial M}{\partial a}\right)\left(\dfrac{\partial M}{\partial c}\right) \\[2ex] \Sigma\left(\dfrac{\partial M}{\partial b}\right)\left(\dfrac{\partial M}{\partial a}\right) & \Sigma\left(\dfrac{\partial M}{\partial b}\right)^2 & \Sigma\left(\dfrac{\partial M}{\partial b}\right)\left(\dfrac{\partial M}{\partial c}\right) \\[2ex] \Sigma\left(\dfrac{\partial M}{\partial c}\right)\left(\dfrac{\partial M}{\partial a}\right) & \Sigma\left(\dfrac{\partial M}{\partial c}\right)\left(\dfrac{\partial M}{\partial b}\right) & \Sigma\left(\dfrac{\partial M}{\partial c}\right)^2 \end{bmatrix}$$

Then the variance-covariance matrix, $\mathbf{\Sigma}$, for the parameters is given by:

$$\mathbf{\Sigma} = \sigma^2 (\mathbf{D'D})^{-1}$$

The variance for a predicted value from the model is equal to $\text{tr } \mathbf{\Sigma\Delta}$, where $\mathbf{\Delta}$ is a matrix of partial derivatives of the model with respect to the parameters in the model, evaluated at the value(s) of the independent variable(s) used to predict the dependent variable. The symbol tr stands for trace, which is the sum of the diagonal elements of a matrix.

This is one of several methods for obtaining parameter estimates in least-squares fitting of nonlinear regression. This technique may converge slowly, may oscillate widely, or fail to converge at all. In such cases other techniques should be investigated. Initial choices for parameter values are important since some of the problems of convergence are due to the poor representation by a Taylor series expansion when initial and actual values are not close.

It is important to seek statistical help early in any program design if you are not absolutely sure of the experimental design and methods to be used. Far too many resources have been wasted because of inappropriate design and analysis.

Index

Library of Congress Cataloging in Publication Data

Everhart, Watson Harry, 1918–
    Principles of fishery science.

    Includes bibliographies and index.
    1. Fishery management.   2. Fish populations.   3. Fish-culture.   I. Youngs,
William D., joint author.   II. Title.
SH328.R69 1981   639.3   80-15603
ISBN 0-8014-1334-6

DATE DUE

DE 14 '00			
FE 21 '94			
MAY 2			
96			
AP 14 '96			
AUG 0 1 1998			

DEMCO 38-297